Advances in Experimental Medicine and Biology

Volume 765

For further volumes:
http://www.springer.com/series/5584

William J. Welch · Fredrik Palm
Duane F. Bruley · David K. Harrison
Editors

Oxygen Transport to Tissue XXXIV

Springer

Editors
William J. Welch
Division of Nephrology and Hypertension
Georgetown University
Washington, DC, USA

Fredrik Palm
Department of Medical Cell Biology
Uppsala University
Uppsala, Sweden

Duane F. Bruley
Synthesizer, Inc.
Ellicott City, MD, USA

David K. Harrison
Institute of Cellular Medicine
Newcastle University
Newcastle upon Tyne, UK

ISSN 0065-2598
ISBN 978-1-4614-4771-9 ISBN 978-1-4614-4989-8 (eBook)
DOI 10.1007/978-1-4614-4989-8
Springer New York Heidelberg Dordrecht London

Library of Congress Control Number: 2012944420

Printed on acid-free paper

Springer is part of Springer Science+Business Media (www.springer.com)

Dedication: Britton Chance, M.D., Ph.D., D.Sc.

Lin Z. Li, Shoko Nioka, and Kyung A. Kang

L.Z. Li (✉)
Department of Radiology, University of Pennsylvania, 250 Anatomy Chemistry, Philadelphia, PA 19014, USA

Johnson Research Foundation, Department of Biochemistry and Biophysics, University of Pennsylvania, Philadelphia, PA 19104, USA
e-mail: linli@mail.med.upenn.edu

S. Nioka
Johnson Research Foundation, Department of Biochemistry and Biophysics, University of Pennsylvania, Philadelphia, PA 19104, USA

K.A. Kang
Chemical Engineering Department, University of Louisville, Louisville, KY 40292, USA

Abstract Professor Britton Chance was one of the most outstanding scientists in the world. He was born on July 24, 1913 in Wilkes-Barre, PA, USA and passed away on November 16th, 2010 in Philadelphia at the age of 97. He has left with us a tremendous legacy that many other extraordinary human beings cannot match. As a scientist and an engineer, he has invented and developed numerous physical instruments and employed them to answer some of the most pressing research questions in biology and medicine, ranging from enzyme kinetics through bioenergetics and electron transport in mitochondria, reactive oxygen species, quantum tunneling in biology, in vivo NMR, to biophotonics for brain functional studies and the detection, diagnosis, and treatment of diseases. He had a high impact on every major research field in which he worked and was regarded as a founding father for mitochondrial bioenergetics, redox sciences, in vivo NMR, and biophotonics. With his keen capacities in electronics, he worked in the MIT Radiation Lab on precision bombing and radar systems that were used in World War II and contributed to the development of the world's first general purpose computer, ENIAC, at the University of Pennsylvania. As an athlete, he won an Olympic gold medal (5.5 m sailing) in Helsinki in 1952 and several World Championships in the late 1950s–1960s. He was a member of the National Academy of Sciences in the USA and an academician in five other countries. He won the National Medal of Sciences in 1974. As an educator, he tirelessly trained thousands of students and researchers and many of them have become established leaders in various fields of scientific research. He had been an ISOTT member since the founding of the Society and was President of ISOTT in 1976. In his 90s, he traveled to China, Singapore, and Taiwan for various research and educational activities and helped local scientists to develop cutting-edge research projects and institutions. In 2008 and 2009 he received the two highest honors for foreign scientists from the Chinese government.

Keywords Olympic gold medal • radiation lab • mitochondria redox • biophotonics, NMR

Selected Photos of Dr. Chance

The Life, Times and Legacy of Dr. Chance in Science

Education

1935	B.S., University of Pennsylvania, Philadelphia, PA
1936	M.S., University of Pennsylvania, Philadelphia, PA
1940	Ph.D., University of Pennsylvania (Physical Chemistry), Philadelphia, PA
1942	Ph.D., Cambridge University (Biology "B", Physiology), England
1952	D.Sc., Cambridge University, England

Honorary Degrees

1962	M.D., Karolinska Institute
1974	D.Sc., Medical College of Ohio at Toledo
1976	D.Sc., Semmelweis University
1977	D.Sc., Hahnemann Medical College
1985	D.Sc., University of Pennsylvania
1990	D.Sc., University of Helsinki
1991	M.D., University of Dusseldorf
1993	M.D., University of Buenos Aires
1995	M.D., University of Copenhagen
1997	M.D., University Degli Studi Di Roma "Tor Vergata"

Academic Career

1940–1941	Acting Director, E.R. Johnson Foundation, University of Pennsylvania
1941–1949	Assistant Professor of Biophysics and Physical Biochemistry, School of Medicine, University of Pennsylvania, Philadelphia, PA
1949–1977	Professor of Biophysics and Physical Biochemistry, Graduate Group of Biophysics, School of Medicine, University of Pennsylvania, Philadelphia, PA
1964–1975	Professor, Eldridge Reeves Johnson Professor of Biophysics, Department of Biophysics, School of Medicine, University of Pennsylvania, Philadelphia, PA
1949–1975	Chairman, Department of Biophysics and Physical Biochemistry, School of Medicine, University of Pennsylvania, Philadelphia, PA
1975–1983	Professor, Biochemistry and Biophysics, School of Medicine, University of Pennsylvania, Philadelphia, PA
1949–1983	Director, E.R. Johnson Foundation, School of Medicine, University of Pennsylvania, Philadelphia, PA
1982–1990	Director, Institute for Structural and Functional Studies, University City Science Center, Philadelphia, PA

1990–1999	Director, Institute for Biophysical and Biomedical Research, University City Science Center, Philadelphia, and Scientific Director to the President, University City Science Center, Philadelphia, PA
1983–1992	Eldridge Reeves Johnson University Professor Emeritus of Biochemistry and Biophysics and Physical Biochemistry, School of Medicine, University of Pennsylvania, Philadelphia, PA
1998–2006	President, Medical Diagnostic Research Foundation (MDRF), Philadelphia, PA
1983–2010	Eldridge Reeves Johnson University Professor Emeritus of Biochemistry and Biophysics and Physical Chemistry and Radiologic Physics, School of Medicine, University of Pennsylvania, Philadelphia, PA
2007–2010	Professor of Department of Biomedical engineering, Drexel University Philadelphia, PA

Postgraduate Training and Fellowship Appointments

1941–1946	Massachusetts Institute of Technology, Radiation Laboratory: (1941–1945) Research Associate (1941–1945) Group Leader (1942–1945) Associate Division Head (1942–1945) Steering Committee (1945–1946) Editorial Board, Radiation Lab Series (1954–1956) Visiting Committee (Biology)
1941	Office of Scientific Research and Development, Investigator
1946–1948	Guggenheim Fellow, Nobel Institute, Stockholm and Moleno Institute, Cambridge
1966	Foreign Fellow, Churchill College, Cambridge

Special Appointments

1948	U.S.N Consultant to Attaché for Research, London
1948	American Red Cross and National Research Council Committee on Blood and Blood Derivatives
1951–1956	National Science Foundation Consultant (Molecular Biology)
1955–1959	Bartol Research Foundation, Visiting Committee
1959–1960	President's Scientific Advisory Committee
1971–1975	Council Member, National Institute on Alcohol Abuse and Alcoholism
1973	Council Member, Working Group on Molecular Control, National Cancer Institute
1983	NIH Panel Member, Biotechnology Research Resources Program, Division of Research Resources

Awards and Honors

1950	Paul Lewis Award in Enzyme Chemistry, USA
	U.S. President's Certificate of Merit, USA
1954	Harvey Lecturer, The New York Academy of Medicine, USA
1956, 1965	Phillips Lecturer, University of Pittsburgh, PA, USA
1957	Pepper Lecturer, University of Pennsylvania, PA, USA
1961	Morlock Award, Institute of Electrical & Electronics Engineers (IEEE), USA
1965	Genootschapps Medaille, Dutch Biochemical Society, The Netherlands
1966	Keilin Lecturer, England
	Harrison Howe Award, Rochester Section, American Chemical Society, USA
	Franklin Medal, Franklin Institute, Philadelphia, USA
	David P. Hackett Lecturer, University of California, USA
1968	Pennsylvania Award for Excellence (Life Sciences), USA
1969	Philadelphia Section Award, American Chemical Society, USA
1970	Nichols Award, New York Section, American Chemical Society, USA
	Heineken Medal, Netherlands Academy of Science and Letters, The Netherlands
	Redfearn Memorial Lecturer, University of Leicester, England
1972	Gairdner Award, Canada
1973	Post-IUB Symposium on Energy Transducing Membrane Function, Dedicated to B. Chance on the Occasion of his 60th Birthday, Sweden
1974	Semmelweis Medal, Hungary
	National Medal of Science, USA
1975	Presidential Lecturer, University of Pennsylvania, PA, USA
1976	35th Richtmeyer Memorial Lecture of American Association of Physics Teachers, USA
	2nd Julius L. Jackson Memorial Lecture, Wayne State University, USA
	ISCO Award for Significant Contributions Field of Biochemical Instruments, USA
	DaCosta Oration, Philadelphia County Medical College, USA
1978	Phillip Morris Lecturer, USA
1979	Rudolf-Lemberg Memorial Lecture Series, Australia
1980	Distinguished Lecturer, Society of General Physiologists International Congress of Physiological Sciences, Hungary
	Plenary Lecturer, First European Bioenergetics Congress, Urbino, Italy
	Hastings Lecturer, Scripps Clinic & Research Foundation, USA
1981	Plenary Lecturer, VII International Biophysics Congress/II Pan American Biochemistry Congress, Mexico

1981	Radvin Lecturer, American College of Physicians and Surgeons, USA
1984	Troy C. Daniels Lectureship Award, University of California, USA
	IEEE Philadelphia Section Award, USA
	Sober Lectureship Award, American Society of Biological Chemists, USA
	Plenary Lecturer, Pan American Association of Biochemical Societies (PAABS) Congress, Argentina
	Invited Speaker, Symposium on Bioscience, Biomedical Imaging Sponsored by the Takeda Science Foundation, Kyoto, Japan
	Plenary Lecturer, International Symposium on Fast Reactions in Biological Systems, Kyoto, Japan
1986	Elizabeth Winston Lanier Award, American Academy of Orthopedic Surgeons Kappa Delta Awards, USA
	Senior Investigator Award, American Heart Association, USA
1987	Gold Medal Distinguished Service to Medicine, College of Physicians, USA
1987	Biological Physics Prize, The American Physical Society, USA
1988	Gold Medal, Society of Magnetic Resonance in Medicine, USA
1989	J. Henry Wilkinson Award, International Society for Clinical Enzymology, USA
1990	The Benjamin Franklin Medal for Distinguished Achievement in the Sciences, American Philosophical Society
1992	Christopher Columbus Discovery Award in Biomedical Research (National Institutes of Health, USA)
1992	John Scott Medal Award, City of Philadelphia
1995	Stellar-Chance Laboratories, University of Pennsylvania, Philadelphia, PA. Dedicated June 28, 1995
1999	American College of Sports Medicine Honor Award, USA
2003	Senior Science Advisor, International Association of Yan Xin Life Science and Technology
2005	International Society for Optical Engineering (SPIE) Lifetime Achievement Award
2005	ICAS Liberty Award, USA
2006	Gold Medal, American Roentgen Ray Society
2006	Distinguished Achievement Award, American Aging Association
2008	Britton Chance Reading Room, The Library of American Philosophical Society. Dedicated in April, 2008
2008	The Friendship Award, China
2008	The Chime Bell Award, Hubei Province, China
2009	The International Science and Technology Cooperation Award, China

Elections to Honorary Societies

National

1954	National Academy of Science, USA
1954	Harvey Society, New York Academy of Medicine
1955	American Academy of Arts and Sciences, Fellow and Member, Rumford Committee
1958	American Philosophical Society
	Wistar Association (1969–Present)
	Secretary (1969)
	Vice-President (1984–1990)
1959	Biophysical Society, Founder and Councillor (1959–1962)
1957	Society of General Physiologists, Councillor (1957–1960)
1966	American Association for the Advancement of Science, Vice President and Chairman, Section of Medical Sciences
1974	Philadelphia College of Physicians, Fellow
1975	Optical Society of America, Fellow
1989–1990	The Oxygen Society, Honorary President

International

1959	Royal Academy of Arts and Sciences, Member
1960	Royal Society of Arts, Benjamin Franklin Fellow
1968	Royal Swedish Academy of Sciences, Foreign Member Medical Sciences
1970	Bavarian Academy of Sciences, Munich, Member
1971	International Federation of Institutes for Advanced Study, Member of Board of Trustees (1971–1982)
1971	Leopoldina Academy, Halle, Germany
1972	International Union of Pure and Applied Biophysics
	Vice President (1972–1975)
	President (1975–1979)
	Honorary Vice President (1979–2010)
1973	International Society of Oxygen Transport to Tissue
	President (1976)
1974	Max-Planck-Gesellschaft zur Förderung der Wissenschaften, Foreign Member
1975	Argentine National Academy of Sciences, Member
1987	National Academy of Lincei, Foreign Member
1981	Royal Society (London), Foreign Member
1988–1990	Society for Free Radical Research (International) (SFRR), President

Short Summary of Scientific Contributions

1. Study of enzyme–substrate kinetics and the first experimental demonstration of the existence of an enzyme–substrate complex using the micro stop-flow device he developed.
2. Development of electronic analog calculators and precision targeting radar/computing systems in the World War II and the contribution to ENIAC, the world's first general purpose computer at the University of Pennsylvania.
3. Development of the dual beam (dual wavelength) spectrometer widely used for studying turbid biological samples.
4. In-depth study of the bioenergetics, redox state, and electron transport in mitochondrial respiration.
5. First discovery of the generation of reactive oxygen species (hydrogen peroxide) in mitochondrial metabolism.
6. First experimental demonstration of electron tunneling phenomena in biological systems.
7. Ground-breaking development of in vivo nuclear magnetic resonance spectroscopy and applications for perfused organs, animals, and human subjects (patients and athletes).
8. Founding the field of biophotonics by developing novel near-infrared spectroscopy and imaging methods including time-resolved spectroscopy (TRS), photon diffusion tomography (PDT), and their biomedical applications to study brain functions (fNIR) and various diseases including breast cancer, etc.
9. Development of mitochondrial metabolic/redox state fluorescence imaging (redox scanning) and its biomedical applications such as tumor biomarkers.
10. Invention of a number of scientific instruments, some of which have been widely used in research and industry throughout the world until the present, including the micro stop-flow device, the dual beam (dual wavelength) spectrometer, NADH fluorometer, redox scanner, RunMan, and a handheld NIR breast cancer detector. These instruments have made it possible for many key developments in biochemistry, biophysics, molecular biology, biomedical research, and clinical practice. In addition, he inspired the development of multiple NMR technologies and made important contributions to the instrumentation development for simultaneous optical and multinuclear NMR spectroscopy/imaging studies.
11. Over 1500 manuscripts published. According to ISI Science Citation Indexes accessed on July 10, 2011, the H index of Dr. Chance was 123. He had six papers cited more than 1,000 times, including two on mitochondrial bioenergetics, two papers on hydrogen peroxide, one on the assay of catalases and peroxidases, and one on the time-resolved measurement of tissue optical properties. A review paper on the metabolism of hydrogen peroxide published in 1979 has the highest citation rate at 3,600 times.
12. Contributions to the development of scientific research communities all over the world. He had trained thousands of students, postdocs, and researchers, with

many of them becoming leaders in scientific research. He devoted himself to a Summer Science Program for Minority High School Students in Philadelphia from 1996 to 2006. He tirelessly promoted academic exchanges between the East and West, and supported and facilitated the development of biomedical photonics in Asia, particularly in China, during the last 10 or so years of his life.

Copyright Acknowledgement All photos are by courtesy of the Johnson Research Foundation, Department of Biochemistry and Biophysics, Perelman School of Medicine, University of Pennsylvania. The assistance from Mary Leonard is appreciated.

References

1. Ohnishi T, Zweier JL (2011) Celebrating Professor Britton Chance (1913–2010), a Founding Father of Redox Sciences. Antioxidants & Redox Signaling 15(11): 2815–2817
2. Dutton PL (2011) Britton Chance. Phys Today, 64(11);65–66
3. Li LZ (2011) Celebrating the life and legacy of Britton Chance. Editorial for a special section in memory of Professor Britton Chance. J Innov Optical Health Sci 4(2):v–vii
4. Kuozomi H (2011) Professor Britton Chance and the early times he lived. J Innov Optical Health Sci 4(3): 215–219
5. Cicco G (2011) Memories of Britton Chance and The Britton Chance Award. J Innov Optical Health Sci 4(4):477–478
6. Li LZ & Nioka S (2011) Britton Chance: One of the most outstanding scientists in the world, Editorial for a special issue in memory of Professor Britton Chance, J Innov Optical Health Sci 4(3):v–viii
7. Li LZ & Fu L (2011) Tao Li Tian Xia. Editorial for a special section in memory of Professor Britton Chance, J Innov Optical Health Sci 4(4):v–vi
8. Li LZ & Fu L (2011) Britton Chance, Biophysical Society, February, 2011
9. Estabrook R, Sies H, and Packer L (2010) Remembering Britton Chance, Oxygen Club of California, December
10. Dutton PL (2010) Retrospective: Britton Chance (1913–2010). Science 330:1641
11. Dutton PL (2010) Britton Chance, Olympian and Biophysics Researcher, Dies at 97. The New York Times, November 28
12. Dutton PL (2010) Scientist specialized in reactions. The Philadelphia Inquirer, November 28
13. Dutton PL (2010) Britton Chance dies at 97; pioneer in study of ultra-fast reactions in human biology, Los Angeles Times, December 5
14. Dutton PL (2010) Dr. Chance, Biochemistry & Biophysics. University of Pennsylvania Almanac, November 23
15. Britton Chance: his life, times and legacy (http://www.med.upenn.edu/biocbiop/chance/index.html).

Preface

The 39th annual ISOTT conference was held on the campus of Georgetown University in the capital of the USA. The urban campus of the Jesuit institution provided an ideal setting for the exchange of ideas and research from this organization.

The program covered many aspects of oxygen transport from air to the cells, organs, and organisms; instrumentation and methods to sense oxygen and clinical evidence. It consisted of 9 invited keynote lectures, 75 abstracts, and attracted 117 registered participants (21 trainees) from 10 countries. Typical to ISOTT tradition, the audience was highly interactive with lively discussions and follow-up comments on each presentation. The highlight of the meeting was the memorial session on one of the bright lights of ISOTT, Britton Chance. In addition to an overview of his career by Jerry Glickson, a poster was organized by Lin Li and Kyung Kang and presented in the Proceedings (first chapter). Robert Balaban, from the NHLBI/NIH, presented the meeting Keynote lecture on mitochondrial energetics. One session was devoted to translational aspects of oxygen delivery and hypoxia.

ISOTT 2011 featured 15 presentations from trainees, many of whom were awarded travel awards and special awards. The society continues to encourage participation and presentations from the next generation of oxygen investigators.

Social events at the meeting included a pleasant dinner and cruise on the Potomac River, a dinner event in downtown DC, and an elegant banquet on the final night of the meeting. The combination of excellent science, feature presentations of trainees, and fun social activities made this a successful meeting. We look forward to next year's meeting in Belgium.

Washington, DC, USA — Bill Welch, PhD

Acknowledgements

Sponsors

The following institutions and corporations kindly provided financial support for ISOTT 2011:

AstraZeneca

GU · HU · MHRI · ORNL · WDCVAMC

GHUCCTS
Georgetown-Howard Universities
Center for Clinical and Translational Science

GEORGETOWN UNIVERSITY MEDICAL CENTER

The Center for Hypertension, Kidney and Vascular Research, Georgetown University
Dr. Christopher S. Wilcox, Director

The Department of Medicine, Georgetown University
Dr. Bruce Luxon, Chairman

genzyme

ADINSTRUMENTS

Organization ISOTT 2011

Organizing Committee

William J. Welch, Georgetown University, Washington, DC, USA
Fredrik Palm, Uppsala University, Uppsala, Sweden
Peter Hansell, Uppsala University, Uppsala, Sweden
Anton Wellstein, Georgetown University, Washington, DC, USA
Mike Espy, National Institutes of Health, Bethesda, MD, USA
Lin Z. Li, University of Pennsylvania, Philadelphia, PA, USA
Duane Bruley, Ellicott City, MD, USA

ISOTT Officers and Executive Committee

William J. Welch (President, Washington, USA)
Oliver Thews (Secretary, Halle, Germany)
Peter E. Keipert (Treasurer, San Diego, USA)
Duane F. Bruley (Knisely Award Committee, Ellicott City, USA)
Martin Wolf (Past President, Zurich, Switzerland)
Sabine van Huffel (President elect, Brussels, Belgium)
Lin Li (Philadelphia, USA)
Michelle A. Puchowicz (Cleveland, USA)
Hiromi Sakai (Tokyo, Japan)
Ludwig Schleinkofer (Munich, Germany)
Ilias Tachtsidis (London, UK)
Clare Elwell (London, UK)
Ursula Wolf (Bern, Switzerland)

Awards

The Melvin H. Knisely Award

The Melvin H. Knisely Award was established in 1983 to honor Dr. Knisely's accomplishments in the field of transport of oxygen and other metabolites and anabolites in the human body. Over the years, he has inspired many young investigators, and this award is to honor his enthusiasm for assisting and encouraging young scientists and engineers in various disciplines. The award is to acknowledge outstanding young investigators. This award was first presented during the banquet of the 1983 annual conference of ISOTT in Ruston, Louisiana. The award includes a Melvin H. Knisely plaque and a cash prize. Award recipients:

1983 Antal G. Hudetz, Hungary
1984 Andras Eke, Hungary
1985 Nathan A. Bush, USA
1986 Karlfried Groebe, Germany
1987 Isumi Shibuya, Japan
1988 Kyung A. Kang, Korea/USA
1989 Sanja Batra, Canada
1990 Stephen J. Cringle, Australia
1991 Paul Okunieff, USA
1992 Hans Degens, The Netherlands
1993 David A. Benaron, USA
1994 Koen van Rossem, Belgium
1995 Clare E. Elwell, UK

1996 Sergei A. Vinogradov, USA
1997 Chris Cooper, UK
1998 Martin Wolf, Switzerland
1999 Huiping Wu, USA
2000 Valentina Quaresima, Italy
2001 Fahmeed Hyder, Bangladesh
2002 Geofrey De Visscher, Belgium
2003 Mohammad Nadeem Khan, USA
2004 Fredrik Palm, Sweden
2005 Nicholas Lintell, Australia
2007 Ilias Tachtsidis, UK
2008 Kazuto Masamoto, Japan
2009 Rossana Occhipinti, USA
2010 Sebastiano Cicco, Italy
2011 Mei Zhang, USA

The Dietrich W. Lübbers Award

The Dietrich W. Lübbers Award was established in honor of Professor Lübbers's long-standing commitment, interest, and contributions to the problems of oxygen transport to tissue and to the society. This award was first presented in 1994 during the annual conference of ISOTT in Istanbul, Turkey. Award recipients:

1994 Michael Dubina, Russia
1995 Philip E. James, UK/USA
1996 Resit Demit, Germany
1997 Juan Carlos Chavez, Peru
1998 Nathan A. Davis, UK
1999 Paola Pichiule, USA
2000 Ian Balcer, USA
2001 Theresa M. Busch, USA
2002 Link K. Korah, USA
2003 James J. Lee, USA
2004 Richard Olson, Sweden
2005 Charlotte Ives, UK
2006 Bin Hong, China/USA
2007 Helga Blockx, Belgium
2008 Joke Vanderhaegen, Belgium
2009 Matthew Bell, UK
2010 Alexander Caicedo Dorado, Belgium
2011 Malou Friederich, Sweden

The Britton Chance Award

The Britton Chance Award was established in honor of Professor Chance's long-standing commitment, interest, and contributions to the science and engineering aspects of oxygen transport to tissue and to the society. This award was first presented in 2004 during the annual conference of ISOTT in Bari, Italy. Award recipients:

2004 Derek Brown, Switzerland
2005 James Lee, USA
2006 Hanzhu Jin, China/USA
2007 Eric Mellon, USA
2008 Jianting Wang, USA
2009 Jessica Spires, USA
2010 Ivo Trajkovic, Switzerland
2011 Alexander Caicedo Dorado, Belgium

The Duane F. Bruley Travel Awards

The Duane F. Bruley Travel Awards were established in 2003 and first presented by ISOTT at the 2004 annual conference in Bari, Italy. This award was created to provide travel funds for student researchers in all aspects of areas of oxygen transport to tissue. The awards signify Dr. Bruley's interest in encouraging and supporting young researchers to maintain the image and quality of research associated with the society. As a cofounder of ISOTT in 1973, Dr. Bruley emphasizes cross-disciplinary research among basic scientists, engineers, medical scientists, and clinicians. His pioneering work constructing mathematical models for oxygen and other anabolite/metabolite transports in the microcirculation, employing computer solutions, were the first to consider system nonlinearities, time dependence, including multidimensional diffusion, convection, and reaction kinetics. It is hoped that receiving the Duane F. Bruley Travel Award will inspire students to excel in their research and will assist in securing future leadership for ISOTT. Award recipients:

2004 Helga Blocks (Belgium), Jennifer Caddick (UK), Antonio Franco (Italy), Charlotte Ives (UK), Nicholas Lintell (Australia), Leonardo Mottola (Italy), Samin Rezania (USA/Iran), Ilias Tachtsidis (UK), Liang Tang (USA/China), Iyichi Sonoro (Japan)
2005 Robert Bradley (UK), Kathy Hsieh (Australia), Harald Oey (Australia), Jan Shah (Australia)
2006 Ben Gooch (UK), Ulf Jensen (Germany), Smruta Koppaka (USA), Daya Singh (UK), Martin Tisdall (UK), Bin Wong (USA), and Kui Xu (USA)
2007 Dominique De Smet (Belgium), Thomas Ingram (UK), Nicola Lai (USA), Andrew Pinder (UK), Joke Vanderhaegen (Belgium)
2008 Sebastiano Cicco (Italy)
2009 Lei Gao (UK), Obinna Ndubuizu (USA), Joke Vanderhaegen (Belgium), Jianting Wang (USA)
2010 Zareen Bashir (UK), Martin Biallas (Switzerland), Takashi Eriguchi (Japan), Jack Honeysett (UK), Catalina Meßmer (USA), Tracy Moroz (UK), Yoshihiro Murata (Japan), Mark Muthalib (Australia)
2011 Patrik Persson (Sweden), Kouichi Yoshihara (Japan), Catherine Hesford (UK), Luke S. Holdsworth (UK), Andreas Metz (Switzerland), Felix Scholkmann (Switzerland), Maria D. Papademetriou (UK).

Contents

The original version of this book was revised. An erratum to this book can be found at DOI 10.1007/978-1-4614-4989-8_53

Contributors

The Georgetown University support staff who helped make ISOTT 2011 included Tolise Miles, Research Administrator, Department of Medicine; Sarah Kamenski, Research administrator; Emily Chan, Division of Nephrology administrator; and Sean Hawkins, administrator to the Executive Vice-President. The meeting would not have been successful without their invaluable service.

Reviewers

The editors would like to thank the following experts, who scientifically reviewed the papers:

Alexander Caicedo Dorado Katholieke Universiteit Leuven, Leuven, Belgium

Daniel Coman Yale University, New Haven, CT, USA

Chris Cooper University of Essex, Colchester, Essex, UK

Clare Elwell University College London, London, UK

Elaine Fisher University of Akron, Akron, OH, UK

Arnab Ghosh University College London, London, UK

Jerry Glickson University of Pennsylvania, Philadelphia, PA, USA

Peter Hansell Uppsala University, Uppsala, Sweden

David Harrison Newcastle University, Newcastle upon Tyne, UK

Kyung Kang University of Louisville, Louisville, KY, USA

Joseph LaManna Case Western Reserve University, Cleveland, OH, USA

Lin Li University of Pennsylvania, Philadelphia, PA, USA

Masaomi Nangaku University of Tokyo, Tokyo, Japan

Paul Okunieff University of Florida, Gainesville, FL, USA

Fredrik Palm Uppsala University, Uppsala, Sweden

Maria Papademetriou University College London, London, UK

Michelle Puchowicz Case Western Reserve University, Cleveland, OH, USA

Joseph Rifkind National Institute on Aging, Baltimore, MD, USA

Felix Scholkmann University of Zurich, Zurich, Switzerland

Ilias Tachtsidis University College London, London, UK

Oliver Thews University of Halle-Wittenberg, Halle, Germany

SabineVanHuffel Katholieke Universiteit Leuven, Leuven, Belgium

Peter Vaupel Technical University of Munich, Munich, Germany

William Welch Georgetown University, Washington, DC, USA

David Wilson University of Pennsylvania, Philadelphia, PA, USA

Martin Wolf University of Zurich, Zurich, Switzerland

Ursula Wolf University of Bern, Bern, Switzerland

Chis Wolff St. Bartholomew's Hospital, London, UK

He Xu University of Pennsylvania, Philadelphia, PA, USA

Also, many thanks to the following technical reviewers:

Laraine Visser-Isles Rotterdam, The Netherlands

Eileen Harrison St. Lorenzen, Italy

Chapter 1
Effects of Experimentally Deviated Mandibular Position on Stress Response

Ai Amemiya, Tomotaka Takeda, Kazunori Nakajima, Keiichi Ishigami, Takeo Tsujii, and Kaoru Sakatani

Abstract The purpose of this study was to investigate the effects of stress on prefrontal cortex (PFC), emotion (using visual analogue scale, VAS, and State-Trait Anxiety Inventory, STAI), and the autonomic nervous system (ANS). Two types of stress were applied: (1) malocclusion-induced physical stress and (2) mental stress induced by an arithmetic task. Malocclusion was induced using an experimentally deviated mandibular device (EDMD) to obtain an experimentally deviated mandibular position (EDMP). A total of 11 healthy volunteers participated in the study. On day 1 they performed a pretrial arithmetic task followed by a 10-min rest, after which they performed a posttrial EDMD + arithmetic task or rest device + arithmetic task. These two tasks were selected at random and assigned at the rate of one per day. Activity in the PFC tended to show an increase in the pretrial arithmetic tasks and rest device + arithmetic task, but a decrease in the EDMD + arithmetic task compared with the rest device + arithmetic task. Heart rate significantly increased during the rest device + arithmetic task, whereas no significant difference was observed during the EDMD + arithmetic task. The EDMD + arithmetic task significantly increased STAI scores ($p = 0.0047$), and the significant decrease in VAS indicated "unpleasant" ($p = 0.035$). These findings suggest that EDMP-induced reduction in the level of PFC activity was a response to discomfort, indicating that EDMP affects systemic function such as that of the ANS as an unpleasant stressor.

Keywords Prefrontal cortex • Near-infrared time-resolve spectroscopy • Physical stress • Mental stress • Experimentally deviated mandibular position

---

A. Amemiya (✉) • T. Takeda • K. Nakajima • K. Ishigami
Department of Sports Dentistry, Tokyo Dental College,
1-2-2 Masago, Mihama-ku, Chiba, 261-8502, Japan
e-mail: aamemiya@tdc.ac.jp

T. Tsujii • K. Sakatani
Division of Optical Brain Engineering, Department of Neurosurgery,
Nihon University School of Medicine, Tokyo, Japan

W.J. Welch et al. (eds.), *Oxygen Transport to Tissue XXXIV*, Advances in Experimental Medicine and Biology 765, DOI 10.1007/978-1-4614-4989-8_1,

1 Introduction

Many studies have investigated the effects of mental stress: reaction of the circulation and central nervous system [4] and activity in prefrontal cortex (PFC) and the autonomic nervous system (ANS) [9, 10] were examined using a mental arithmetic task. Similarly, many studies have investigated physical stress: immunoglobulin A (IgA) antibody levels in saliva [5] and hypothalamic–pituitary–adrenal response [8] during restriction stress loading were investigated in animals; reaction of the ANS to cold stress has also been investigated [3]. The relationship between mental and physical stress, however, remains to be clarified. In an earlier study, we reported on experimentally deviated mandibular position (EDMP) and arithmetic-induced elevated levels of activity in the PFC using near-infrared spectroscopy (NIRS). However, with NIRS, the average path length is unknown and assumed to be a constant value. In practice, a certain value obtained by experimental measurements is substituted in each subject. Near-infrared time-resolve spectroscopy (TRS) is an emerging method which enables the absolute hemoglobin concentration in tissue to be evaluated.

The purpose of this study was to investigate the effects of malocclusion-induced physical stress and arithmetic task-induced mental stress on PFC activity, emotion, and the ANS.

2 Materials and Methods

A total of 11 healthy male volunteers participated in the study (mean age, 24.2 ± 4.3 years). In order to avoid the influence of environmental stress, the participants were seated in a comfortable chair in an air-conditioned room with temperature and humidity maintained at approximately 23 °C and 50 %, respectively. The study was conducted in accordance with the Principles of the Declaration of Helsinki, and the protocol was approved by the Ethics Committee of Tokyo Dental College (Ethical Clearance NO.164). Written informed consent was obtained from all participants.

On day 1 the participants were required to perform a pretrial arithmetic task. After a 10-min rest, they then performed a posttrial EDMD + arithmetic or rest device + arithmetic task at random (Fig. 1.1). Activity in the PFC was measured by TRS (TRS-20, Hamamatsu Photonics K.K.). The TRS probe was placed at Fp2. The location of the probe was determined according to the international 10–20 system. Heart rate (HR) was monitored simultaneously with PFC activity by placing a pulse oximeter (WristOx, Nonin, USA) on the participant's left earlobe to measure pulse waves.

We used the 100-mm visual analogue scale (VAS) and State-Trait Anxiety Inventory (STAI) to assess psychological stress levels.

We employed a mental arithmetic task as a psychological stressor. Participants were required to mentally subtract 2 digits from 4-digit numbers displayed on a personal computer (example: 3425–79) and input the answers using a keyboard. They were instructed to make their calculations as accurately and as quickly as

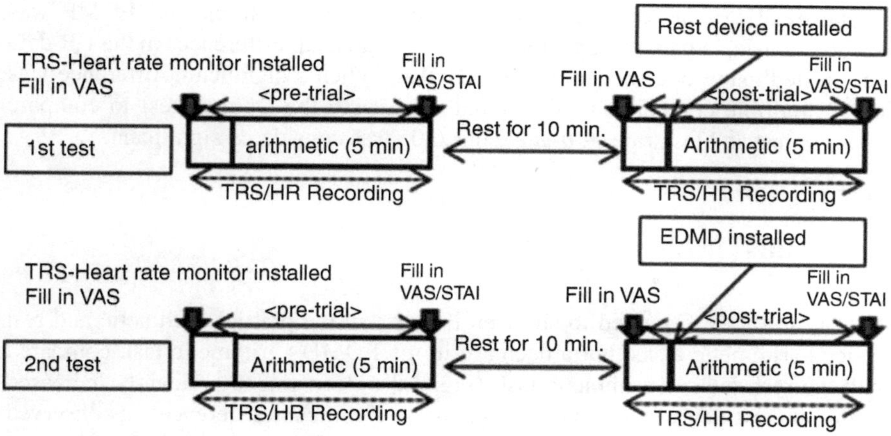

Fig. 1.1 Experimental protocol for rest device + arithmetic tasks and EDMD + arithmetic tasks to elicit TRS signals from participants. Trial consisted of baseline of 1 min; each task took 5 min. Rest device and EDMD were adopted at random

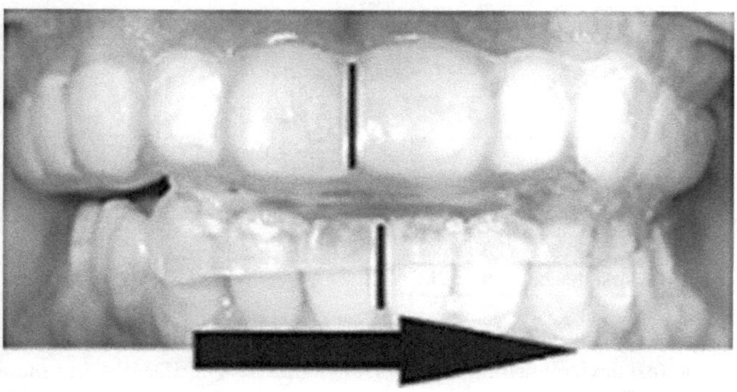

Fig. 1.2 In EDMP + arithmetic task, mandible was deviated toward the non-preferred chewing side using EDMD

possible. This mental arithmetic task has previously been used to investigate mental stress-induced PFC activity [11].

The mandible was in the mandibular rest position during the arithmetic tasks and rest device + arithmetic task. In the EDMP + arithmetic task, it was deviated toward the non-preferred chewing side using the experimentally deviated mandibular device (EDMD) (Fig. 1.2). The EDMD shifted the mandible until the cusps of the upper and lower canine teeth on the non-preferred chewing side touched each other. To minimize the influence of the EDMD itself on measurement, a resilient 3-mm Drufosoft sheet (Dreve Dentamid GmbH, Germany) was used to cover the crowns in the upper jaw and about 1/3 of the crowns in the lower jaw. The mandible was fixed in the deviated position [7]. The rest device was made of the same type of sheet, but allowed free movement.

Statistical evaluation of changes between the rest position and EDMP was performed using a paired Student's *t*-test. Only statistical differences in the HR data were tested using one-way measures ANOVA. When a significant difference was found, multiple comparisons were conducted using the Dunnett test to compare values at rest and task rate. A *p*-value of <0.05 was considered significant.

3 Results

Activity in the PFC tended to show an increase in the pretrial arithmetic and rest device + arithmetic tasks, but a decrease in the EDMD + arithmetic task compared with the rest device + arithmetic task (Fig. 1.3). Heart rate significantly increased during the rest device + arithmetic task, but no significant difference was observed during the EDMD + arithmetic task (Fig. 1.4). The EDMD + arithmetic task significantly increased STAI scores ($p = 0.0047$) (Fig. 1.5), and the significant decrease in VAS indicated "unpleasant" ($p = 0.035$) (Fig. 1.6).

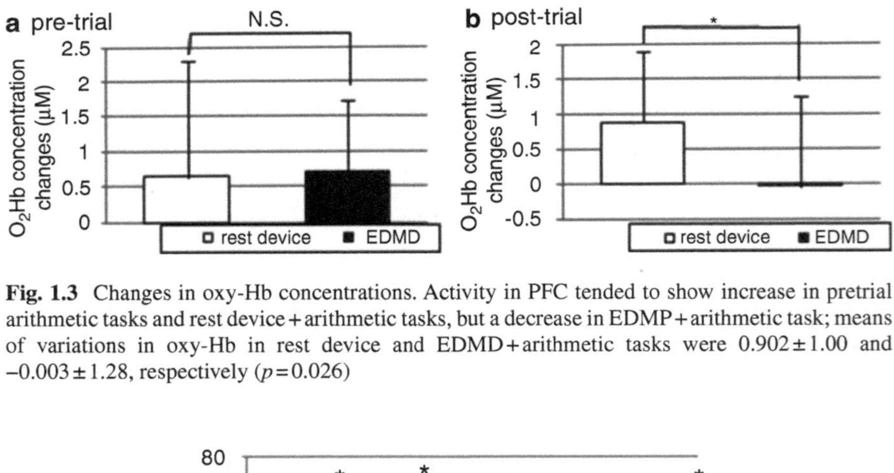

Fig. 1.3 Changes in oxy-Hb concentrations. Activity in PFC tended to show increase in pretrial arithmetic tasks and rest device + arithmetic tasks, but a decrease in EDMP + arithmetic task; means of variations in oxy-Hb in rest device and EDMD + arithmetic tasks were 0.902 ± 1.00 and -0.003 ± 1.28, respectively ($p = 0.026$)

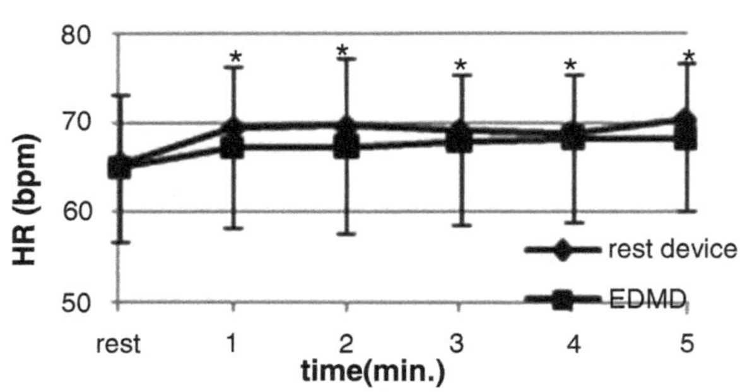

Fig. 1.4 Changes in HR concentrations. HR significantly increased in rest device + arithmetic task ($p < 0.05$), but no significant difference was observed in EDMD + arithmetic task

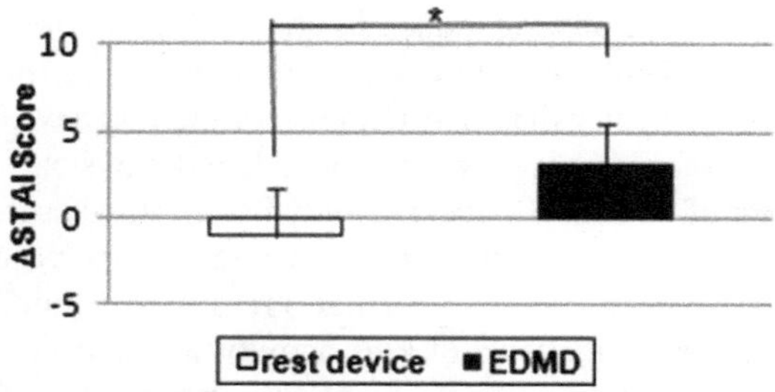

Fig. 1.5 Changes in STAI scores. EDMD + arithmetic task significantly increased STAI scores; means of variations in STAI scores in rest device and EDMD + arithmetic tasks were -1.00 ± 2.79 and 3.09 ± 2.34, respectively ($p = 0.0047$)

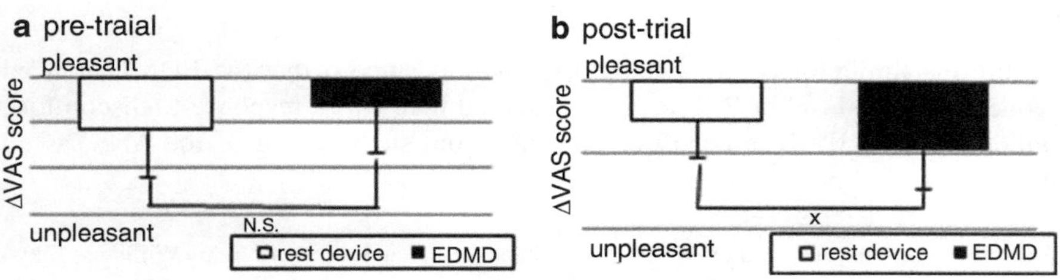

Fig. 1.6 Changes in VAS. The significant decrease in VAS indicated "unpleasant" at posttrial; means of variations in VAS in rest device and EDMD + arithmetic tasks were $-105 \pm 126$and -190 ± 117, respectively ($p = 0.035$)

4 Discussion

The results showed that activity in the PFC tended to show an increase in the arithmetic task, but when the EDMD was installed, activity in the PFC showed a significant decrease compared with activity when the rest device was used. Although HR significantly increased in the rest device + arithmetic task, no significant increase was observed in the EDMD + arithmetic task. The EDMD + arithmetic task significantly increased STAI scores, and the significant decrease in VAS indicated "unpleasant." In a previous study, Tanida et al. [10] found PFC activation during the mental arithmetic task, which reflected general cognitive operations such as attention and emotion. In the present experiment, decreased activity in the PFC during the EDMD + arithmetic task suggests that the EDMD disturbed general cognitive operations. This assumption is consistent with the findings that occlusal disharmony induced spatial memory impairment in mice [2], occlusal disharmony increased stress response [12], occlusal interference affected the ANS [1], and clenching and malocclusal were involved in emotion and/or pain-related neural processing in the brain [6].

Mental and physical stresses can induce sympathetic excitation and cause a rapid increase in HR and blood pressure. However, in this experiment, the EDMD-induced increase in HR was no greater than that with use of the rest device. This may have been due to an interaction between the mental and physical stresses applied. However, to the best of our knowledge, no study has clarified such an interaction.

The EDMP induced a significant decrease in VAS and increase in STAI scores, indicating an increase in the level of mental stress. These findings suggest that EDMP-induced decrease in the level of PFC activity was a response to discomfort, indicating that EDMP affects systemic function, such as that of the ANS, as an unpleasant stressor.

5 Conclusions

Within the limitations of this study, the findings suggest that the EDMP-induced reduction in the level of PFC activity observed here was a response to discomfort, indicating that EDMP affects systemic function, such as that of the ANS, as an unpleasant stressor.

Acknowledgments The authors would like to thank Associate Professor Jeremy Williams, Tokyo Dental College, for his assistance with the English of the manuscript. This research was partly supported by the Japan Science and Technology Agency under the Strategic Promotion of Innovative Research and Development Program and a Grant-in-Aid from the Ministry of Education, Culture, Sports, Sciences and Technology of Japan (B23300247).

References

1. Iwayama K, Ono Y, Komasa Y (2007) Influence of occlusal interference on the autonomic nervous system. Shika Igaku 70:81–90
2. Kubo KY, Yamada Y, Iinuma M, Iwaku F, Tamura Y, Watanabe K, Nakamura H, Onozuka M (2007) Occlusal disharmony induces spatial memory impairment and hippocampal neuron degeneration via stress in SAMP8 mice. Neurosci Lett 414:188–191
3. Lind L, Johansson K, Hall J (2002) The effects of mental stress and the cold pressure test on flow-mediated vasodilation. Blood Press 11:22–27
4. Liu X, Iwanaga K, Koda S (2011) Circulatory and central nervous system responses to different types of mental stress. Ind Health 49:265–273
5. Muneta Y, Yoshikawa T, Minagawa Y, Shibahara T, Maeda R, Omata Y (2010) Salivary IgA as a useful non-invasive marker for restraint stress in pigs. J Vet Med Sci 72:1295–1300
6. Otsuka T, Watanabe K, Hirano Y, Kubo K, Miyake S, Sato S, Sasaguri K (2009) Effects of andibular deviation on brain activation during clenching: an fMRI preliminary study. Cranio 27:88–93
7. Takayama K (1993) Study on the relation between stomatognathic system and the systemic condition-influence of load time at change of mandibular position on upright posture. J Jpn Prosthodont Soc 37:582–596

8. Tan Z, Nagata S (2002) Superimposed cold stress-induced hypothalamic-pituitary-adrenal response during long-duration restraint stress. J UOEH 24:361–373
9. Tanida M, Katsuyama M, Sakatani K (2007) Relation between mental stress-induced prefrontal cortex activity and skin conditions: a near-infrared spectroscopy study. Brain Res 1184:210–216
10. Tanida M, Sakatani K, Takano R, Tagai K (2004) Relation between asymmetry of prefrontal cortex activities and the autonomic nervous system during a mental arithmetic task: near infrared spectroscopy study. Neurosci Lett 369:69–74
11. Wang J, Rao H, Wetmore GS, Furlan PM, Korczykowski M, Dinges DF, Detre JA (2005) Perfusion functional MRI reveals cerebral blood flow pattern under psychological stress. Proc Natl Acad Sci U S A 102:17804–17809
12. Yoshihara T, Taneichi R, Yawaka Y (2009) Occlusal disharmony increases stress response in rats. Neurosci Lett 452:181–184

Chapter 2
Kidney EPO Expression During Chronic Hypoxia in Aged Mice

Girriso F. Benderro and Joseph C. LaManna

Abstract In order to maintain normal cellular function, mammalian tissue oxygen concentrations must be tightly regulated within a narrow physiological range. The hormone erythropoietin (EPO) is essential for maintenance of tissue oxygen supply by stimulating red blood cell production and promoting their survival. In this study we compared the effects of 290 Torr atmospheric pressure on the kidney EPO protein levels in young (4-month-old) and aged (24-month-old) C57BL/6 mice. The mice were sacrificed after being anesthetized, and kidney samples were collected and processed by Western blot analysis. Relatively low basal expression of EPO during normoxia in young mice showed significant upregulation in hypoxia and stayed upregulated throughout the hypoxic period (threefold compared to normoxic control), showing a slight decline toward the third week. Whereas, a relatively higher normoxic basal EPO protein level in aged mice did not show significant increase until seventh day of hypoxia, but showed significant upregulation in prolonged hypoxia. Hence, we confirmed that there is a progressively increased accumulation of EPO during chronic hypoxia in young and aged mouse kidney, and the EPO upregulation during hypoxia showed a similarity with the pattern of increase in hematocrit, which we have reported previously.

Keywords Erythropoietin • Aging • Mouse kidney • Hypobaric hypoxia • Hematocrit

---

G.F. Benderro
Department of Anatomy, Case Western Reserve University, Cleveland, OH 44106, USA

J.C. LaManna (✉)
Department of Physiology and Biophysics, Case Western Reserve University, 10900 Euclid Avenue, Robbins E 611, Cleveland, OH 44106-4970, USA
e-mail: joseph.lamanna@case.edu

W.J. Welch et al. (eds.), *Oxygen Transport to Tissue XXXIV*, Advances in Experimental Medicine and Biology 765, DOI 10.1007/978-1-4614-4989-8_2,

1 Introduction

In order to maintain normal cellular function, mammalian tissue oxygen concentrations must be tightly regulated within a narrow physiological range [1–5]. One of the adaptive physiological changes in response to chronic hypoxia is polycythemia (increase in hematocrit or packed red blood cell (RBC) volume) [5]. The hypoxic milieu is sensed by the kidney peritubular capillary endothelial cells which increase expression of hypoxia inducible factor-1α (HIF-1α) to enhance erythropoietin (EPO) production [6–8]. The hormone EPO is essential for maintenance of tissue oxygen homeostasis by stimulating RBCs production and promoting their survival [7, 8].

Kidneys are the major sites of EPO production in adults. EPO-expressing cells in the kidney have been identified as interstitial cells in the peritubular space in the cortex and medulla [6]. The liver produces EPO essentially in fetal life and has the potential to produce EPO and actually hepatocytes secrete a small quantity of EPO in adult life [4]. There is also paracrine EPO production in brain [2, 9] and in reproductive organs such as the uterus and testes [10].

There are reports from mice, rats, rabbits, and human studies indicating attenuated induction of HIF-1α in response to hypoxia which may result in reduction of expression of the genes with hypoxia response elements (such as EPO) in old age [11, 12]. It is also shown that many of the HIF-1α-regulated gene protein levels are also upregulated in aged rats and mouse, brain and skeletal muscles, during hypoxia by HIF-1α-independent mechanisms [11, 13]. In this study we also report upregulated EPO expression in the kidneys of aged mice during normoxia and hypoxia.

2 Methods

2.1 *Exposure to Chronic Hypobaric Hypoxia*

Male C57BL6 mice were purchased from the National Institute for Aging (NIA) and maintained at the Case Western Reserve University Animal Care Facility for a few days before the experiment. Animal housing and handling met Institutional Animal Care and Use Committee standards. Mice that were exposed to hypoxia were placed in hypobaric chambers for a period up to 3 weeks, at a constant pressure of 0.4 ATM (290 Torr, equivalent to 8 % normobaric oxygen). The littermate normoxic control mice were housed in the same room next to the hypobaric chamber to ensure identical ambient conditions.

2.2 *Tissue Collection and Preparation of Whole Cell Lysates*

Mice were weighed, anesthetized, and sacrificed. Blood samples were obtained for hematocrit determination. Kidneys were dissected and immediately frozen in liquid nitrogen and stored at −80 °C until further processing. Frozen kidney samples were dissected and homogenized in ice-cold lysis buffer (150 mM NaCl, 50 mM Tris–HCl pH 7.4, 1 mM EDTA, 1% Triton X-100, 0.5% sodium deoxycholate, 0.1% SDS, in ddH_2O) containing EDTA-free protease inhibitor tablet (Complete Mini, Roche Diagnostics, Indianapolis, IN). Homogenates were kept on ice for an hour and then centrifuged at 14,000 × *g* for 30 min at 4 °C. Protein contents in the supernatants were determined by a Bradford protein assay (Bio-Rad, Hercules, CA) with bovine serum albumin (BSA) as a standard.

2.3 *Western Blot Analysis*

Proteins from the whole cell lysates were denatured in Laemmli loading buffer, separated by using SDS gel electrophoresis and transferred to nitrocellulose membranes (Bio-Rad) by standard procedures. The membranes were incubated in 5 % skimmed milk in tris-buffered saline with 0.1 % Tween (TBS-T) for 1 h to block nonspecific binding. Then, the membranes were incubated overnight in the same blocking solution with the primary antibodies. The specific primary antibodies of interest were: EPO (1:500; Santa Cruz Biotechnology, Santa Cruz, CA) and β-actin (1:1,000; Santa Cruz, CA). The membranes were washed three times each for 15 min with TBS-T, followed by incubation with the appropriate horseradish peroxidase-conjugated secondary antibodies (Invitrogen, Camarillo, CA). After a series of three washes each for 15 min with TBS-T, immunoreactive protein bands were visualized using an enhanced chemiluminescence detection system (SuperSignal ECL Kit, Thermo Scientific, IL) and subsequent exposure of the membrane to Hyperfilm (Thermo Scientific, IL). Computer-aided Sigma Scan was used to quantify densitometry of the protein bands and normalized to that of β-actin (optical density ratio).

2.4 *Statistical Analysis*

Quantitative data are expressed as mean ± standard deviation (SD). Statistical analysis was carried out using Origin Lab Data Analysis and Graphing software version 8.0. Statistical comparisons were performed by one-way ANOVA followed by Tukey's comparison. In all cases, $p < 0.05$ was considered statistically significant.

3 Results

We quantified the kidney EPO protein level in young and aged mice to investigate whether the EPO production correlates with our previously reported hematocrit record [11]. As shown in Figs. 2.1 and 2.2, a relatively low constitutive expression of EPO during normoxia in young mice showed a significantly robust expression starting from the first day of hypoxia and stayed upregulated throughout the hypoxic

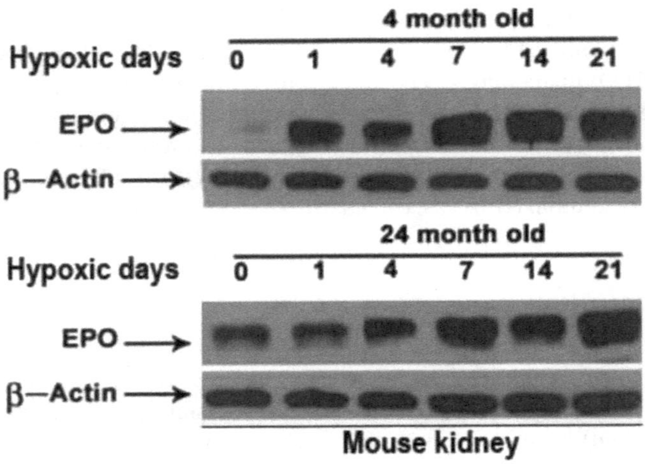

Fig. 2.1 EPO expression in the kidneys. Representative immunoblot showing rapid significant accumulation of EPO protein in young mice from the first day of hypoxia and upregulated basal level of EPO protein in aged mice in normoxia, which shows a significant increase only after the first week of hypoxia

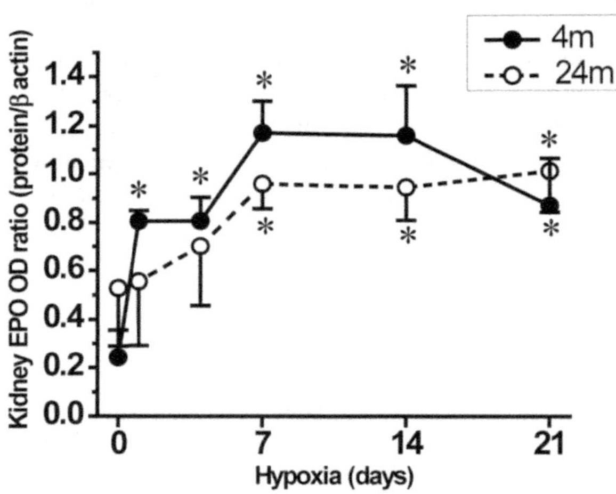

Fig. 2.2 Optical density (OD) ratio of EPO normalized to β-actin, during normoxic control (0), and 1, 4, 7, 14 and 21 days of hypoxia. Values are mean ± SD. $n = 5$ mice per time point. $^*p < 0.05$ compared with corresponding age normoxic control value

period (threefold compared to normoxic control), showing a slight decline toward the third week. Whereas, a relatively higher basal EPO protein level in aged mice did not show significant upregulation until the seventh day of hypoxia and showed steady upregulation in the second and third week.

In young mice hematocrit was significantly elevated from 46 ± 1 ($n = 19$) in normoxia to 58 ± 4 ($n = 10$) after 4 days and by 1 week (66 ± 2, $n = 10$) of hypoxia.

It continued to be higher at 2 weeks (78 ± 4, n = 10) and 3 weeks (81 ± 3, n = 16) of hypoxia, whereas in aged mice relatively reduced normoxic hematocrit (40 ± 1, n = 11) was only slightly elevated to 43 ± 1 (n = 10) after 4 days and significantly increased to 48 ± 2 (n = 11) at 1 week of hypoxia. It continued to be higher (61 ± 4, n = 10) at 2 and at 3 weeks (74 ± 5, n = 10) of hypoxia The differences in the hematocrit values between young and aged mice in normoxic control as well as throughout the end of third week of hypoxic exposure were significant ($p < 0.05$).

4 Discussion

Exposure of animals to chronic mild hypoxia results in systemic and brain adaptational changes that allow the animals to acclimatize to the environment [14]. The immediate response of oxygen deprivation (hypoxia) to mammalian organs includes a reversible increase in blood flow. In contrast, long-term responses to hypoxia involve physiological changes, metabolic regulation, and vascular remodeling [5]. Decrease in body weight, reduction in metabolism and body temperature, and increase in hematocrit are some of the systemic responses for adaptation to chronic hypoxia [5, 15]. These adaptations allow mammals to reside permanently at higher altitudes of up to 4–5 km [5]. However, in this and previous studies [1, 11, 13] we found significant differences in some of these adaptive mechanisms with aging.

Polycythemia is one of the major systemic responses of an animal to chronic hypoxia [5]. Upregulation of EPO in mammalian tissue during hypoxia has been reported [7, 16]. In this study we also confirmed that there is a progressively increased accumulation of EPO during chronic hypoxia in young and aged mouse kidney, and the pattern of EPO upregulation during hypoxia correlates with the pattern of increase in hematocrit. Elevated constitutive basal expression of EPO in normoxia in aged mice and corresponding reduced hematocrit may imply defects in the RBC lifecycle such as decreased EPO-bone marrow activity or increased destruction of the RBC in peripheral circulation. Of note, the hematocrit value depends on the balance between production and destruction of the RBC [8]. High levels of EPO observed in normoxia and hypoxia in aging mouse kidney under the condition where there is generalized attenuation of HIF-1α [1, 3] implies the presence of HIF-1α-independent EPO induction pathway, as reported from studies on brain [11–13].

Acknowledgment This study was supported by NIH R01 NS038632.

References

1. Chavez JC, LaManna JC (2003) Hypoxia-inducible factor-1alpha accumulation in the rat brain in response to hypoxia and ischemia is attenuated during aging. Adv Exp Med Biol 510:337–341
2. Chikuma M, Masuda S, Kobayashi T et al (2000) Tissue-specific regulation of erythropoietin production in the murine kidney, brain, and uterus. Am J Physiol Endocrinol Metab 279:E1242–E1248

3. Katschinski DM (2006) Is there a molecular connection between hypoxia and aging? Exp Gerontol 41:482–484
4. Koury ST, Bondurant MC, Koury MJ et al (1991) Localization of cells producing erythropoietin in murine liver by in situ hybridization. Blood 77:2497–2503
5. LaManna JC, Chavez JC, Pichiule P (2004) Structural and functional adaptation to hypoxia in the rat brain. J Exp Biol 207:3163–3169
6. Bachmann S, Le HM, Eckardt KU (1993) Co-localization of erythropoietin mRNA and ecto-5¢-nucleotidase immunoreactivity in peritubular cells of rat renal cortex indicates that fibroblasts produce erythropoietin. J Histochem Cytochem 41:335–341
7. Fisher JW (2003) Erythropoietin: physiology and pharmacology update. Exp Biol Med (Maywood) 228:1–14
8. Kendall RG (2001) Erythropoietin. Clin Lab Haematol 23:71–80
9. Rankin EB, Biju MP, Liu Q et al (2007) Hypoxia-inducible factor-2 (HIF-2) regulates hepatic erythropoietin in vivo. J Clin Invest 117:1068–1077
10. Yasuda Y, Masuda S, Chikuma M et al (1998) Estrogen-dependent production of erythropoietin in uterus and its implication in uterine angiogenesis. J Biol Chem 273:25381–25387
11. Benderro GF, LaManna JC (2011) Hypoxia-induced angiogenesis is delayed in aging mouse brain. Brain Res 1389:50–60
12. Chavez JC, Baranova O, Lin J et al (2006) The transcriptional activator hypoxia inducible factor 2 (HIF-2/EPAS-1) regulates the oxygen-dependent expression of erythropoietin in cortical astrocytes. J Neurosci 26:9471–9481
13. Ndubuizu OI, Tsipis CP, Li A et al (2010) Hypoxia-inducible factor-1 (HIF-1)-independent microvascular angiogenesis in the aged rat brain. Brain Res 1366:101–109
14. Lenfant C, Sullivan K (1971) Adaptation to high altitude. N Engl J Med 284:1298–1309
15. Mortola JP, Matsuoka T, Saiki C et al (1994) Metabolism and ventilation in hypoxic rats: effect of body mass. Respir Physiol 97:225–234
16. Ribatti D, Vacca A, Roccaro AM (2003) Erythropoietin as an angiogenic factor. Eur J Clin Invest 33:891–896

Chapter 3
Nature's "Silver Bullet" for Anticoagulation: Mechanism of Zymogen Protein C to Activated Protein C

Duane F. Bruley and Michael B. Streiff

Abstract We have defined the Zymogen Protein C (ZPC) to Activated Protein C (APC) process as the "silver bullet" of blood anticoagulation. This definition suggests that the anticoagulation activity occurs when and where it is needed, resulting in local anticoagulation without enhanced bleeding. It is important for man to be able to manufacture an inexpensive ZPC product or to find a substitute drug to duplicate one of God's natural anticoagulant/antithrombotic processes, in vivo, in human blood. After intense research and at great expense scientists have not been able to produce a safe anticoagulant. All products that are now being used can cause bleeding even if dosing is carefully monitored. In fact many professionals in the health care and the pharmaceutical industries define an anticoagulant as a drug that "does" cause bleeding. This results in a large financial burden that has been placed on the health care industry because of necessary emergency treatments for dangerous occurrences. In addition, many patients are dying annually due to internal and external bleeds created or enhanced by presently administered anticoagulants. Since there are no safe drugs available it is necessary to use the existing products when a medical condition calls for an anticoagulant. This paper will discuss the ZPC process and why its mechanistic design is one of nature's unique defenses against unwanted blood clotting. The prevention and lysis of clots allows normal blood flow and therefore results in the required tissue oxygenation for cell function and survival. If clinical research is carried out with great care it could uncover other uses of ZPC that will allow safer medical procedures, in addition to its use with standard PC deficiency cases. An important example might be for some brain surgeries where the use of existing anticoagulants is unsafe because of potential bleeds.

D.F. Bruley (✉)
Synthesizer, Inc., 2773 Westminster Road, Ellicott City, MD 21043, USA
e-mail: Bruley@umbc.edu

M.B. Streiff
Division of Hematology, Department of Medicine, Anticoagulation Management Service, Johns Hopkins Medical Institutions, Baltimore, MD 21205, USA

W.J. Welch et al. (eds.), *Oxygen Transport to Tissue XXXIV*, Advances in Experimental Medicine and Biology 765, DOI 10.1007/978-1-4614-4989-8_3,

Clinical research could reveal an efficacious ZPC level (for instance, 125, 150, or 200% of normal) that would prevent dangerous clotting situations from occurring without unnecessary bleeding.

Keyword Protein C

1 Introduction

International Society on Oxygen Transport to Tissue (ISOTT) research focuses on all processes of normal and pathological oxygen transport within the blood and tissue [1, 2]. Foremost of interest is blood hemostasis to ensure optimal tissue oxygenation. This includes the coagulation cascade, particularly the significance of natural anticoagulants and substitutes for them in cases of pathology. The phenomenon of "blood sludging" (agglutination) was first observed by Dr. Melvin H. Knisely via his quartz rod crystal optical technology [3]. Our initial studies included mathematical modeling of oxygen transport in the microcirculation along with polarographic microelectrode oxygen sensing [4–6]. This work was complemented with studies on anti-adhesive drugs to enhance tissue oxygenation [7].

We have defined the in vivo Zymogen Protein C (ZPC) to Activated Protein C (APC) process as the "silver bullet" of blood anticoagulation. This definition suggests that the anticoagulation mechanism occurs when and where it is needed, resulting in local anticoagulation without enhanced bleeding.

Historically, PC was not so named until Stenflo isolated the protein from bovine plasma and labeled it Protein C because it was the third peak in the chromatograph elution [8]. Originally, APC was coined autoprothrombin IIA (APIIA) because it was believed to be derived from prothrombin [9]. Twelve years later, it was shown that APIIA's precursor was not only distinct from prothrombin but also it inhibited the factor Xa-mediated activation of prothrombin as well as facilitating fibrinolysis [10]. Shortly after Stenflo isolated the bovine PC, Seegers verified it to be identical to APIIA [11]. Three years later, Kisiel was able to purify human PC [12].

ZPC is the pivotal anticoagulant and antithrombotic in the human blood coagulation cascade [13]. PC is a glycoprotein with a molecular weight of 62,000 Da. Human ZPC is synthesized in the liver as a single chain precursor and circulates in the blood primarily as a two chain inactive zymogen until it is activated by proteolytic cleavage. The protein is a serine protease that requires vitamin K for normal biosynthesis [13]. It is a member of the vitamin K-dependent (VKD) family also consisting of coagulation proteins such as factors VII, IX, proteins S and Z, and prothrombin.

ZPC is a trace protein in human blood with a concentration of 4 μg/mL [12, 14, 15]. Serious problems can occur when the PC level in the blood is lowered. For example, patients deficient in PC are at risk of deep vein thrombosis (DVT) [16] and other clotting complications, resulting in tissue oxygen deprivation, some of which can be life threatening. When these blood clots break away from the surface of the vein and enter the blood stream, they will induce strokes, heart attacks, and pulmonary embolisms. Venous thrombosis is one of the most frequent complications in medical patients [17]. It is the most common cause of death in patients undergoing major orthopedic operations. In the United States it has been estimated that 300,000 hospitalizations and 50,000 deaths occur [17], and this amounts to millions and possibly billions of dollars in medical expenses annually.

At the present time heparin and Coumadin are used to treat ZPC deficiency and other hypercoagulable conditions. The disadvantage of these drugs is that both have dangerous side effects. Excessive internal bleeding is a major problem that can result from their use, possibly causing a stroke or major organ failure. Additionally, Coumadin-induced skin necrosis and heparin-induced skin necrosis can lead to amputation of extremities and death. Also, pregnant women cannot use Coumadin.

Previous animal testing [18–21] and clinical trials indicate that PC is an effective anticoagulant/antithrombotic for many medical indications without harmful side effects. This unique feature is a function of the mechanistic behavior of PC in the body. Differing from all other anticoagulants, PC circulates the blood in an inactive form and is activated only at the site where it is needed and at the time when it is needed. Also, ZPC can be used at greater than normal blood concentrations without bleeding complications.

PC has many clinical applications. Not only can it be used to treat genetically deficient patients, but it can also be used to treat septic shock [22], hip and knee replacement patients, Coumadin-induced skin necrosis patients, heparin-induced thrombocytopenia, patients doing fibrinolytic therapy, and patients undergoing angioplasty or suffering from unstable angina, etc. [23]. Additionally, research has shown that the use of safe anticoagulants could lower the rate of strokes in the USA from 80,000 to 40,000 per year, reduce patient complications, and save the medical industry an estimated $600 million per year [24].

ZPC concentrate has been shown to be successful for the prevention and treatment of thrombosis in individuals with inherited and acquired PC deficiency and to avoid the problems associated with fresh and frozen plasma administration [25–28]. When one considers that PC is the only known anticoagulant/antithrombotic without bleeding side effects and 1 in 300 people are hereditary PC deficient [17], it is easy to see the enormous benefit of having inexpensive PC available to medical patients. Additionally, there are innumerable patients with acquired PC deficiency.

2 The ZPC to APC Process and APC Function

Although very complex, three main pathways are involved in regulating coagulation [29]. One pathway that utilizes heparin-like molecules and antithrombin III results in the inhibition of coagulation proteases. A second pathway, referenced as lipoprotein associated coagulation inhibitor, or extrinsic pathway inhibitor, blocks the activity of factor VIIa–tissue factor complex [30]. The third and most important pathway involves APC which neutralizes factors Va and VIIIa [13]. These major pathways function together to inhibit both the proteases and regulatory proteins (cofactors) of the clotting system. It has been shown through clinical studies that antithrombin III, protein C, and factor S deficiencies all exhibit thrombotic complications [16, 31].

Although inefficient, PC can be activated by thrombin alone. This activation process can be enhanced by a factor of at least 1,000 via a complex between thrombin and a membrane protein called thrombomodulin on the surface of endothelial cells [13, 32]. The activated PC is a potent serine protease that regulates blood coagulation by forming a complex with Protein S (PS) on both endothelial and platelet surfaces. This deactivates factors Va and VIIIa, thus preventing generation of the enzymes factors Xa and thrombin [33, 34].

Cofactor PS circulates in the blood as a free agent and in complex with C4bBP which is a regulatory protein of the complement system. The PS–C4bBP complex is not functional as a cofactor for activated PC in factor Va inactivation and therefore downregulates the effectiveness of PC in the inactivation of Va and VIIIa. Va is required to produce thrombin. Once thrombin is produced, it activates factor I (fibrinogen) to form fibrin, which synthesizes a soft clot. VIIIa is then required to solidify the clot. So by inactivating Va and VIIIa, you down regulate the clotting process.

A component of the process that complements local anticoagulation includes the relative half-lives of ZPC vs. APC. ZPC circulates in the body with an approximate 8 h half-life and as needed, is activated to APC with a half-life of approximately 10 min [13]. The immediate use of APC along with its short half-life minimizes unwanted bleeding.

3 ZPC Production

ZPC is currently obtained from human blood plasma [35–38], cell culture [39], and transgenic animal milk [40]. Human blood plasma is a limited source of PC and only approximately 12 kg PC could be produced if the total blood plasma in the USA was used for PC production. Because of the difficulty of the synthesis of the correct biological activity, few mammalian cells can be used and the production is very expensive because of high substrate cost and low yield. Genetic engineering of animals is an alternative method to produce large quantities of human proteins. The Virginia Polytechnic Institute (VPI) and the American Red Cross (ARC) have conducted research to produce ZPC at concentrations as high as 1 mg/mL in transgenic swine milk40 with 95 % PC purity. However, the total recovery is very low. Almost half of the PC is lost in the step for PC separation from caseins by PEG precipitation [41].

The two existing protein C products (Ceprotin, Baxter International; Xigris, Eli Lilly) are so expensive that they are rarely used even in cases where patient survival is in question. The zymogen could be used for prophylactic treatment of protein C deficiency, as well as other disease states or in medical procedures. There are at least four significant examples of perioperative procedures utilizing ZPC to prevent clotting without enhancing bleeding [42–45]. These examples establish a need to produce inexpensive ZPC. The methods might include isolation of human PC from blood and from transgenic animals [40], or via recombinant DNA technology utilizing cell cultures. Future work utilizing molecular design might produce anticoagulant molecules that can mimic the APC/ZPC mechanism. Isolation of ZPC from Cohn Fraction IV-1 of blood plasma is complicated by the fact that ZPC is a trace protein in blood and there are other VKD homologous blood factors. One possibility of less expensive production is to develop a process utilizing immobilized metal affinity chromatography (IMAC) instead of the presently used immuno affinity chromatography technology [35–37, 46, 47]. Another major bioprocess engineering problem is the administration of ZPC. Since ZPC is a parenteral drug, more patient compliant administration technologies will have to be developed for direct injection into the blood stream or through subcutaneous depots.

4 Conclusion

Protein C deficiency is a hidden disease. Therefore patients are unaware of their condition until a potentially deadly situation occurs. This is also true for patients with protein S deficiency, antithrombin III deficiency, and Factor V Leiden. For these cases, it is important that research be continued to find an effective anticoagulant that does not cause bleeding. This is a difficult task because the protein C mechanism is very complex. However, it is nature's pivotal anticoagulant/antithrombotic process and it does not cause dangerous bleeds. Through research and development, investigators should design a process to recover inexpensive ZPC or discover another anticoagulant that mimics the ZPC/APC mechanism.

Acknowledgments The authors would like to express their appreciation to E. Eileen Thiessen for the preparation of the power point slide presentation and the production of this paper.

References

1. Bruley DF (2012) The history of ISOTT. In: Wolf M, Bucher HU, Rudin M et al (eds) Oxygen transport to tissue XXXIII, Adv Exp Med Biol, Springer, New York 737:1–9
2. Bruley DF (2008) ISOTT: roots, founding and beyond. In: Kang KA, Harrison DK, Bruley DF (eds) Oxygen transport to tissue XXIX, Adv Exp Med Biol. Springer, New York, 614:1–8
3. Goro FW (1948) Blood sludge. Life Mag 24(22):49–59
4. Thews G (1960) Oxygen diffusion in the brain. A contribution to the question of the oxygen supply of the organs. Pflugers Archiv 271:197–226
5. Reneau DD, Bruley DF, Knisely MH (1967) A mathematical simulation of oxygen release, diffusion and consumption in the capillaries and tissue of the human brain. In: Hershey D (ed) Chemical engineering in medicine and biology. Plenum, New York, pp 135–241
6. Artigue RS, Bruley DF (1983) The transport of oxygen, glucose, carbon dioxide and lactic acid in the human brain: mathematical models. In: Oxygen transport to tissue IV, Adv Exp Med Biol. Plenum 159:1–4

7. Bicher HI, Bruley DF, Knisely MH (1973) Anti-adhesive drugs and tissue oxygenation. In: Bruley DF, Bicher HI (eds) Adv Exp Med Biol. Plenum 37:657–667
8. Stenflo J (1976) Á new vitamin K-dependent protein. J Biol Chem 251:355–363
9. Mammen EF, Thomas WR, Seegers WH (1960) Activation of purified prothrombin to autoprothrombin II (platelet cofactors II or autoprothrombin IIA). Thromb Diath Haemorrh 5:218–249
10. Marcianik E (1972) Inhibitor of human blood coagulation elicited by thrombin. J Lab Clin Med 79:924–934
11. Seegers WH, Novoa E, Henry RL et al (1976) Relationship of "new" vitamin K-dependent Protein C and "old" autoprothrombin II-a. Thromb Res 8(5):543–552
12. Kisiel W (1979) Human plasma protein C: isolation, characterization and mechanism of activation by a thrombin. J Clin Invest 4:761–769
13. Esmon CT (1989) The roles of protein C and thrombomodulin in the regulation of blood coagulation. J Biol Chem 264(9):4743–4746
14. Fernlund P, Stenflo J (1982) Ámino acid sequence of the light chain of bovine protein C. J Biol Chem 257(20):12170–12179
15. Kisiel W, Carnfield WM, Ericsson LH et al (1977) Ánticoagulant properties of bovine plasma protein C following activation by thrombin. Biochemistry 16(26):5824–5831
16. Clouse LH, Comp PC (1986) The regulation of hemostasis: the protein C system. N Engl J Med 314(20):1298–1304
17. Bertina RM (1988) Protein C and related proteins: biochemical and clinical aspects (contemporary issues in haemostasis and thrombosis). Churchill Livingstone, New York
18. Colucci M (1984) Influence of protein C activation on blood coagulation and fibrinolysis in squirrel monkey. J Clin Invest 74:200–204
19. Comp P, Esmon C (1981) Generation of fibrinolytic activity by infusion of activated protein C into dogs. J Clin Invest 68(5):1221–1228
20. Gruber A, Griffin JH, Harker LA et al (1989) Inhibition of platelet-dependent thrombosis formation by human activated protein C in a primate model. Blood 73:639–642
21. Okijima K, Koga S, Kaji M et al (1990) Effect of protein C and activated protein C on coagulation and fibrinolysis on normal human subjects. Thromb Haemostasis 3:48–53
22. Sharma GR, Grinnell BW et al (2008) Activated protein C modulates chemokine response and tissue injury in experimental sepsis. In: Kang KA, Harrison DK, Bruley DF (eds) Oxygen transport to tissue XXIX, Adv Exp Med Biol 614:83–91
23. Bruley DF, Drohan WN (1990) Protein C and related anticoagulants. Advances in applied biotechnology series, volume 11. Gulf Publishing Company (Portfolio Publishing Company), Houston, TX.
24. Winslow R (1995) Increased use of blood-thinning drugs could prevent 40,000 strokes a year, Wall St. J Sep 8
25. Vukovich T, Auerger K, Weil J et al (1988) Replacement therapy for a homozygous protein C deficiency-state using a concentrate of human protein C and S. Br J Haematol 70(4):435–440
26. Dreyfus M, Magny JF, Bridey F et al (1991) Treatment of homologous protein C deficiency and neonatal purpura fulminans with a purified protein C concentrate. N Engl J Med 325(22):1565–1568
27. Manco-Johnson M, Nuss R (1992) Protein C concentrate prevents peripartum thrombosis. Am J Hematol 40(1):69–70
28. Marlar RA, Montgomery RR, Broakman AW (1989) Diagnosis and treatment of homozygous protein C deficiency Report of the working party on homozygous protein C deficiency of the subcommittee on protein C and protein S, International Committee on Thrombosis and Haemostasis. J Paediatrics 114(4 Pt 1):528–534
29. Esmon CT (1990) Regulation of coagulation: the nature of the problem. In: Bruley DF, Drohan WN (eds) Protein C and related coagulants. Gulf, Houston, TX, p 3
30. Rapaport SI (1989) Inhibition of factor VIIa/tissue factor-induced blood coagulation: with particular emphasis upon a factor Xa-dependent inhibitory mechanism, Blood 73(2):359–365
31. Rosenberg R, Marcum JA (1985) Heparin-like molecules and thrombotic disease. ASAIO J 8:215

32. Zushi M, Gomi K, Yamamoto S et al (1989) The last three consecutive epidermal growth factor-like structures of human thrombomodulin comprise the minimum functional domain for protein C-activating cofactor activity and anticoagulant activity. J Biol Chem 264(18):10351–10353
33. Walker FJ, Sexton PW, Esmon CT (1979) The inhibition of blood coagulation by activated protein C through the selective inactivation of activated factor V. Biochim Biophys Acta 571(2):333–342
34. Fulcher CA, Gardiner JE, Griffin JH et al (1984) Proteolytic inactivation of human factor VIII procoagulant protein by activated human protein C and its analogy with factor V. Blood 63(2):486–489
35. Wu H, Bruley DF, Kang KA (1998) Protein C separation from human blood plasma Cohn fraction IV-1 using immobilized metal affinity chromatography. In: Hudetz AG, Bruley DF (eds) Oxygen transport to tissue XX, Adv Exp Med Biol, Plenum, New York, 454:697–704
36. Wu H, Bruley DF (1999) Homologous blood protein separation using immobilized metal affinity chromatography: protein c separation from prothrombin with application to the separation of factor IX and prothrombin. Biotechnol Prog 15(5):928–931
37. Wu H, Bruley DF (2002) Chelator, metal ion and buffer studies for protein c separation. Comp Biochem Physiol A Mol Integr Physiol 132(1):213–220
38. Tadepalli SS, Bruley DF, Kang KA et al (1997) Immobilized metal affinity chromatography process identification and scale-up for protein C production. In: Harrison DK, Delpy DT (eds) Oxygen transport to tissue XIX, Adv Exp Med Biol, Plenum, New York, 428:31–43
39. Grinnell B. Executive Director Biotechnology Research, Eli Lilly and Company
40. Velander WH, Johnson JL, Page RL et al (1992) High-level expression of a heterologous protein in the milk of transgenic swine using the cDNA encoding human protein C. Proc Natl Acad Sci U S A 89(24):12003–12007
41. Drohan WH, Wilkins TD, Lattime E et al (1994) A scalable method for the purification of recombinant human protein C from the milk of transgenic swine. Adv Bioprocess Eng, Kluwer Academic Publishers, Netherlands, 501–507
42. Bruley DF (2009) Zymogen protein c concentrate for safer heterozygote surgery, I am a guinea pig! Oxygen Transport to Tissue XXX, Adv Exp Med Biol, Springer, New York, 645:115–121
43. Bruley DF, Mears SC, Streiff MB (2010) Safer surgery using zymogen protein C Concentrate. In: Takahashi EJ, Bruley DF (eds) Oxygen transport to tissue XXXI, Adv Exp Med Biol, Springer, New York, 662:439–445
44. Bruley DF, Jagannath SB, Streiff MB (2011) Zymogen protein C to prevent clotting without bleeding during invasive medical procedures. In: LaManna JC, Puchowicz MA, Xu K et al. (eds) Oxygen transport to tissue XXXII, Adv Exp Med Biol, Springer, New York, 701:91–97
45. Thakkar SC, Streiff MB, Bruley DF et al (2010) Perioperative use of protein c concentrate for protein c deficiency in THA. Clinical orthopaedics and related research. Springer, NY, 468(7):1986–1990
46. Thiessen EE, Bruley DF (2003) Theoretical studies of IMAC interfacial phenomena for the purification of protein C. Oxygen transport to tissue XXV, Adv Exp Med Biol 540:183
47. Lee J, Thiessen EE, Bruley DF (2005) Anticoagulant/anti-inflammatory/antithrombotic protein C production-metal ion/protein interfacial interaction in immobilized metal affinity chromatography. Oxygen transport to tissue XXVI, Adv Exp Med Biol, Springer, NY, 566:381–387

Chapter 4
Canonical Correlation Analysis in the Study of Cerebral and Peripheral Haemodynamics Interrelations with Systemic Variables in Neonates Supported on ECMO

Alexander Caicedo, Maria D. Papademetriou, Clare E. Elwell, Aparna Hoskote, Martin J. Elliott, Sabine Van Huffel, and Ilias Tachtsidis

Abstract Neonates supported on extracorporeal membrane oxygenation (ECMO) are at high risk of brain injury due to haemodynamic instability. In order to monitor cerebral and peripheral (muscle) haemodynamic and oxygenation changes in this population we used a dual-channel near-infrared spectroscopy (NIRS) system. In addition, to assess interrelations between NIRS and systemic variables, collected simultaneously, canonical correlation analysis (CCA) was employed. CCA can quantify the relationship between a set of variables and assess levels of dependency. In four out of five patients, systemic variables were found to be less inter-related with cerebral rather than peripheral NIRS measurements. Moreover, during ECMO flow manipulations, we found that the interrelation between the systemic and the NIRS cerebral/peripheral variables changed. The CCA method presented here can be used to assess differences between NIRS cerebral and NIRS peripheral responses due to systemic variations which may be indicative of physiological differences in the mechanisms that regulate oxygenation and/or haemodynamics of the brain and the muscle.

Keyword ECMO

The original version of this chapter was revised. An erratum to this chapter can be found at DOI 10.1007/978-1-4614-4989-8_53

A. Caicedo • S. Van Huffel
ESAT/SCD, Department of Electrical Engineering & IBBT Future Health Department, Katholieke Universiteit Leuven, Leuven, Belgium

M.D. Papademetriou • C.E. Elwell • I. Tachtsidis (✉)
Department of Medical Physics and Bioengineering, University College London, 18 Dickensons Lane, London, WC1E 6BT, UK
e-mail: iliastac@medphys.ucl.ac.uk

A. Hoskote • M.J. Elliott
Great Ormond Street Hospital for Children, London, UK

W.J. Welch et al. (eds.), *Oxygen Transport to Tissue XXXIV*, Advances in Experimental Medicine and Biology 765, DOI 10.1007/978-1-4614-4989-8_4,

1 Introduction

Extracorporeal membrane oxygenation (ECMO) is a life support system for patients with intractable cardio-respiratory failure. Neonates supported on ECMO often suffer from periods of haemodynamic instability, hypoxia and/or hypercapnia. In addition, the ECMO procedure itself may cause physiological changes due to ligation of the major neck vessels, heparinization and haemodilution, which can cause alterations in cerebral blood flow and potentially disrupt autoregulation [1]. Consequently, ECMO patients have increased risk for brain injury with reported abnormal neuroimaging ranging from 28 to 52%, depending on the imaging technique used [2].

Several studies have described changes in the cerebral haemodynamics before, during and after ECMO procedure. Liem et al. [1] reported that mean arterial blood pressure (MABP), arterial oxygen saturation (SaO_2) and partial pressures of oxygen and CO_2 measured transcutaneously were some of the variables that better explained changes in cerebral total haemoglobin (HbT) measured by NIRS. Ejike et al. [3] reported that the regional cerebral oxygenation presented a negative correlation with arterial partial pressure of CO_2 (pCO_2) and no significant correlation with changes in ECMO flow. Papademetriou et al. [4] used dual-channel NIRS system during ECMO flow changes and reported the presence of low frequency oscillations (<0.1 Hz) in peripheral oxyhaemoglobin (HbO_2), which are not present in cerebral HbO_2, demonstrating differences between cerebral and peripheral haemodynamics in this patient group.

Several studies have investigated the relationship between spontaneous changes in MABP and cerebral NIRS signals as assessment of brain autoregulation [5–7]. Brady et al. [6] investigated the correlation between NIRS and MABP in paediatric patients undergoing cardiac surgery with cardiopulmonary bypass for correction of congenital heart defects. They found an association between hypotension during cardiopulmonary bypass and impairment of autoregulation. We have also previously [7] studied the relation between MABP and haemoglobin difference ($HbD = HbO_2 - HHb$, oxy minus reduced haemoglobin) and tissue oxygenation index ($TOI = HbO_2/HbO_2 + HHb$) by means of correlation, coherence and partial coherence analysis, and its use in clinical outcome prediction; although higher values were found in the population with adverse clinical outcome, indicating a stronger relation between MABP and HbD/TOI, no strong evidence was established. However, ECMO is a complex procedure and study of the interrelation of haemodynamic variables, only, with MABP may not be sufficient.

In this study we describe the use of canonical correlation analysis (CCA) to investigate the differences between the interrelations in cerebral and peripheral NIRS measurements with the systemic variables in ECMO patients. In our analysis the systemic variables were defined as the independent dataset, while the cerebral and peripheral NIRS measurements were defined as dependent variables.

2 Methods

CCA is a statistical method that analyzes the interrelation between variables in multidimensional datasets. CCA can be seen as an extension to normal correlation analysis, in which the proximity between two multidimensional datasets, instead of vectors, is analyzed by means of canonical angles [8]. CCA determines how strongly the variables in both datasets are related. It is also possible to determine which and how many of the independent variables explain most of the variation in the dependent dataset.

Measurements from five subjects (ranging from 1 to 1,825 days) on veno-arterial (VA) ECMO procedure were used in this study. A dual-channel near-infrared system (NIRO 200, Hamamatsu Photonics KK) was used to measure the changes in HbO_2, HHb and TOI using spatially resolved spectroscopy. From these signals HbD and total haemoglobin changes ($HbT = HbO_2 + HHb$) were calculated and used, together, with TOI for further analysis. NIRS data were collected at a frequency of 6 Hz. Channel 1 was placed on the forehead in order to assess cerebral NIRS changes, while channel 2 was placed on the calf to assess peripheral NIRS changes. A full set of systemic data including MABP, central venous pressure (CVP), end-tidal carbon dioxide pressure ($EtCO_2$), heart rate (HR), respiration rate (RR), core and skin temperatures and SaO_2 were continuously measured in real time at the bedside (IntelliVue MP70, Philips Medical). All signals were down-sampled to 1 Hz and artefacts were removed manually by means of interpolation. Figure 4.1 shows an example of the systemic and NIRS measurements from one neonate. Measurements were done during stepwise changes in the ECMO flow; the flow was reduced from baseline (100% ECMO flow) in steps of 10%, approximately every 10 min, until 70% of the baseline ECMO flow was reached, afterwards the flow was increased back to baseline following the same profile. In cases where the patients could not accommodate a 30% reduction in ECMO flow it was only reduced by a total of 20%. The interrelations between the set of peripheral and cerebral NIRS changes with the systemic variables were studied using two different approaches. The first approach used the complete measurement period and the ECMO flow when available as a parameter in the analysis. In the second approach, the signals were segmented in epochs of constant ECMO flow and the methods were applied separately to each epoch. In addition, in order to normalize the results to be comparable between patients, we estimated the ratio between the percentage of variance in the peripheral NIRS explained by the systemic variables and the percentage of variance of the cerebral NIRS explained by the systemic variables. We call this index the peripheral to cerebral haemodynamic ratio (PCHR). PCHR can be used to quantify the differences in the interrelations between both cerebral and peripheral circulation mechanisms versus systemic variations. PCHR values lower than 1 indicate that variations in the systemic variables are more likely to be reflected in the muscle than in the brain.

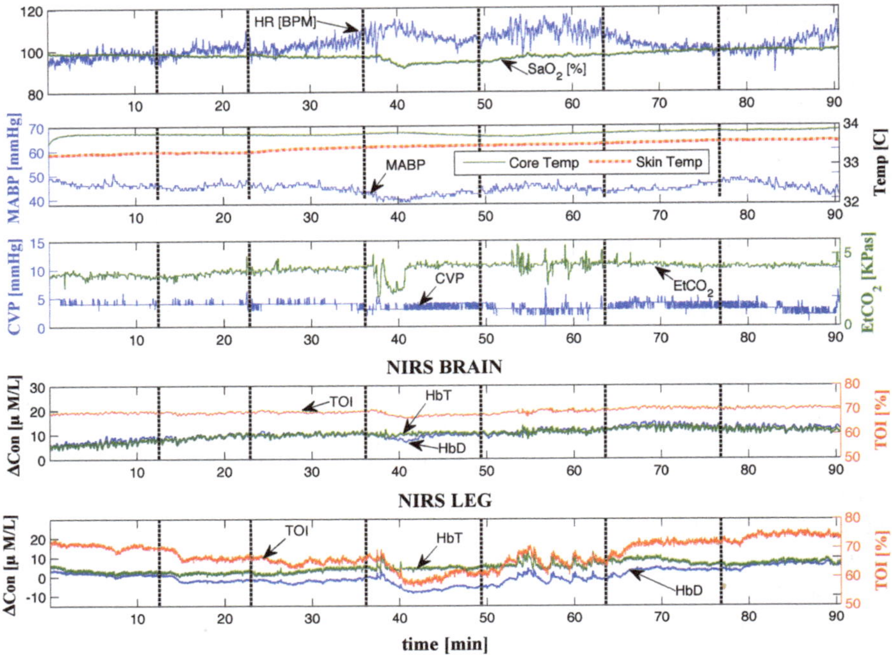

Fig. 4.1 Systemic and haemodynamics measurements recorded from patient 3. The *dashed vertical lines* represent the changes in ECMO flow

3 Results

Table 4.1 shows PCHR for different ECMO flows and for the full measurement period. Results for the full measurement period show that only patient 4 presented a PCHR > 1, indicating that the variations in the systemic variables were more likely to be reflected in the brain NIRS measurements rather than the peripheral in that patient; in the other patients the peripheral NIRS changes are more likely to be affected by variations in the systemic variables. In addition, patient 3, in contrast with the other patients, was the only one who, at every ECMO flow rate, consistently presented PCHR < 1 and as it happened was the only patient in our group that the clinicians were confident to reduce the flow down to 70% from baseline. Table 4.2 shows which systemic variables contributed more to the changes in cerebral/peripheral NIRS variables, when analyzing the complete measurement period. HR and skin temperature were the systemic variables that most affected the cerebral and/or peripheral haemodynamic variables in the patients. MABP, CVP, skin temperature, SaO_2 and ECMO affected the NIRS variables to a lesser extent.

Table 4.1 Ratio between the percentage of the variance in the cerebral haemodynamics and the peripheral haemodynamics explained by variations in the systemic variables (PCHR), for different ECMO flow percentages and the full measurement period

		ECMO flow percentage							
	Age (days)	100%	90%	80%	70%	80%	90%	100%	Full period
Patient 1	180	0.56	0.76	1.19	–	–	1.59	1.00	0.85
Patient 2	1	0.68	1.55	1.61	–	–	0.55	–	0.82
Patient 3	2	0.91	0.95	0.49	0.89	0.91	0.47	0.38	0.91
Patient 4	4	0.66	1.68	0.73	–	–	0.71	1.00	1.21
Patient 5	1,825	0.75	1.42	–	–	–	–	1.33	0.82

A PCHR lower than 1 indicates that the variations in systemic variables are more likely to be reflected in the muscle than in the brain. Spaces marked with a dash (–) indicate that it was not possible to perform the analysis due to the lack of measurements

Table 4.2 Systemic variables that presented a correlation higher than 0.5, in absolute value, with cerebral/peripheral haemodynamic variables represented by a C&P sign, respectively

Patient	HR	MABP	CVP	CoreT	SkinT	SaO_2	RR	$EtCO_2$	Flow
1	C	C&P	C		P				–
2	C&P	C	C	C	C&P	P	C&P	–	C&P
3	P	P		C&P	C&P	C&P	–	C	C&P
4	C&P		P	–	C&P	–			C&P
5	C&P	C	C&P		C&P	C&P			–

The blank spaces indicate no correlation, and a dash (–) indicates that the parameter was not used in the analysis due to lack of measurements or distortion by artefacts

4 Discussion

PCHR < 1, but close to 1, were observed in four out of five patients when analyzing the full measurement period. Earlier studies on neonates supported on ECMO indicate that autoregulation may be disrupted [1]; this can be the cause of PCHR values close to 1 as the brain haemodynamic and oxygenation changes will respond passively to systemic variations. In addition, when reducing ECMO flows it is expected that the peripheral circulation will be more affected by systemic changes than the brain circulation. At baseline level (100% ECMO flow) all patients reported a PCHR < 1; when reducing the flow to 90% three out of five patients reported PCHR > 1, indicating that systemic changes were more reflected in the cerebral circulation than in the peripheral circulation, while patient 5 could not accommodate extra reductions in the flow and was returned to baseline level. At 80% flow two out of four patients presented PCHR > 1; furthermore, in this stage, three of the four patients could not accommodate more reductions in the flow and were returned to baseline level. Only patient 3 was able to accommodate a reduction to 70% in the ECMO flow and was the only patient who consistently presented PCHR < 1. When returning ECMO flow to baseline level, three out of four patients presented

PCHR ≥ 1. These results suggest that at low ECMO flows cerebral circulation is more vulnerable to changes in systemic variables and this effect is apparent in the most vulnerable patients.

In the population studied HR and skin temperature were the variables that most affected the cerebral and peripheral NIRS signals. Out of five patients, MABP was correlated with cerebral and peripheral haemodynamic variables in three and two patients, respectively. Out of four patients, SaO_2 was correlated with cerebral and peripheral haemodynamic variables in three and two patients, respectively. Out of four patients $EtCO_2$ was correlated with the cerebral haemodynamic variables in one patient. Furthermore, ECMO flow presented a strong correlation with cerebral and peripheral NIRS variables in all three patients with flow measurements. The lack of homogeneity in relation to the systemic variables that affect the cerebral or peripheral circulation in our study suggests differences in the clinical condition of each patient; however, due to the small patient numbers it is difficult to present a clinical hypothesis. Interestingly, and in contrast with these results, Tisdall et al. [9] in healthy adults found that changes in SaO_2 and $EtCO_2$ make a large contribution to changes in the cerebral NIRS TOI signal. In addition, CCA indicated that variation in the skin temperature strongly affects the cerebral and peripheral NIRS changes; conversely, Harper et al. [10] reported that a change in core temperature in adults can produce changes in blood flow due to changes in blood viscosity and metabolic rate (among other reasons). Whilst, Davis et al. [11] reported that changes in temperature produce changes in skin blood flow that can have a significant impact in the NIRS derived signals; therefore, special care should be taken when analyzing these results.

Several factors should be taken into account before interpreting the results provided by CCA. Among the limitations, the length of the signal under analysis and the presence of noise, non-linearities and nonstationarity can be cited. In order to overcome some of these problems, and compare the results between patients, normalization such as the PCHR ratio should be used; otherwise, the results should be interpreted carefully. Cerebral and peripheral haemodynamic and oxygenation changes are affected by multiple factors; hence underlying the necessity of measuring several systemic variables in order to obtain a general idea of the mechanisms affecting them. CCA is a useful tool to investigate this problem as it helps to assess and quantify the interrelation between a multiple set of variables, simultaneously.

Acknowledgments A.C.D. was supported by the Flemish fund for research FWO, and author I.T. is a Wellcome Trust Research Fellow (088429/Z/09/Z).

References

1. Liem K, Hopman J, Oesenburg B et al (1995) Cerebral oxygenation and haemodynamics during induction of extracorporeal membrane oxygenation as investigated by near infrared spectrophotometry. Pediatrics 93:555–561
2. Bulas D, Glass P (2005) Neonatal ECMO: neuroimaging and neurodevelopmental outcome. Semin Perinatol 29:58–65

3. Ejike J, Schenkeman K, Seidel K et al (2006) Cerebral oxygenation in neonatal and pediatric patients during veno-arterial extracorporeal life support. Pediatr Crit Care Med 7:154–158
4. Papademetriou MD, Tachtsidis I, Leung TS et al (2010) Cerebral and peripheral tissue oxygenation in children supported on ECMO for Cardio-respiratory failure. Adv Exp Med Biol 662:447–453
5. Wong FY, Leung TS, Austin T et al (2008) Impaired autoregulation in preterm infants identified by using spatially resolved spectroscopy. Pediatrics 121(3):e604–e611
6. Brady KM, Mytar JO, Lee JK et al (2010) Monitoring cerebral blood flow pressure autoregulation in pediatric patients during cardiac surgery. Stroke 41:1957–1962
7. Caicedo A, De Smet D, Vanderhaegen J et al (2011) Impaired cerebral autoregulation using near-infrared spectroscopy and its relation to clinical outcomes in premature infants. Adv Exp Med Biol 701:233–239
8. Hotelling H (1936) Relations between two sets of variates. Biometrika 8:321–377
9. Tisdal MM, Taylor C, Tashtsidis I et al (2009) The effect on cerebral tissue oxygenation index of changes in the concentrations of inspired oxygen and end-tidal carbon dioxide in healthy adult volunteers. Anesth Anal 109(3):906–913
10. Harper AM, Jennet S (1990) Cerebral blood flow and metabolism. Manchester University Press, Manchester, p 20
11. Davis SL, Fadel PJ, Cui J et al (2006) Skin blood flow influences near-infrared spectroscopy-derived measurements of tissue oxygenation during heat stress. J Appl Physiol 100:221–224

Chapter 5
Blood Oxygen Level Dependent Magnetization Transfer (BOLDMT) Effect

Kejia Cai, Mohammad Haris, Anup Singh, Lin Z. Li, and Ravinder Reddy

Abstract A few studies have reported that magnetization transfer (MT) preparation interacts with blood oxygen level dependent (BOLD) contrast used for functional magnetic resonance imaging (MRI). However, the mechanism is still not well established. This study shows that blood oxygenation level itself affects MT contrast. MT ratio (MTR) decreases with increased blood oxygenation, which is demonstrated by ex vivo and in vivo experiments. Oxygenated blood shows less MTR contrast compared to deoxygenated blood sample; and higher levels of oxygen inhalation decrease tissue MTR in vivo especially in brain tumor region. The percentage reduction of MTR due to hyperoxia inhalation, referred to as the blood oxygen dependent magnetization transfer (BOLDMT) effect, correlates well with tissue oxygen extraction, which is highest in well-vascularized tumor rim, followed by inner tumor, gray matter (GM), and white matter (WM) normal tissue. Simulations and experiments demonstrate that BOLDMT effect induced with hyperoxia inhalation may be generated by decreased tissue T_1 due to increased O_2 dissolution and increased tissue T_2 due to reduced deoxyhemoglobin (dHb) concentration. Compared to regular T_2* weighted BOLD contrast, BOLDMT has higher insensitivity to B_0 inhomogeneities. BOLDMT may potentially serve as a reliable and novel biomarker for tumor oxygen extraction.

Keywords Blood oxygen level dependent • Magnetization transfer • Tumor oxygen extraction • Hyperoxia inhalation

K. Cai (✉) • M. Haris • A. Singh • L.Z. Li • R. Reddy
Department of Radiology, Center for Magnetic Resonance and Optical Imaging, University of Pennsylvania, B1 Stellar-Chance Laboratories, 422 Curie Boulevard, Philadelphia, PA 19014, USA
e-mail: kcflying@gmail.com

W.J. Welch et al. (eds.), *Oxygen Transport to Tissue XXXIV*, Advances in Experimental Medicine and Biology 765, DOI 10.1007/978-1-4614-4989-8_5,

1 Introduction

Conventional magnetization transfer (MT) in magnetic resonance imaging (MRI) provides a unique contrast associated to bound protons [1, 2]. Saturating macromolecular bound protons with off-resonance radio frequency (RF) pulses creates MRI detectable signal change from the bulk water protons due to magnetization transfer between the macromolecular protons and bulk water. Blood oxygenation level dependent (BOLD) imaging [3], typically based on T_2* weighted gradient echo signals, provides contrast from paramagnetic deoxyhemoglobin (dHb) in the blood. Increases in oxygenated hemoglobin concentration leads to increased T_2* or decreased R_2* ($1/T_2$*), higher BOLD-MRI signals. This concept has been widely used in MRI of oxygenation in different tissues including brain [3], kidney [4], and tumors [5]. However, regular T_2* weighted BOLD MRI is prone to artifacts associated with B_0 field inhomogeneity and fluctuations [6, 7].

MT and BOLD contrasts could interact with each other. Zhou et al. investigated the MT effect in rat brain at a range of arterial pCO_2 levels [8] and showed an altered MT response in the brain due to varied arterial pCO_2. In another study, functional MRI with MT preparation has been proposed for quantification of arterial cerebral blood volume (CBVa) alternation [9]. However, the fundamental mechanism on how MT contrast is affected by functional activation and blood oxygenation is still not clear.

Our hypothesis is that blood oxygenation level itself could directly affect MT contrast. Blood oxygen level dependent magnetization transfer (BOLDMT) effect is demonstrated in this study with stationary ex vivo blood experiments and tumor rats subjected to different percentages of oxygen inhalation. The advantage of BOLDMT over regular BOLD MRI is demonstrated.

2 Methods

Fresh rat blood samples with heparin are bubbled with 100% N_2 or 100% O_2 for 10 min at a rate of 2 cc/min and imaged on a 7 T Siemens whole-body MRI scanner with a 35 mm-diameter custom-built solenoid coil. A custom pulse sequence was written to provide a frequency selective saturation pulse followed by a segmented RF spoiled gradient echo (GRE) readout (128 segments per shot for every 10 s). Sequence parameters used were: slice thickness = 5 mm, flip angle = 10°, readout TR = 12.2 ms, TE = 6 ms, in-plane resolution = 270 × 270 μm². Z spectra between −100 p.p.m. and +100 p.p.m. were collected using ten Hanning-windowed rectangular pulses (100 ms each with 200 μs inter-delay) at the average B_1 of 250 Hz.

To develop intracranial tumors, rat gliosarcoma cells (9 L) were injected into Syngeneic female Fisher rats (F344/NCR, 4–6 weeks old) weighing 130–150 g, as reported previously [10, 11]. At 4–6 weeks after tumor implantation, brain tumor-bearing rats (n = 5) were scanned at a Varian 9.4 T horizontal bore small animal scanner and placed in a 35 mm diameter commercial quadrature proton coil. *Z* spectra of an axial slice of rat brain were acquired from rats subjected to different levels of inhalation oxygen (8, 21, 40, 60, 80, 100% diluted with N_2), at a fixed flow rate of 35 cc/min. After any specific gas had been given for 3 min, to allow blood oxygenation to reach a steady state, *Z* spectra (100, -/+20, -/+10, -/+5, -/+4, -/+3.5, -/+3, -/+2.5 … 0 p.p.m.) were collected using a similar customized sequence as mentioned except using 1 s continuous rectangle saturation pulse with 250 Hz B_1, slice thickness = 2 mm, flip angle = 15°, readout TR = 6.2 ms, TE = 2.9 ms, and 64 segment acquisition per shot for every 4 s (saturation pre-delay time = 3 s). Following *Z* spectra acquisition, T_2*-weighted images were acquired using a gradient echo sequence at various TE (3, 5, 7.5, 10, 15, and 20 ms), from which R_2* maps were generated. B_0 field map of the same slice was also generated with short TEs (2.5, 3.0, and 3.5 ms) [12]. All animal experiments were performed according to an approved institutional animal care and use committee protocol.

Magnetization transfer ratio (MTR) contrast depends on various physiological parameters including macromolecular concentration, bound proton exchange rate, tissue T_1 and T_2. MT effect as a function of bulk water T_1 (1–2.5 s) and T_2 (20–80 ms) was simulated using the Bloch equations modified for two-pool chemical exchange model (Bloch-McConnell equations) [13]. Other parameters are exchange rate = 10 Hz, free and bound proton concentration = 110 M and 4 M, respectively, bound proton offset = −2.4 p.p.m., bound proton $T_1/T_2 = 1$ s/10 μs. Saturation pulse parameters are the same as those in the blood experiments at 7 T. MTR as a function of field fluctuation was simulated by shifting B_0 static field between ±1 p.p.m.

Image processing was performed using MATLAB. MTR spectrum is the plot of $(1 - MT_{\text{“on”}}/MT_{\text{“off”}})$ as a function of the saturation offset, where $MT_{\text{“on”}}$ and $MT_{\text{“off”}}$ are the signals with and without saturation, respectively. Specifically for generating MTR maps, +20 p.p.m. signal was used for $MT_{\text{“on”}}$ and 100 p.p.m. for $MT_{\text{“off”}}$ since it preserves over 99% of signal. BOLDMT contrast represented the percentage MTR change induced by hyperoxia inhalation compared to the baseline under normoxia. Regular BOLD contrast is generally calculated as the relative changes of T_2* weighted signal or R_2* ($1/T_2$*).

3 Results

Blood oxygen saturation reaches 100 and ~80% after being bubbled with 100% O_2 and 100% N_2, respectively, under these experimental conditions. Signal at a given saturation offset from fully oxygenated blood is generally higher than deoxygenated blood, presumably due to regular BOLD effect (Fig. 5.1a). MTR spectrum (Fig. 5.1b) is broader in deoxygenated blood. As a result, MTR at +20 p.p.m. of deoxygenized blood is (20.1 ± 2.6%) higher (Fig. 5.1c).

Z spectra of a brain tumor-bearing rat subject to different oxygen inhalations are shown in Fig. 5.2a. MTR spectral line-width of tumor increases as the inhalation

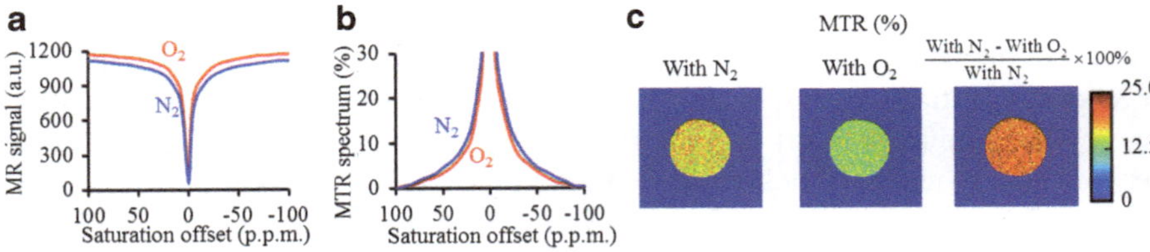

Fig. 5.1 Ex vivo experiments on stationary blood samples. *Z* spectra (**a**) and MTR spectra (**b**) of blood samples bubbled with 100 % O_2 (*red*) or 100 % N_2 (*blue*). Corresponding MTR maps at +20 p.p.m. and percentage MTR change are shown in (**c**)

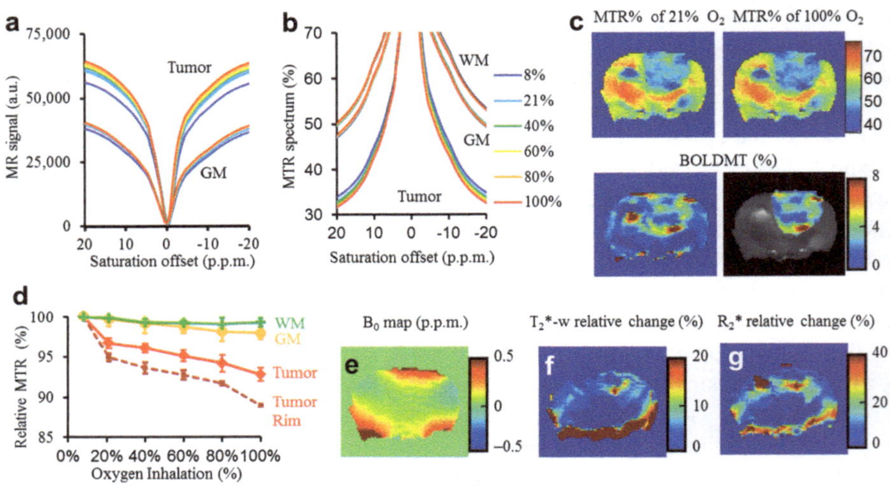

Fig. 5.2 Representative *Z* spectra (**a**) and MTR spectra (**b**) from tumor and normal brain tissues under different levels of oxygen inhalations. (**c**) +20 p.p.m. MTR (*top*) and BOLDMT (*bottom*) maps of the same slice. Tumor BOLDMT contrast (*bottom right*) was color-coded over the anatomical image in gray scale. (**d**) Relative MTR contrast (mean ± standard deviation, $n = 5$) at different oxygen inhalations. (**e**) B_0 field map. (**f** and **g**) Relative T_2*-weighted (TE = 10 ms) and R_2* changes due to hyperoxia inhalation

oxygen decreases, while the line-widths of gray matter (GM) and white matter (WM) are much less sensitive to oxygen inhalation (Fig. 5.2b). Relative MTR contrast at +20 p.p.m. under different oxygen inhalations are plotted in Fig. 5.2d, showing tumor rim and tumor has >4 times MTR change induced by hyperoxia inhalation comparing to the normal brain tissues. In the +20 p.p.m. MTR maps (Fig. 5.2c, top), lowest MTR contrast was seen from regions with high vessel density (mostly tumor rim), followed by tumor, GM, and WM. Tumor MTR decreases slightly under hyperoxia inhalation. BOLDMT map (Fig. 5.2c, bottom) shows highest contrast from tumor rim followed by tumor, GM, and WM. However, T_2*-weighted and R_2* contrast maps (Fig. 5.2 f, g) are highly modulated by B_0 field (Fig. 5.2e) with over 0.5 p.p.m. variation.

4 Discussions

Our experimental data in phantom and in vivo have demonstrated the changes in MTR due to changes in blood oxygenation. Increased supply of oxygen reduces blood dHb concentration, which primarily raises T_2 and T_2* [14–16]. Additionally, excess O_2 dissolves into tissue, primarily reducing the T_1 of tissues [15, 17]. Theoretical simulation demonstrated that both decreasing bulk water T_1 and increasing T_2 reduce MTR contrast (Fig. 5.3a, b).

MTR contrast is insensitive to field inhomogeneity with <5% of MTR variation under 1 p.p.m. field fluctuation (Fig. 5.3c), since MT preparation pulse has a large frequency offset away from the center water resonance. In comparison, regular BOLD contrast is highly sensitive to B_0 homogeneity [7, 18] as demonstrated in vivo (Fig. 5.2f, g).

In general, tumors have higher metabolism compared to normal tissue. Increased oxygen inhalation may provide more oxygen to tumor tissue and increase the tissue oxygen extraction. Tumor BOLDMT contrast map due to hyperoxia inhalation matches the expected oxygen extraction ratio of brain tumor by showing the highest magnitude in tumor rim and then inner tumor tissue, followed by GM and WM, suggesting that BOLDMT could be used as an indicator of tumor oxygen extraction with hyperoxia inhalation. Oxygen-sensitive electrodes [19–21] and histological staining of hypoxia biomarker, such as EF5 [22], may help us to establish the correlation between BOLDMT map and tumor oxygen extraction ratio in the future work. Future research on the contributions of blood flow and saturation pre-delay time on the in vivo BOLDMT effect would help us to fully understand and optimize BOLDMT contrast.

Imaging techniques for tumor oxygenation, characterized by hypoxia, have been reviewed [23]. So far, there is no reliable method for MRI of tumor oxygenation. BOLDMT effect based on hyperoxia inhalation might be used in clinical studies for imaging pathologies with altered oxygenation, such as tumor, Alzheimer's disease, and stroke.

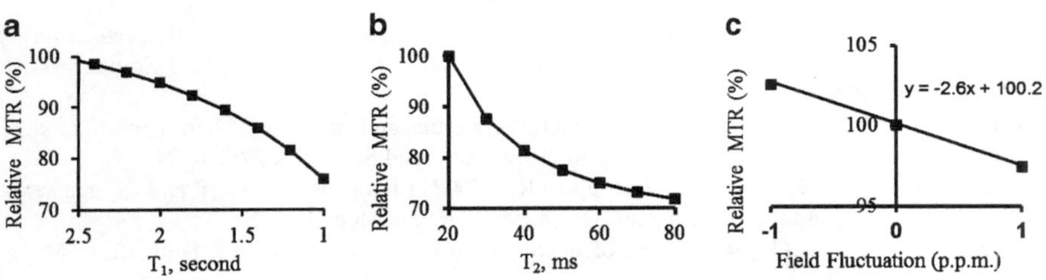

Fig. 5.3 Relative MTR at +20 p.p.m. as a function of bulk water T_1 (**a**) and T_2 (**b**) and static B_0 field fluctuation (**c**)

Acknowledgments The authors thank Drs. Harish Poptani, Ranjit Ittyerah, and Damodar Reddy for their help with the animal model; Matt Fenty, Weixia Liu, Steve Pickup for their technical assistance in using small animal research scanners; Kalli Grasley and Prianka Waghray for literature review. This work was performed at an NIH supported resource with funding from P41RR02305.

References

1. Grossman RI, Gomori JM, Ramer KN et al (1994) Magnetization transfer: theory and clinical applications in neuroradiology. Radiographics 14:279–290
2. Henkelman RM, Stanisz GJ, Graham SJ (2001) Magnetization transfer in MRI: a review. NMR Biomed 14:57–64
3. Ogawa S, Lee TM, Kay AR et al (1990) Brain magnetic resonance imaging with contrast dependent on blood oxygenation. Proc Natl Acad Sci U S A 87:9868–9872
4. Prasad PV, Edelman RR, Epstein FH (1996) Noninvasive evaluation of intrarenal oxygenation with BOLD MRI. Circulation 94:3271–3275
5. Baudelet C, Gallez B (2002) How does blood oxygen level-dependent (BOLD) contrast correlate with oxygen partial pressure (pO_2) inside tumors? Magn Reson Med 48:980–986
6. Jezzard P, Song AW (1996) Technical foundations and pitfalls of clinical fMRI. Neuroimage 4:S63–S75
7. Port JD, Pomper MG (2000) Quantification and minimization of magnetic susceptibility artifacts on GRE images. J Comput Assist Tomogr 24:958–964
8. Zhou J, Payen JF, van Zijl PC (2005) The interaction between magnetization transfer and blood-oxygen-level-dependent effects. Magn Reson Med 53:356–366
9. Kim T, Hendrich K, Kim SG (2008) Functional MRI with magnetization transfer effects: determination of BOLD and arterial blood volume changes. Magn Reson Med 60:1518–1523
10. Kim S, Pickup S, Hsu O et al (2008) Diffusion tensor MRI in rat models of invasive and well-demarcated brain tumors. NMR Biomed 21:208–216
11. Wang S, Kim S, Chawla S et al (2009) Differentiation between glioblastomas and solitary brain metastases using diffusion tensor imaging. Neuroimage 44:653–660
12. Haris M, Cai K, Singh A et al (2011) In vivo mapping of brain myo-inositol. Neuroimage 54:2079–2085
13. Woessner DE, Zhang S, Merritt ME et al (2005) Numerical solution of the Bloch equations provides insights into the optimum design of PARACEST agents for MRI. Magn Reson Med 53:790–799
14. Tadamura E, Hatabu H, Li W et al (1997) Effect of oxygen inhalation on relaxation times in various tissues. J Magn Reson Imaging 7:220–225
15. Uematsu H, Takahashi M, Hatabu H et al (2007) Changes in T1 and T2 observed in brain magnetic resonance imaging with delivery of high concentrations of oxygen. J Comput Assist Tomogr 31:662–665
16. Pauling L, Coryell CD (1936) The magnetic properties and structure of hemoglobin, oxyhemoglobin and carbonmonoxyhemoglobin. Proc Natl Acad Sci U S A 22:210–216
17. Silvennoinen MJ, Kettunen MI, Kauppinen RA (2003) Effects of hematocrit and oxygen saturation level on blood spin–lattice relaxation. Magn Reson Med 49:568–571
18. Jezzard P, Clare S (1999) Sources of distortion in functional MRI data. Hum Brain Mapp 8:80–85
19. Vaupel P, Schlenger K, Knoop C et al (1991) Oxygenation of human tumors: evaluation of tissue oxygen distribution in breast cancers by computerized O_2 tension measurements. Cancer Res 51:3316–3322

20. Kavanagh MC, Sun A, Hu Q et al (1996) Comparing techniques of measuring tumor hypoxia in different murine tumors: Eppendorf pO_2 Histograph, [3H]misonidazole binding and paired survival assay. Radiat Res 145:491–500
21. Mason RP, Hunjan S, Constantinescu A et al (2003) Tumor oximetry: comparison of 19F MR EPI and electrodes. Adv Exp Med Biol 530:19–27
22. Gillies RJ, Raghunand N, Karczmar GS et al (2002) MRI of the tumor microenvironment. J Magn Reson Imaging 16:430–450
23. Krohn KA, Link JM, Mason RP (2008) Molecular imaging of hypoxia. J Nucl Med 49(Suppl 2):129S–148S

Chapter 6
Characterizing Prostate Tumor Mouse Xenografts with CEST and MT-MRI and Redox Scanning

Kejia Cai, He N. Xu, Anup Singh, Mohammad Haris, Ravinder Reddy, and Lin Z. Li

Abstract The main goal of this study was to use multimodality imaging methods to reveal the heterogeneity in prostate cancer and seek the correlation between the characteristic heterogeneity and tumor aggressiveness. Here we report the preliminary data on chemical exchange saturation transfer (CEST) and magnetization transfer (MT) magnetic resonance imaging (MRI) and redox scanning [cryogenic NADH/Fp (reduced nicotinamide adenine dinucleotide/oxidized flavoproteins) fluorescence imaging] of two aggressive human prostate tumor lines (DU-145 and PC-3) xenografted in athymic nude mice. The results obtained by these methods appeared to be consistent, with all showing a higher level of heterogeneity in DU-145 tumors than in PC-3 tumors. DU-145 tumors showed CEST maps with both positive and negative areas while PC-3 CEST maps were relatively homogeneous. The mean CEST value for PC-3, 23.0±2.1 %, is at a significantly higher level ($p<0.05$) than DU-145 (1.9±6.7 %) at the peak of the CEST asymmetric curve (+2 ppm). Fp redox ratio (Fp/(NADH+Fp)) images exhibited localized highly oxidized regions in DU-145 tumors, whereas PC-3 tumors appeared to be less heterogeneous. These results suggest a possible role of metabolism in tumor progression. More studies, including an indolent prostate tumor line and with larger sample size, will be performed in the future to identify the biomarkers for prostate tumor aggressiveness.

Keywords Mitochondrial redox state • Fluorescence imaging • Metabolism • Tumor aggressiveness

---

Kejia Cai and He N. Xu contributed equally.

K. Cai • H.N. Xu • A. Singh • M. Haris • R. Reddy • L.Z. Li (✉)
Department of Radiology, University of Pennsylvania,
250 Anatomy Chemistry, Philadelphia, PA 19014, USA
e-mail: linli@mail.med.upenn.edu

W.J. Welch et al. (eds.), *Oxygen Transport to Tissue XXXIV*, Advances in Experimental Medicine and Biology 765, DOI 10.1007/978-1-4614-4989-8_6,

1 Introduction

Prostate cancer tops the most common cancers and is the second leading cause of cancer death in males in the USA (Center for Disease Control and Prevention http://www.cdc.gov/cancer/dcpc/data/men.htm). Globally, the highest incidence rates were recorded primarily in the developed countries of Oceania, Europe, and North America in 2008 [1]. Metastasis is the primary cause of death of prostate cancer patients. Predicting prostate tumor aggressiveness (metastatic potential) to assist treatment strategies is of major importance. Our ultimate goal for this study was to identify effective metabolic/functional imaging biomarkers for the prediction of prostate tumor metastatic potential. Previously, we showed that multimodality imaging methods (NMR and optical) are beneficial in predicting tumor aggressiveness in five lines of human melanoma xenografts spanning a full range of metastatic potential [2, 3]. In this paper, we report the preliminary data on the characterization of prostate tumor xenografts using chemical exchange saturation transfer (CEST) and magnetization transfer (MT) magnetic resonance imaging (MRI) and the redox scanning methods.

Conventional MT-MRI provides a unique contrast associated to bound water protons [4, 5]. After saturating macromolecular bound protons with off-resonance radio frequency pulses, the macromolecular proton spins transfer saturation to the bulk water, creating the MRI-detectable signal change from the bulk water. CEST-MRI that utilizes proton chemical exchange effects specifically from –OH, –NH, and $–NH_2$ groups of metabolites has been recently explored to map the tissue pH [6, 7], glycosaminglycans in cartilage [8], as well as liver glycogen [9]. The redox scanning of snap-frozen tissue [10, 11] gives the in vivo mitochondrial redox state by imaging the intrinsic fluorescences of NADH and Fp including FAD (flavin adenine dinucleotide), providing sensitive indicators of tumor mitochondrial metabolic states as shown in our previous studies [2, 3, 12, 13].

2 Methods

Two classical cell lines of human prostate cancer, PC-3 and DU-145, were chosen for the initial experiments, which have high and moderate invasive potential, respectively [14]. All animal experiments were performed according to a protocol approved by the University of Pennsylvania Institutional Animal Care and Use Committee (IACUC). The propagated cells were implanted subcutaneously into the upper thighs of athymic nude mice (~7 million cells/site).

Mice bearing DU-145 ($n=3$) and PC-3 ($n=4$) prostate tumors (5 weeks growth period) were under anesthesia using isofluorane and scanned at a Varian 9.4-T horizontal MRI scanner. CEST Z-spectra were collected from tumor central cross sections using a custom-programmed sequence, with a frequency selective rectangle saturation pulse ($B_1=250$ Hz, 1 s), followed by a segmented RF spoiled gradient echo readout. Sequence parameters were: field of view 35×35 mm^2, slice thickness

2 mm, flip angle 15°, readout TR 6.2 ms, TE 2.9 ms, matrix size 128×128, and number of averages 2. One saturation pulse followed with 64 segment acquisition was repeated every 4 s. Saturation frequency was arrayed from −5 to 5 ppm (part per million with respect to the water resonance) with 0.25 ppm step size. In addition, images with saturation at ±10, ±20, and ±100 ppm were acquired. MT "on" and "off" images are images with +20 and +100 ppm saturation offsets, respectively. Saturation with +100 ppm preserves over 99 % of signal based on our observation. B_0 and B_1 maps were generated for correction CEST contrast [15].

For redox scanning, mice under anesthesia underwent snap-freezing procedures and tumors were harvested and embedded as previously reported [13, 16, 17]. Redox scanning of four DU-145 and three PC-3 tumors was performed at various tissue depths. The nominal concentrations of tissue NADH and Fp were calibrated to corresponding reference standards.

All image processing and data analysis was performed using Matlab. Z-spectrum was plotted as the percentage of saturated signal (averaged within tumor regions and across all tumors in each line) as a function of saturation offset in ppm. The MTR asymmetric curve or the CEST contrast as a function of saturation offset was generated by subtracting the reference signal at the negative side of the Z-spectrum from their positive counterparts and then normalized by the reference signal. CEST contrast at a specific chemical shift (2 ppm) and MT ratio (MTR or $1-\mathrm{MT}_{\text{“on”}}/\mathrm{MT}_{\text{“off”}}$) were also obtained on a pixel basis and then color mapped and overlaid onto the anatomical images acquired when MT was turned "on."

3 Results

DU-145 shows higher levels of signal compared to PC-3 across the entire Z-spectrum (Fig. 6.1a), indicating lower level of magnetization transfer effects from DU-145. The MTR asymmetric curve of PC-3 (Fig. 6.1b) shows positive CEST contrast, while contrast from DU-145 is mostly negative with large variation. DU-145 tumors

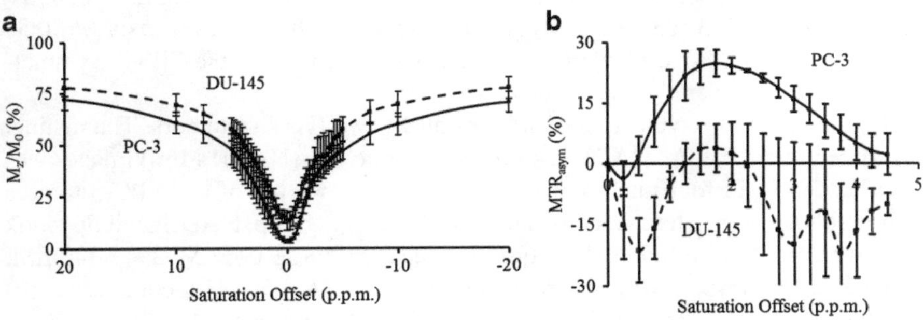

Fig. 6.1 Z-spectra (**a**) and MTR asymmetric curves (**b**) of DU-145 (*dashed*) and PC-3 (*solid*) prostate tumors. Data are averaged within tumor regions and across all tumors for each tumor line

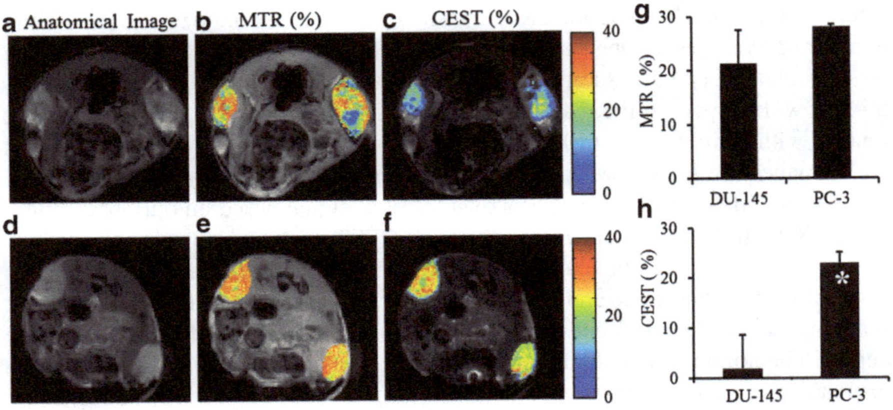

Fig. 6.2 Anatomy images (**a**, **d**), +20 ppm MTR (**b**, **e**), and +2 ppm CEST (**c**, **f**) maps of a typical DU-145 tumor (*top*) and a typical PC-3 tumor (*bottom*). CEST contrast of PC-3 ($n=4$) is significantly higher than DU-145 ($n=3$) (**h**, $p<0.05$), while their MTR contrasts are similar (**g**, $p>0.05$)

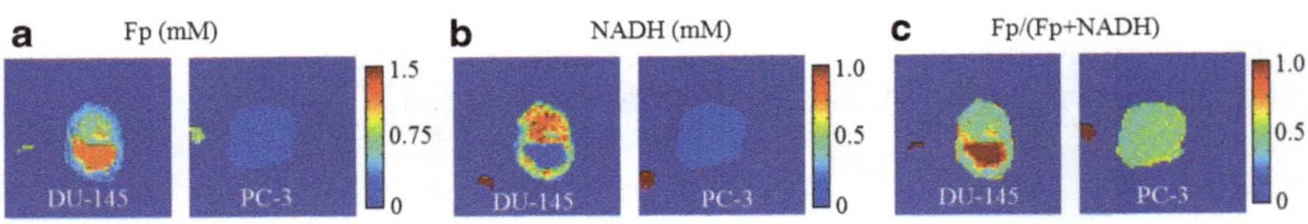

Fig. 6.3 Typical redox images (Fp, NADH, and Fp redox ratio) of DU-145 and PC-3 xenografts. The *small round spots* outside the tumors are the reference standards

showed a heterogeneous MTR and CEST distribution while PC-3 maps were relatively homogeneous (Fig. 6.2a–f). Extremely low MTR regions seen from DU-145 indicate possible tumor necrosis. Although there is no significant difference in MTR mean values (Fig. 6.2g) between DU-145 and PC-3, the mean CEST contrast (Fig. 6.2h) for PC-3 (23.0±2.1 %) is at a significantly higher level ($p<0.05$) than DU-145 (1.9±6.7 %) at +2 ppm of chemical shift, the peak of the CEST asymmetric curve.

Figure 6.3 shows typical redox images of the two types of tumors. The distinct heterogeneity in the Fp, NADH, and Fp redox ratio (Fp/(NADH+Fp)) images was clearly seen in all four tumors of DU-145, whereas two out of three PC-3 tumors appeared to be less heterogeneous. The global averages of Fp, NADH, and Fp redox ratio for DU-145 (4 tumors, 15 sections scanned) are 582±196 μM, 390±140 μM, and 0.59±0.09, respectively; whereas those for PC-3 (3 tumors, 11 sections scanned) are 553±303 μM, 350±170 μM, and 0.58±0.11, respectively. These values obtained by averaging the mean from each tissue section of the same type of tumor are not significantly different between the two lines.

The results obtained by the three imaging methods appeared to be consistent, with all showing a higher level of heterogeneity in the DU-145 tumors than that in the PC-3 tumors.

4 Discussion and Conclusions

MTR contrast originates from bound water protons that are associated with relatively stationary structural tissue proteins and lipids. The typical exchange rate for these bound protons is less than 20 Hz, while the CEST effect of macromolecule, peptides, and free amino acids is due to exchangeable protons with an exchange rate (k) from 10 to several kHz [18, 19]. In order to observe the CEST effect, k should be within a slow to intermediate regime compared to the chemical shift ($\Delta\omega$), i.e., under the condition of $k \leq \Delta\omega$. Additionally, at any given saturation amplitude, a maximum CEST effect is observed when $k \sim 2\pi B_1$ in Hz [20]. That is why saturation with B_1 as low as 50 Hz has been used for CEST imaging of amide proton transfer (APT) with $k \sim 30$ Hz [7]. Using higher B_1 decreases CEST contribution from APT and activates small molecule metabolites and mobile peptides. In this study, CEST contrast created by using 250 Hz B_1 may be mostly due to free amino acids with an exchange rate of ~1.5 kHz (or $2\pi \times 250$ Hz). CEST contrast could be used to reflect tissue metabolism involving free amino acids (such as alanine and glycine) and other organic molecule (such as creatine). Further investigation is needed to find the specific small molecule metabolites that contribute to CEST-MRI contrast in prostate cancers.

Our previous quantitative studies on the metastatic potential of melanoma [2, 3] and breast tumor [13] xenografts indicated that a high level of heterogeneity in mitochondrial redox state is associated with tumor aggression, and the aggressive tumors have localized highly oxidized regions while the indolent ones are usually more homogeneous and do not have localized oxidized regions. Although we have not found data showing direct measurement of the metastatic potential of these two lines in vivo, published data suggest that the DU-145 line is less invasive than PC-3 in vitro [14, 21]. The observed heterogeneity in the MRI and redox state imaging results seems to support that the intra-tumor heterogeneity of the DU-145 tumors was more distinct than that of PC-3 tumors. These results prompt further examination to elucidate whether such heterogeneity difference is due to the intrinsic difference in cancer biology. Global average of the redox indices in this study shows no significant difference between the two prostate tumor lines, suggesting that more specific analysis on redox state of tumor heterogeneity is required, which will be conducted in the near future.

More studies including an indolent prostate tumor line (such as LnCap) and with larger sample size will be carried out to identify the molecular imaging biomarkers for prostate tumor aggressiveness. We would also be interested in measuring tissue oxygenation to know whether these tumors were under acute or chronic hypoxia which, according to a recent in vitro study [22], may increase the invasiveness or cause cell death. Histological staining of vasculatures, cell death, or hypoxia (HIF) may help us to gain more understanding of our observations from the MRI and the redox scanning.

As a conclusion, CEST- and MT-MRI and redox scanning were successfully employed to reveal the heterogeneity in prostate tumor xenografts. Data showed more heterogeneity in the DU-145 than in the PC-3 xenografts. The CEST and MT-MRI are non-invasive and potentially translatable to the clinic while the redox scanning can be applied to the biopsy specimen.

Acknowledgments The authors thank Drs. Weixia Liu and Steve Pickup for their technical assistance with animal MRI scanners. This work was supported by an NIH research resource (P41RR02305, R. Reddy) and the SAIR grant 2U24-CA083105 (J.D. Glickson and L. Chodosh).

References

1. Jemal A, Bray F, Center MM et al (2011) Global cancer statistics. CA Cancer J Clin 61:69
2. Li LZ, Zhou R, Xu HN et al (2009) Quantitative magnetic resonance and optical imaging biomarkers of melanoma metastatic potential. Proc Natl Acad Sci U S A 106:6608–6613
3. Li LZ, Zhou R, Zhong T et al (2007) Predicting melanoma metastatic potential by optical and magnetic resonance imaging. Adv Exp Med Biol 599:67–78
4. Grossman RI, Gomori JM, Ramer KN et al (1994) Magnetization transfer: theory and clinical applications in neuroradiology. Radiographics 14:279–290
5. Henkelman RM, Stanisz GJ, Graham SJ (2001) Magnetization transfer in MRI: a review. NMR Biomed 14:57–64
6. Sun PZ, Zhou J, Sun W et al (2006) Detection of the ischemic penumbra using pH-weighted MRI. J Cereb Blood Flow Metab 27:1129
7. Zhou J, Payen JF, Wilson DA et al (2003) Using the amide proton signals of intracellular proteins and peptides to detect pH effects in MRI. Nat Med 9:1085–1090
8. Ling W, Regatte RR, Navon G et al (2008) Assessment of glycosaminoglycan concentration in vivo by chemical exchange-dependent saturation transfer (gagCEST). Proc Natl Acad Sci U S A 105:2266–2270
9. van Zijl PCM, Jones CK, Ren J et al (2007) MRI detection of glycogen in vivo by using chemical exchange saturation transfer imaging (glycoCEST). Proc Natl Acad Sci 104:4359–4364
10. Chance B, Schoener B, Oshino R et al (1979) Oxidation–reduction ratio studies of mitochondria in freeze-trapped samples. NADH and flavoprotein fluorescence signals. J Biol Chem 254:4764–4771
11. Quistorff B, Haselgrove JC, Chance B (1985) High spatial resolution readout of 3-D metabolic organ structure: An automated, low-temperature redox ratio-scanning instrument. Anal Biochem 148:389–400
12. Xu HN, Nioka S, Chance B et al (2011) Heterogeneity of mitochondrial redox state in premalignant pancreas in a PTEN null transgenic mouse model. Adv Exp Med Biol 201:207–213
13. Xu HN, Nioka S, Glickson JD et al (2010) Quantitative mitochondrial redox imaging of breast cancer metastatic potential. J Biomed Opt 15:036010
14. Pulukuri SM, Gondi CS, Lakka SS et al (2005) RNA interference-directed knockdown of urokinase plasminogen activator and urokinase plasminogen activator receptor inhibits prostate cancer cell invasion, survival, and tumorigenicity in vivo. J Biol Chem 280:36529–36540
15. Haris M, Cai K, Singh A et al (2010) In vivo mapping of brain myo-inositol. Neuroimage 54:2079–2085

16. Xu HN, Wu B, Nioka S et al (2009) Calibration of redox scanning for tissue samples. Proc SPIE 7174:71742F1–71742F8
17. Xu HN, Wu B, Nioka S et al (2009) Quantitative redox scanning of tissue samples using a calibration procedure. J Innov Opt Health Sci 2:375–385
18. van Zijl PCM, Zhou J, Mori N et al (2003) Mechanism of magnetization transfer during on-resonance water saturation. A new approach to detect mobile proteins, peptides, and lipids. Magn Reson Med 49:440
19. van Zijl PCM, Yadav NN (2011) Chemical exchange saturation transfer (CEST): what is in a name and what isn't? Magn Reson Med 65:927
20. Woessner DE, Zhang S, Merritt ME et al (2005) Numerical solution of the Bloch equations provides insights into the optimum design of PARACEST agents for MRI. Magn Reson Med 53:790–799
21. Laniado ME, Lalani EN, Fraser SP et al (1997) Expression and functional analysis of voltage-activated Na+ channels in human prostate cancer cell lines and their contribution to invasion in vitro. Am J Pathol 150:1213–1221
22. Dai Y, Bae K, Siemann DW (2011) Impact of hypoxia on the metastatic potential of human prostate cancer cells. Int J Radiat Oncol 81:521–528

Chapter 7
In Vitro Sirius Red Collagen Assay Measures the Pattern Shift from Soluble to Deposited Collagen

Chun Chen, Shanmin Yang, Mei Zhang, Zhenhuan Zhang, Bingrong Zhang, Deping Han, Jun Ma, Xiaohui Wang, Jingshen Hong, Yansong Guo, Paul Okunieff, and Lurong Zhang

Abstract In this study, we compared two in vitro collagen production assays ([^{3}H]-proline incorporation and Sirius Red) for their ability to determine the pattern shift from soluble to deposited collagen. The effect of the antifibrotic agent, triptolide (TPL), on collagen production was also studied. The results showed that: (1) 48 h after NIH 3T3 (murine embryo fibroblast) and HFL-1(human fetal lung fibroblast) were exposed to transforming growth factor-beta 1 (TGF-β), there was an increase in soluble collagen in the culture medium; (2) on day 4, soluble collagen declined, whereas deposited collagen increased; (3) Sirius Red was easier to use than [^{3}H]-proline incorporation and more consistently reflected the collagen pattern shift from soluble to deposited; (4) the in vitro Sirius Red assay took less time than the in vivo assay to determine the effect of TPL. Our results suggest that: (a) the newly synthesized soluble collagen can sensitively evaluate an agent's capacity for collagen production and (b) Sirius Red is more useful than [^{3}H]-proline because it is easier to use, more convenient, less time consuming, and does not require radioactive material.

Keyword Collagen

C. Chen • S. Yang • M. Zhang • Z. Zhang • B. Zhang • D. Han • J. Ma • X. Wang • J. Hong
Y. Guo • P. Okunieff • L. Zhang (✉)
Department of Radiation Oncology, UF Shands Cancer Center, University of Florida, 2033 Mowry Road, Gainesville, 100385, FL 32610, USA
e-mail: lurongzhang@ufl.edu

W.J. Welch et al. (eds.), *Oxygen Transport to Tissue XXXIV*, Advances in Experimental Medicine and Biology 765, DOI 10.1007/978-1-4614-4989-8_7,

1 Introduction

Ionizing radiation (IR) often leads to fibrosis, which impairs the given organ's functions, reduces the host's quality of life, and can even lead to death [1, 2]. The underlying mechanism for IR-induced fibrosis is still largely unknown and the effort to reduce IR-induced fibrosis has progressed slowly with unsatisfactory results. In part, this lack of progress is due to the difficulties associated with using an in vitro system to mimic the long-term in vivo pathological process. Fibrosis characteristically deposits a large amount collagen in the organs; collagen production is a transition from a single alpha-chain procollagen secreted from cells to 3 alpha chains, which is the final mature and deposited form [3–6]. We believe that blocking procollagen production from cells early in the process is the key to reducing fibrosis. In this study, the two most frequently used in vitro collagen production assays, [^{3}H]-proline incorporation [7] and the Sirius Red method [8], were compared. In addition, the simultaneous measurement of collagen in both media and cell lysates revealed the transition from procollagen to collagen. Finally, the effect of the antifibrotic agent, triptolide (TPL), was used to determine the ability of Sirius Red to screen effective agents.

2 Methods

For the Sirius Red assay, 12-well plates were seeded with 20×10^4 fibroblast cells in 1 ml Iscove's minimal essential medium (IMEM) with 10 % newborn calf serum (NCS); thereafter, the cells were grown to 70–80 % confluence. To stimulate the production of collagen, we used 0, 2.5, or 5 ng/ml doses of transforming growth factor-beta (TGF-β). Both culture media and cell lysate were collected on days 1, 2, 4, and 6 after TGF-β treatment. To prove that collagen was stimulated by TGF-β, interleukin-1 alpha (IL-1α), which did not strongly stimulate collagen production, was used as negative cytokine control.

For the [^{3}H]-proline incorporation assay, 0.5 μCi/well [^{3}H]-proline cells were added to cells cultured similar to those above 24 h after being seeded. On days 1, 2, 4, and 6 after TGF-β treatment, the media and cell lysates were collected with 0.1 mg/ml pepsin in 0.5 M acetic acid. Since [^{3}H]-proline was incorporated into many proteins during biosynthesis in order to distinguish the collagen from other proteins, the samples were divided into two groups in which: (1) total incorporated [^{3}H]-proline was counted and (2) collagenase was used to digest the collagen. After incubation at 37 °C for 3 h, the incorporated [^{3}H]-proline was precipitated with 10 % trichloric acid. The pellets were washed three times with phosphate buffer and dissolved in 0.2 N NaOH; a scintillation counter read the proteins labeled with [^{3}H]-proline. The following calculation was used: collagen level = total [^{3}H]-proline protein − noncollagenous protein.

We also compared the Sirius Red and [^{3}H]-proline incorporation assays side-by-side to determine not only which assay more sensitively distinguished the alterations caused by TGF-β but also which assay was easier to perform. Since collagen production closely correlated with the cell number, a thiazolyl blue tetrazolium bromide (MTT) assay was used according to standard protocol [9] to determine if the cell number was similar in all groups. MTT determined the cell number when cells were seeded and during the experiment when cells were treated with different factors. The transition of procollagen in media to deposited collagen was examined. In all experiments, the culture media and cell lysates were collected on days 1, 2, 4, and 6 after treatment and analyzed for the procollagen in media and deposited collagen in cell lysates as a function of time.

Finally, two experiments were performed to determine the effect of TPL on the reduction of collagen production. (1) In vivo study: the whole chest of C57BL/6 mice (female, 8 weeks old) received 15 Gy at a dose rate of 1.8 Gy/min with a cesium-137 source. Mice were then divided into two groups for treatment with vehicle alone (saline) or TPL (0.25 mg/kg, intravenous, twice a week, for 3 months). Nonirradiated mice were also treated with saline as an aging control or TPL as agent toxicity control. Eight months after lung IR, mice were euthanized, and lung hydroxyproline levels, representing the deposited collagen, were measured with a biochemistry method [10] and compared among the four groups. The protocol was approved by Institutional Animal Use Committee. (2) In vitro study: NIH 3T3 or HFL-1 cells (ATCC) were treated with 5 ng/ml of TPL 4 h before 5 ng/ml of mouse or human TGF-β (mTGF-β or hTGF-β) was added. The conditional media was collected on days 1 and 2 after treatment for procollagen production.

All data were statistically analyzed using either *t*-test or analysis of variance. $P<0.05$ was considered significant.

3 Results

Both Sirius Red and [^{3}H]-proline incorporation measured the alteration of TGF-β-induced collagen production. Figure 7.1a, b shows that mTGF-β stimulated collagen production in the cell culture medium and cell lysate of NIH 3T3 in a dose-dependent manner. IL-1a did not achieve this result, indicating that the two assays were able to specifically measure the collagen stimulated by mTGF-β. To prove this result was not mouse NIH 3T3-cell specific, human fetal lung fibroblast-1 (HFL-1) cells were treated with human TGF-β on days 1, 2, 4, and 6; a similar pattern of increased collagen production was seen in TGF-β-treated cells but not in IL-1α-treated cells (Fig. 7.2a, b). To exclude the possibility that the difference was not due to cell number, the MTT assay was used to measure the number of cells in the same plating and treatment conditions. The results (Figs. 7.1c and 7.2c) showed that the cell numbers in each group at the same time were similar, indicating that the

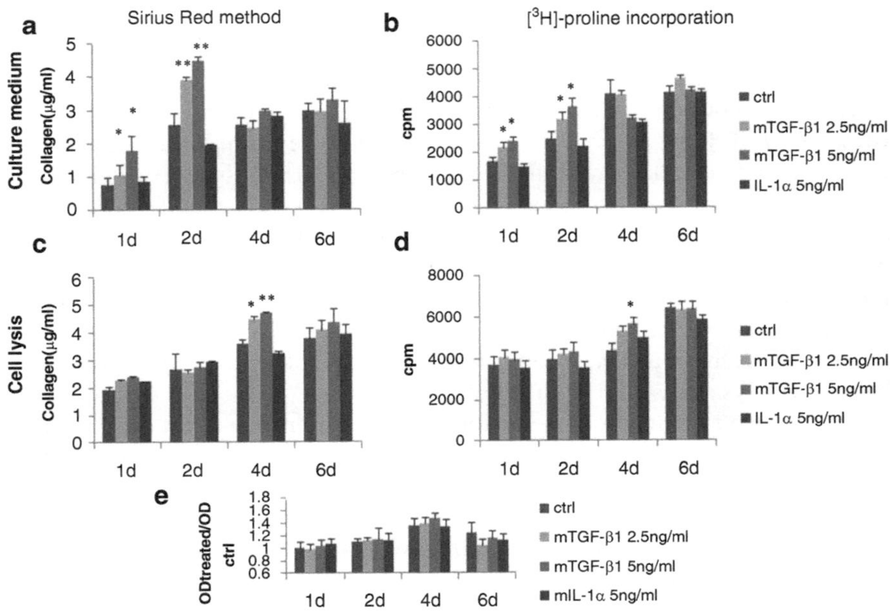

Fig. 7.1 Collagen concentrations in the cell culture medium and cell lysate of NIH 3T3 induced by mTGF-β. (**a**) and (**c**) used Sirius Red, (**b**) and (**d**) used [^{3}H]-proline incorporation, and (**e**) is the OD values of treated groups divided by OD value of control. *$P<0.05$, **$P<0.01$, *t*-test

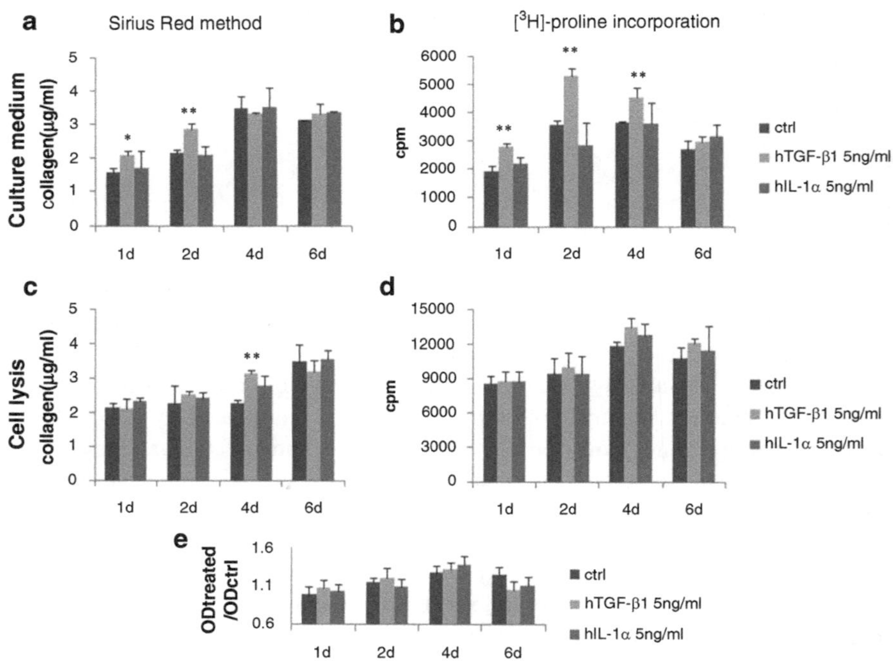

Fig. 7.2 Collagen concentrations in the cell culture medium and cell lysate of HFL-1 induced by hTGF-β. (**a**) and (**c**) used Sirius Red, (**b**) and (**d**) used [^{3}H]-proline incorporation, and (**e**) is the OD values of treated groups divided by OD value of control. *$P<0.05$, **$P<0.01$, *t*-test

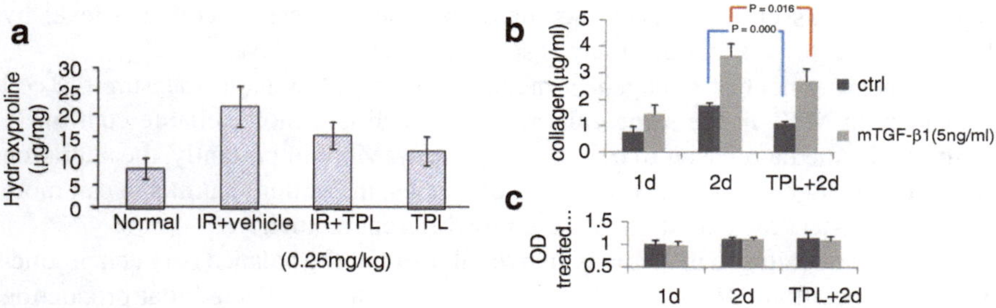

Fig. 7.3 TPL reduced collagen level in IR lung tissue, and TGF-β induced fibroblasts. (**a**) Effect of TPL on the deposition of collagen in IR lungs of C57BL/6 mice at 8 months. (**b**) Effect of TPL on the soluble collagen in the cell culture medium of NIH 3T3 induced by mTGF-β. (**c**) The cell viability of NIH 3T3 was assayed by MTT method

differences among the groups were due to the cells' ability to produce collagen. While the two assays had the same capacity to measure collagen production, the [^{3}H]-proline incorporation assay took longer and involved more procedures than the Sirius Red assay.

Measurement of collagen in both cell media and cell lysates detected the transition of procollagen to collagen. Figures 7.1 and 7.2 show that both assays detected the same pattern of collagen production. (1) In media: on day 1 TGF-β stimulated increased procollagen, which continued to increase on day 2 in a dose-dependent manner. However, on days 4 and 6, the assays revealed no difference. (2) In cell lysates: the stimulatory effect of TGF-β on collagen production was not seen until day 4. This time-shift pattern from an early increase in the media to a later increase in the lysates is consistent with the collagen production progress from a soluble alpha single chain to a deposited insoluble triple helix.

To determine an in vitro collagen assay's assessment of an antifibrotic agent, Sirius Red was used to measure the collagen level in day 2 media from TPL-treated (5 ng/ml) NIH 3T3 fibroblast cells. TPL inhibited the mTGF-β-induced collagen production increase (Fig. 7.3b), while the cell numbers were similar in all groups (Fig. 7.3c). This in vitro result was consistent with the in vivo result (Fig. 7.3a) that TPL inhibited radiation-induced pulmonary fibrosis, as evidenced by the reduced levels of lung hydroxyproline 8 months after irradiation. Thus, Sirius Red could be used to determine antifibrotic agents.

4 Discussion

Through this comparison of two collagen assays, we concluded that Sirius Red is superior to [^{3}H]-proline incorporation because it is easier to use, more convenient, less time consuming, and does not require radioactive material. Unlike [^{3}H]-proline, which incorporates all proteins that contain proline, Sirius Red has better specificity; it mainly stains collagen types I and III, the major forms found in the human body, and negatively stains other extracellular matrixes. The increase in collagen level by TGF-β further proves that the assay is specific for collagen level.

The measurement of collagen in media seems to allow us to measure the cell number with MTT in the same culture well, which is a more reliable control. In addition, the media is easier to collect and to assay. More importantly, the alteration magnitude is higher than the assessment of lysates, indicating that media can more sensitively reflect the alterations induced by different factors.

The level of collagen in media is an overall result of the balance between production and deposition. On days 1–2, the increased collagen reflected that production exceeded deposition; on day 4, the single-chain collagen increased by TGF-β became an insoluble triple helix and deposited as a matrix into the cell layer; on day 6, the effect of TGF-β was reduced, and the production and deposition tended to be flat and similar for all groups. Taken together, measuring media on day 2 with the Sirius Red assay is the best method to study the effect of collagen production.

TPL is a known inhibitor of fibrosis, as evidenced by the in vivo reduction of hydroxyproline levels in lungs treated with TPL. However, these data can only be obtained 8 months after irradiation. With the in vitro Sirius Red assay and 2 day media, we see the same tendency, which suggests that this assay is likely to be a fast, simple, and reliable tool for screening antifibrotic agents.

5 Conclusion

As compared to [^{3}H]-proline incorporation, Sirius Red is the superior in vitro assay to study the effect of collagen production; it allows for faster and easier screening of antifibrotic agents.

Acknowledgments This project is supported in part by U19 AI067733, RC1AI078519, RC2-AI-087580, RC1-AI081274 (NIAID/NIH), and Shands Cancer Center startup funds (University of Florida). We thank Kate Casey-Sawicki for editing this manuscript and Dr. Chihray Liu and his group for setting up the physical dosimetry for accurate radiation.

References

1. Travis EL (2007) Genetic susceptibility to late normal tissue injury. Semin Radiat Oncol 17:149–155
2. McDonald S, Rubin P, Phillips TL et al (1995) Injury to the lung from cancer therapy: clinical syndromes, measurable endpoints, and potential scoring systems. Int J Radiat Oncol Biol Phys 31:1187–1203
3. Leonidas DD, Chavali GB, Jardine AM et al (2001) Binding of phosphate and pyrophosphate ions at the active site of human angiogenin as revealed by X-ray crystallography. Protein Sci 10:1669–1676
4. Subramanian E, Ramachandran GN (2001) Nat Struct Biol 8:489–491

5. Hulmes DJ (2002) Building collagen molecules, fibrils, and suprafibrillar structures. J Struct Biol 137:2–10
6. Hulmes DJ (1992) The collagen superfamily—diverse structures and assemblies. Essays Biochem 27:49–67
7. Estimation of collagen biosynthesis via [3H]-proline incorporation. http://www.graingerlab.org/assets/grl01/protocols/Method_3Hproline.pdf. 13 July 2011
8. Tullberg-Reinert H, Jundt G (1999) In situ measurement of collagen synthesis by human bone cells with a sirius red-based colorimetric microassay: effects of transforming growth factor beta2 and ascorbic acid 2-phosphate. Histochem Cell Biol 112:271–276
9. Vybrant® MTT cell proliferation assay kit. http://www.invitrogen.com/site/us/en/home/References/protocols/cell-culture/mtt-assay-protocol/vybrant-mtt-cell-proliferation-assay-kit.html. 13 July 2011
10. Carlson CG. Determination of hydroxyproline content as a measure of fibrosis in nondystrophic and dystrophic skeletal muscle. http://www.treat-nmd.eu/downloads/file/sops/dmd/MDX/DMD_M.1.2_006.pdf. 13 July 2011

Chapter 8
Intravoxel Incoherent Motion MR Imaging of the Kidney: Pilot Study

Per Eckerbom, Peter Hansell, Tomas Bjerner, Fredrik Palm, Jan Weis, and Per Liss

Abstract MR examinations (Achieva 3 T, Philips, Best, The Netherlands) were performed at five different occasions in a healthy volunteer (male 60 years) and in one renal cancer patient (male 78 years) with normal renal function (creatinine 88 μmol/L). Intravoxel incoherent motion (IVIM) coefficients $D+D^*$ were measured using respiratory-triggered diffusion-weighted spin-echo echo-planar imaging. Perfusion data of the patient were acquired using a saturation-recovery gradient-echo sequence and with the bolus of Gd-BOPTA (Multihance). $D+D^*$ were computed by monoexponential fitting of MR signal intensity attenuation versus b for $b=0$, 50, 100, 150 s/mm^2. Perfusion parameters were evaluated with "NordicICE" software. The map of $D+D^*$ was compared qualitatively with the perfusion map computed from the Gd scan. $D+D^*$ values of the cortex and medulla were in the range 2.3–2.7 and $1.1–1.6\times10^{-3}$ mm^2/s, respectively. In conclusion, in this pilot study a good qualitative relation between IVIM variables $D+D^*$ and renal perfusion has been found.

Keywords Magnetic resonance • Kidney • Blood flow

P. Eckerbom • T. Bjerner • J. Weis • P. Liss (✉)
Department of Radiology, Center for Medical Imaging, Uppsala University Hospital, Uppsala University, Uppsala, 75263, Sweden
e-mail: per.liss@radiol.uu.se

P. Hansell • F. Palm
Department of Medical Cell Biology, Uppsala University, Uppsala, Sweden

W.J. Welch et al. (eds.), *Oxygen Transport to Tissue XXXIV*, Advances in Experimental Medicine and Biology 765, DOI 10.1007/978-1-4614-4989-8_8,

1 Introduction

Due to the risks associated with contrast media and impaired renal function (contrast-induced nephropathy and nephrogenic systemic fibrosis) there is a need of noninvasive methods for imaging of renal disease. Intravoxel incoherent motion (IVIM) magnetic resonance imaging (MRI) combines the effects of capillary perfusion and water diffusion in the extracellular extravascular space. Quantitative parameters that can simultaneously be measured by IVIM MRI are molecular diffusion D, volume fraction f of the diffusion linked to microcirculation of the blood, and perfusion-related diffusion D^* [1]. Relation between these parameters defines LeBihan's equation for diffusion-weighted signal intensity (SI) [1]: $\mathrm{SI}(b)/\mathrm{SI}_0 = (1-f)\times\exp(-b\times D) + f\times\exp(-b\times(D+D^*))$. The aim of the present study was twofold: (1) to determine $D+D^*$ values of the normal human kidney (cortex and medulla) and (2) to investigate whether there is a relation between $D+D^*$ and renal perfusion measured by conventional contrast-enhanced MRI.

2 Methods

A 3 T MR scanner (Achieva, Philips, Best, The Netherlands) was used in this study. Examinations were performed at five different occasions in a healthy volunteer (male 60 years) (no contrast given) and in one renal cancer patient (male 78 years) with normal renal function (creatinine 88 μmol/L). IVIM coefficients $D+D^*$ were measured using coronal respiratory-triggered diffusion-weighted spin-echo echo-planar imaging [2] (FOV = 267 × 260 mm, resolution 1.8 × 1.8 × 5 mm, pixel size 1.4 × 1.4 mm. TR/TE = 1,800/59 ms, diff. grad. timing 29.2/14 ms, EPI factor = 81, b = 0, 50, 100, 150, 200, 300, 500, 700 s/mm^2). Total scan time was approximately 7 min. Perfusion data of the patient were acquired using saturation-recovery gradient-echo sequence (FOV = 450 × 373 mm, resolution 2.3 × 2.3 × 8 mm, TR/TE = 280/0.98 ms, flip angle 12°) [3] and with the bolus of 7 ml Gd-BOPTA (Multihance). $D+D^*$ was computed by monoexponential fitting of MR signal attenuation $\mathrm{SI}(b)/\mathrm{SI}_0$ in the range $0 \leq b \leq 150$ s/mm^2 assuming that at this interval dominates signal decrease $f\times\exp(-b\times(D+D^*))$. Perfusion parameters were evaluated with commercial "NordicICE" software [4]. The map of $D+D^*$ was compared qualitatively with the perfusion map computed from the Gd scan.

3 Results

$D+D^*$ values of the cortex and medulla were in the range 2.3–2.7 and 1.1–1.6 × 10^{-3} mm^2/s, respectively. Figure 8.1 shows images of the regional blood flow (perfusion map) and diffusion $D+D^*$ map.

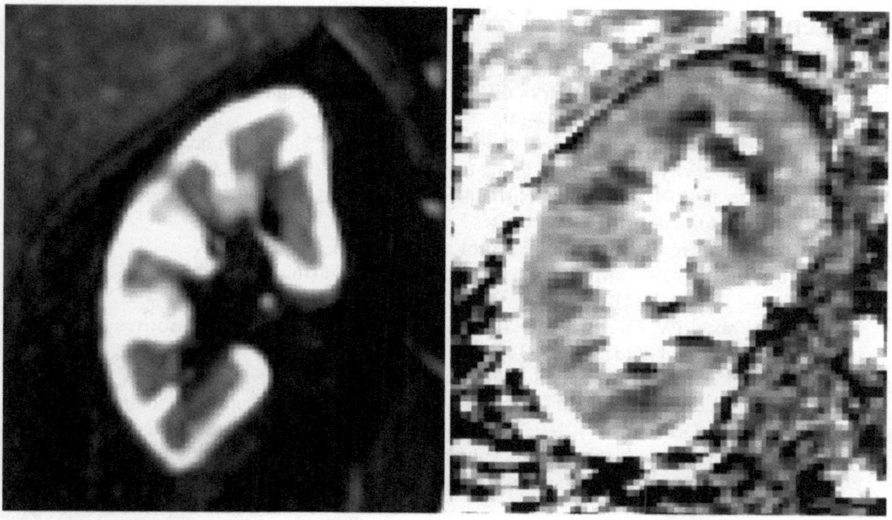

Fig. 8.1 *Left*—perfusion map showing image of the regional renal blood flow computed from gadolinium-enhanced gradient-echo images. *Right*—diffusion map $D+D^*$ computed from diffusion-weighted images with $b=0$, 50, 100, 150 s/mm^2. This map represents "perfusion-weighted" diffusion. Clearly, the highest $D+D^*$ values are restricted almost exclusively to the cortex. Images *left* and *right* are from two different human subjects due to technical reasons. The *central* bright regions in the $D+D^*$ map (*right*) are artifacts due to low signal-to-noise ratio in diffusion-weighted images

4 Discussion

IVIM MRI has previously been shown to be useful in determining renal perfusion [5] although some studies do not support this approach [6]. It should be noted that Wirestam et al. have found a reasonable agreement between regional blood flow measured by conventional MRI techniques and IVIM variables in the brain [7]. To the best of our knowledge, relation between IVIM parameters and regional blood flow in the kidney has not been yet studied. Our results for $D+D^*$ (cortex: 2.3–2.7, medulla: $1.1–1.6\times10^{-3}$ mm^2/s) are in a good agreement with published values [8]. $D+D^*$ maps (Fig. 8.1) seem to be in qualitatively good agreement with the regional renal blood flow images computed from the Gd contrast-enhanced images. The highest $D+D^*$ values are restricted almost exclusively to the cortex. Drawback of our study is lack of sufficient number of patients.

In conclusion, in this pilot study a good qualitative relation between IVIM variables $D+D^*$ and renal perfusion has been found. The finding will be validated in a larger cohort of patients and will include relation between regional renal blood flow and $D+D^*$, D^*, f.

Acknowledgments The work was supported by the Swedish Society for Medical Research, the Swedish Diabetes Foundation, and the Swedish Research Council.

References

1. Le Bihan D et al (1988) Separation of diffusion and perfusion in intravoxel incoherent motion MR imaging. Radiology 168(2):497–505
2. Thoeny HC et al (2005) Diffusion-weighted MR imaging of kidneys in healthy volunteers and patients with parenchymal diseases: initial experience. Radiology 235(3):911–917
3. Michaely HJ et al (2008) Temporal constraints in renal perfusion imaging with a 2-compartment model. Invest Radiol 43(2):120–128
4. NordicNeuroLab, NordicICE, NordicNeuroLab AS (2011). Bergen, Norway
5. Pickens DR III et al (1992) Magnetic resonance perfusion/diffusion imaging of the excised dog kidney. Invest Radiol 27(4):287–292
6. Muller MF, Prasad PV, Edelman RR (1998) Can the IVIM model be used for renal perfusion imaging? Eur J Radiol 26(3):297–303
7. Wirestam R et al (2001) Perfusion-related parameters in intravoxel incoherent motion MR imaging compared with CBV and CBF measured by dynamic susceptibility-contrast MR technique. Acta Radiol 42(2):123–128
8. Yamada I et al (1999) Diffusion coefficients in abdominal organs and hepatic lesions: evaluation with intravoxel incoherent motion echo-planar MR imaging. Radiology 210(3):617–623

Chapter 9
Changes in Gastric Mucosa, Submucosa, and Muscularis IC pH May Herald Irreversible Tissue Injury

Elaine M. Fisher, Sheau Huey Chiu, and Joseph C. LaManna

Abstract Previously we noted an abrupt rise in gastric intracellular pH (IC pH) and bicarbonate buffering between 15 and 30 min of cardiac arrest which we termed agonal alkalinization, failure of pH regulation. Agonal alkalinization may represent the transition point between reversible and irreversible injury. We asked the question, what is the sequence of change in IC pH within the gastric layers, mucosa, submucosa, and muscularis, and which layer is most sensitive? This research explored changes in IC pH within the stomach layers, mucosa, submucosa, and muscularis, at 0, 5, 15, 30, and 40 min, under three conditions, normoxia (control), ischemia (cardiac arrest), and eucapnic hypoxia (12 % oxygen). The mucosa was the most alkalotic gastric layer at baseline. Ischemia and hypoxia at 40″ produced different layer responses with the mucosa and submucosa the most sensitive layers during ischemia and the muscularis during hypoxia. Further study to examine the mechanism of changes between gastric layers using spatial-temporal techniques may assist in understanding the transition to irreversible injury.

Keywords pH • Mucosa

E.M. Fisher • S.H. Chiu
The University of Akron, College of Nursing, Akron, OH, USA

J.C. LaManna (✉)
Department of Anatomy, School of Medicine, Case Western Reserve University, Cleveland, OH, USA

Department of Physiology & Biophysics, School of Medicine, Case Western Reserve University, SOM E 611/10900 Euclid Avenue, Cleveland, OH 44106-4970, USA
e-mail: joseph.lamanna@case.edu

W.J. Welch et al. (eds.), *Oxygen Transport to Tissue XXXIV*, Advances in Experimental Medicine and Biology 765, DOI 10.1007/978-1-4614-4989-8_9,

1 Introduction

The stomach is sensitive to changes in oxygenation and has been targeted as an early monitoring site for compromised blood flow and anaerobic metabolism with one biomarker of anaerobic metabolism being the pH. Pathological conditions causing gastrointestinal hypoxia or ischemia disturb intracellular pH (IC pH) leading to alterations in enzymatic processes and membrane integrity. If unchecked, acid–base imbalance will lead to organ dysfunction and death. Measuring extracellular pH using litmus paper or by pH probe provides an estimate of site-specific pH but does not provide insight into the mechanism of change within tissue.

In previous work we identified an abrupt rise in stomach IC pH and bicarbonate between 15 and 30 min of cardiac arrest, a rise that was similarly noted in sublingual and rectal tissue but at a different time interval, between 30 and 40 min. We defined this reversal in IC pH trend as agonal alkalinization, which we believe represents a loss of pH regulatory function. Failure to regulate IC pH triggers cell injury and death. Agonal alkalization may represent the transition point between reversible and irreversible injury. Thus, understanding the sequence of change in IC pH within gastric layers as cells transition from reversible to irreversible injury may provide clues toward pathogenesis, biomarker discovery, and novel therapeutics and timing of treatment interventions. The purpose of this research was to explore changes in IC pH within the gastric layers, mucosa, submucosa, and muscularis, during normoxia, ischemia (cardiac arrest), and eucapnic hypoxia (12 % oxygen; 35–45 mmHg pCO_2) at 0, 5, 15, 30, and 40 min, and to determine the gastric layer with earliest sensitivity to ischemia and hypoxia.

2 Methods

2.1 *Animal Preparation*

The protocol was approved by the Institutional Animal Care and Use Committee, The University of Akron, in compliance with the *Guide for the Care and Use of Laboratory Animals*. Ninety-one male Wistar rats (375–487 g) were fasted overnight but continued to have free access to water. Rats were randomized to group based on condition (normoxia, ischemia, eucapnic hypoxia) and five (5) time-to-death intervals, 0, 5, 15, 30, and 40 min. Rats were euthanized at each time-to-death interval to allow for IC pH analysis thus, there were 6–8 rats per group.

Rats were placed in an anesthesia box suffused with 3 % isoflurane and anesthesia was maintained using 0.75–1.5 % isoflurane/30 % O_2 via nose cone until an endotracheal tube was placed. Animals were mechanically ventilated using standard nomograms (Inspira Advanced Ventilator, Harvard Apparatus, MA). Tail artery cannulation was performed and the catheter was connected to a transducer leveled at the mid-chest region to monitor MAP (Grass Instruments, Rockland, MA) and for arterial blood sampling. The right external jugular vein was cannulated and used for dye/drug infusion. Catheters were intermittently flushed with 0.9 % normal saline/heparin 5 IU/mL to maintain patency. As part of the larger study, pCO_2 electrodes (MI-720, Microelectrodes Inc., NH) were placed in the sublingual space and rectum. In addition, a midline 5-cm incision was made and a pCO_2 microelectrode was surgically placed in the stomach. A small catheter was placed in the abdominal cavity and externalized to the outer wall for intraperitoneal (IP) dye injection. The abdomen was closed in two layers using 4–0 suture. Body temperature was maintained at 37.0 °C (±1 °C) rectally using a temperature controller with animal heating pad (TR-100; Fine Science Tools, CA).

Post-surgery, 3 mL of a 2 % neutral red (NR) dye solution was slowly infused (30 min) to prevent hypotension. An additional 2 mL of NR dye was instilled into the abdominal cavity via the IP catheter to ensure adequate dye uptake by GI structures. Rats were gently rotated to distribute the dye to abdominal structures. At the appropriate time-to-death interval, rats were euthanized by decapitation. The stomach was removed, flash frozen in liquid nitrogen, and stored at −80 °C.

2.2 *Study Models*

All animals were surgically instrumented, received a 30″ infusion of NR dye, and an IP infusion of NR. Time 0 began at the end of dye infusion. For the *normoxic (control) model*, rats were ventilated using standard algorithms with oxygen delivered between 21 and 30 % to maintain the paO_2 greater than 80 mmHg. For the *ischemic model*, while anesthetized, cardiac arrest was induced by rapid, venous injection of vecuronium (0.10–0.20 mg/kg) followed by a potassium chloride bolus (0.5 M/L; 0.12 mL/100 g of body weight). Time 0 began when MAP reached 0 on the computer display. Temperature was maintained at 37 °C. The *eucapnic hypoxic model* was produced by administering 12 % oxygen to the mechanically ventilated rat. The percent of oxygen was varied to keep the paO_2 within a range of 40–50 mmHg, sufficient for survival of the animal. The respiratory rate on the mechanical ventilator was adjusted to maintain the $paCO_2$ within normal range, 35–45 mmHg. We controlled pCO_2 based on empirical evidence that hypocapnia increases intraluminal pH and partially masks mucosal acidosis. Bicarbonate (HCO_3^-) was administered as needed to keep the arterial HCO_3^- greater than 22 mEq/L. Evidence suggests fluctuations in arterial HCO_3^- affect the luminal pH and tissue pCO_2, other measured variables [1].

2.3 *IC pH Analysis*

IC pH was determined by a reflectance histophotometric imaging technique [2, 3]. This process was adapted for use in gastric tissue and has been previously reported [4].

Grayscale tissue images were used to identify the area of interest (AOI) and these AOIs were identified within the pixel images. Tissues were sliced four times, 50 μm apart. For each tissue slice, four AOIs were identified within each layer. We avoided measuring AOIs too close to layer borders to ensure accurate layer pH measurement.

2.4 Data Analysis

Prior to data analyses, descriptive statistics were computed and decisions about processing the raw data were made. Histograms of IC pH values were constructed by condition at each time point. These graphs provided good visual distribution of the data but did not provide an analytic platform for quantitatively comparing differences in IC pH distribution. Histograms were skewed to the left toward a more acidic IC pH. A paired *t*-test comparing mean to median IC pH values resulted in the mean underestimating IC pH, thus the median value was used in all data analyses ($p < .001$). IC pH pixels outside of the physiologic dye range of 6–8 units were eliminated before calculating the median. The IC pH under normoxic/control condition did not vary at any time point (0–40″). Thus, all control points were collapsed to produce an average median value for each layer. This value is reported as Time 0 (baseline). Analysis of variance (ANOVA) to compare condition-by-time-by-layer with Bonferroni post hoc pair-wise comparisons was used to evaluate interaction effects. The level of significance was set at $p < 0.05$. All IC pH data are reported as mean ± SE.

3 Results

Figure 9.1 depicts the average IC pH value of gastric layers at baseline. The mucosa was the most alkalotic layer and the submucosa was the most acidic layer at baseline ($p < 0.001$). Whole tissue IC pH for this study was 7.05 ± 0.08 (mean ± SE).

Ischemia produced greater changes in gastric layer IC pH than control or eucapnic hypoxia ($p < 0.001$). The layer IC pH relationship changed as oxygen delivery decreased over time for both the hypoxic and ischemic group, most notably at 40″ (Fig. 9.2). A significant interaction effect occurred between condition, time, and layer ($p < 0.03$). During ischemia, the IC pH for all layers was similar to baseline up to 30 min. A decrease in mucosal IC pH occurred at 5″ ischemia. Between 15 and 30″ ischemia, mucosal IC pH reversed direction and increased approximately 0.2 pH units, with significance above baseline reached at 40″ ($p < 0.001$). The change in mucosal IC pH between 15 and 30″ coincides with the time interval for whole tissue agonal alkalinization. For the submucosa, the IC pH was constant up to 30″ and increase by approximately 0.3 pH units at 40″ ($p < 0.001$). The IC pH for the muscularis layer varied with an upward trend beginning at 30″. Thus during ischemia the mucosa was the first layer to change.

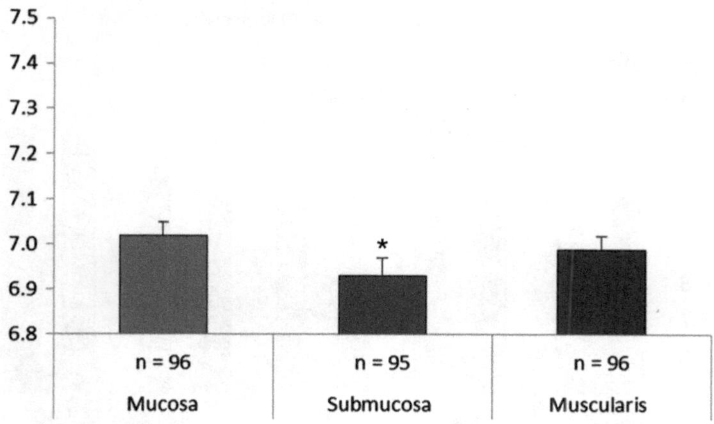

Fig 9.1 Baseline gastric IC pH by layer (mean ± SE). *$p < 0.001$ in comparison to other layers

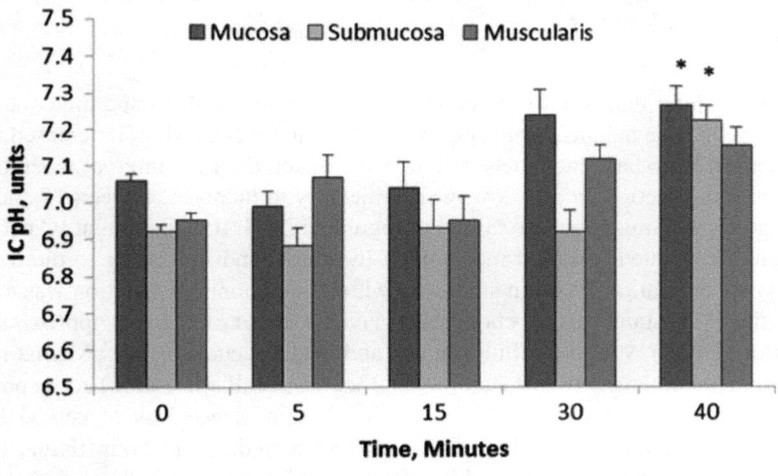

Fig. 9.2 Ischemic changes in the gastric layer IC pH over time. *$p < 0.001$ when layer is compared to Time 0, baseline

Hypoxia produced a limited range of change in IC pH when compared to ischemia (Fig. 9.3). At 40″, only the muscularis reached statistical significance from baseline, rising 0.15 pH units ($p = 0.03$). Of note, the muscularis IC pH was the same value as that of the muscularis at 40″ ischemia. Increases in IC pH between 30 and 40″ were detected for both the mucosa and submucosa. The muscularis was the most sensitive layer to reduced oxygen delivery from hypoxia.

The difference in IC pH between whole tissue and layers under normoxic condition is not surprising since layers of the stomach are functionally different in tissue perfusion and metabolism. Layer sensitivity between hypoxia and ischemia differed at 40″ suggesting the mechanism underlying changes in IC pH may vary based on the severity and/or cause of reduction in oxygen.

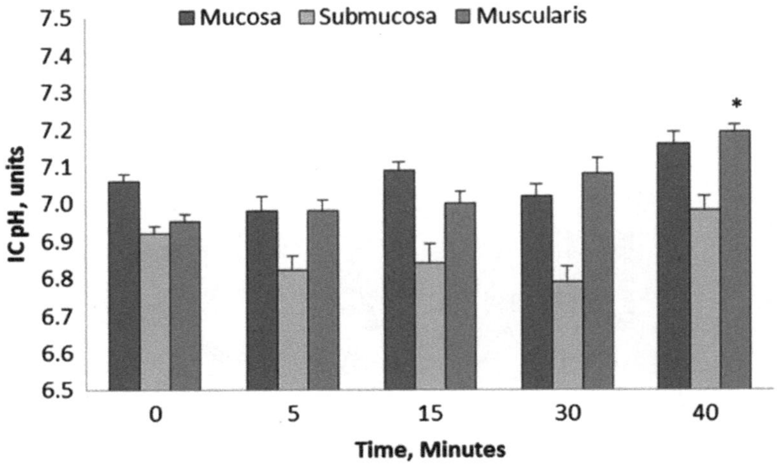

Fig. 9.3 Eucapnic hypoxic changes in gastric layer IC pH over time. $*p<0.001$ when layer is compared to Time 0, baseline

A strong interrelationship exists between energy metabolism and the control of cellular acid–base balance. Between 15 and 30 min, mucosal IC pH increased, with significance from baseline reached at 40″. We linked the time interval whereby IC pH reversed direction from a downward trajectory to increase between 15 and 30″ with agonal alkalinization, the failure to regulate IC pH. It is likely that IC pH dysregulation is related to failed metabolism by mitochondria leading to the loss of transporter regulation. In brain studies in which mitochondrial function was evaluated using TTC staining, mitochondria survived without oxygen for approximately 30 min before irreversible cellular injury and death occurred [5]. LaManna noted changes in the intensity of NR staining in brain tissue (alkalinization) corresponded with failure of the tissue to take up the TTC stain, hence loss of cell viability (Personal communication, 2010). Consistent with findings in brain tissue, Guan et al. studied redox biomarkers and IC pH measures in mucosal cells of the jejunum and found 40–50″ ischemia followed by reperfusion did not restore normal structure or functioning of the majority of mucosal cells, concluding irreversible injury had occurred [6]. While we suspect irreversible injury occurred during agonal alkalinization, we cannot validate these changes.

The limited dynamic range in layer IC pH during hypoxia indicates hypoxia is better tolerated than ischemia. In the broader study, IC pH and tissue pCO_2 were simultaneously measured. Hypoxia protocol prescribed a paO_2 less than 50 mmHg (range 35–51 mmHg) with controlled $paCO_2$ (range 38–46 mmHg). A small, nonsignificant increase in gastric pCO_2 occurred at 40″ hypoxia which would not be considered clinically significant. The stable IC pH during hypoxia may be explained by the presence of intact compensatory mechanism whereby lowering the IC pH triggered afferents to increase blood flow, thus flushing CO_2 from the tissue and stabilizing IC pH [7]. To evaluate whether tissues were more sensitive to the effects of ischemia or hypoxia, Dubin et al [8] compared the pCO_2 gradient (tissue

pCO_2–arterial pCO_2) under sham, hypoxic hypoxia with preserved blood flow, and ischemic conditions. No change in pCO_2 was found at 90″ hypoxia with preserved blood flow, concluding blood flow not hypoxia is the more influential factor in injury. Nevertheless, we have shown that IC pH changes are detectable at 40″ hypoxia. Tracking changes in IC pH over longer time intervals may be useful to define the time course during which irreversible injury develops.

4 Conclusion

Our study quantified IC pH within the gastric mucosa, submucosa, and muscularis under normoxic, ischemic, and hypoxic conditions. The differential pattern of IC pH layer response between ischemia and hypoxia indicates the cause of reduced oxygen delivery is important, with ischemia producing greater pH dysregulation. Further study is needed to examine changes in IC pH between the layers, especially at layer borders, using spatial-temporal analysis methods, cluster analysis, and edge detection. If identifiable IC pH clusters/patterns are found, this knowledge would allow us to generate predictive models mathematically and pose hypotheses to be tested in future translational research.

Acknowledgments Supported by K01 NR009787-01 from NIH:NINR & Delta Omega Chapter, Sigma Theta Tau International, Inc. to first author. Special thanks to Miss Sujin Kim for her assistance with data processing and analysis.

References

1. Jakob SM, Parviainen I, Ruokonen E, Kogan A, Takala J (2008) Tonometry revisited: perfusion-related, metabolic, and respiratory components of gastric mucosal acidosis in acute cardiorespiratory failure. Shock 29(5):543–548
2. LaManna JC, McCracken KA, Whittingham TS, Lust WD (1986) Determination of intracellular pH by color film histophotometry of frozen in situ rat brain. Adv Exp Med Biol 200:253–259
3. Hoxworth JM, Xu K, Zhou Y, Lust WD, LaManna JC (1999) Cerebral metabolic profile, selective neuron loss, and survival of acute and chronic hyperglycemic rats following cardiac arrest and resuscitation. Brain Res 821(2):467–79
4. Fisher E, LaManna J (2006) Intracellular pH in gastric and rectal tissue post cardiac arrest. Adv Exp Med Biol 578:6–11
5. Benedek A, Moricz K, Juranyi Z, Gigler G, Levay G, Harsing LG Jr, Matyus P, Szenasi G, Albert M (2006) Use of TTC staining for the evaluation of tissue injury in the early phases of reperfusion after focal cerebral ischemia in rats. Brain Res 1116(1):159–165
6. Guan Y, Worrell RT, Pritts TA, Montrose MH (2009) Intestinal ischemia-reperfusion injury: reversible and irreversible damage imaged in vivo. Am J Physiol Gastrointest Liver Physiol 297(1):G187–G196
7. Akiba Y, Kaunitz JD (1999) Regulation of intracellular pH and blood flow in rat duodenal epithelium in vivo. Am J Physiol 276(1 Pt 1):G293–G302
8. Dubin A, Murias G, Estenssoro E, Canales H, Badie J, Pozo M, Sottile JP, Baran M, Palizas F, Laporte M (2002) Intramucosal-arterial PCO2 gap fails to reflect intestinal dysoxia in hypoxic hypoxia. Crit Care 6(6):514–520

Chapter 10
Normobaric Hyperoxia Does Not Change Optical Scattering or Pathlength but Does Increase Oxidised Cytochrome *c* Oxidase Concentration in Patients with Brain Injury

Arnab Ghosh, Ilias Tachtsidis, Christina Kolyva, David Highton, Clare Elwell, and Martin Smith

Abstract We report the use of a novel hybrid near-infrared spectrometer for the measurement of optical scattering, pathlength and chromophore concentration in critically ill patients with brain injury. Ten mechanically ventilated patients with acute brain injury were studied. In addition to standard neurointensive care monitoring, middle cerebral artery flow velocity, brain lactate–pyruvate ratio (LPR) and brain tissue oxygen tension were monitored. The patients were subjected to graded normobaric hyperoxia (NBH), with the inspired fraction of oxygen increased from baseline to 60% then 100%. NBH induced significant changes in the concentrations of oxyhaemoglobin, deoxyhaemoglobin and oxidised–reduced cytochrome *c* oxidase; these were accompanied by a corresponding reduction in brain LPR and increase in brain tissue oxygen tension. No significant change in optical scattering or pathlength was observed. These results suggest that the measurement of chromophore concentration in the injured brain is not confounded by

---

The original version of this chapter was revised. An erratum to this chapter can be found at DOI 10.1007/978-1-4614-4989-8_53

A. Ghosh (✉)
Institute of Neurology, University College London, 4 Brookfield Road, Queen Square, London E9 5AH, UK

Neurocritical Care, University College Hospitals, Queen Square, London, UK
e-mail: arnab.ghosh@ucl.ac.uk

I. Tachtsidis • C. Kolyva • C. Elwell
Medical Physics and Bioengineering, University College London, Malet Place, London, UK

D. Highton
Neurocritical Care, University College Hospitals, Queen Square, London, UK

M. Smith
Neurocritical Care, University College Hospitals, Queen Square, London, UK

Medical Physics and Bioengineering, University College London, Malet Place, London, UK

W.J. Welch et al. (eds.), *Oxygen Transport to Tissue XXXIV*, Advances in Experimental Medicine and Biology 765, DOI 10.1007/978-1-4614-4989-8_10,

changes in optical scattering or pathlength and that NBH induces an increase in cerebral aerobic metabolism.

Keywords Hyperoxia • Optical scattering

1 Introduction

The identification and avoidance of cerebral hypoxia/ischaemia is a central tenet of contemporary neurocritical care, and near-infrared spectroscopy (NIRS) has a potential monitoring role in this regard. However, whilst a plethora of studies describe the use of "cerebral oximeter" devices—which express absolute scaled concentrations of oxyhaemoglobin and deoxyhaemoglobin ([HbO_2] and [HHb], respectively) in the form of a combined regional haemoglobin oxygen saturation—few studies outside the context of cardiopulmonary bypass link brain desaturation with neurological outcome or support the routine clinical use of such devices [1].

Cytochrome *c* oxidase (CCO) is the terminal electron acceptor in the mitochondrial electron transport chain and responsible for 95% of cellular oxygen utilisation. Like haemoglobin, the oxidised and reduced forms of CCO have characteristic absorption spectra within the NIR band; these can be measured as the difference spectrum of oxidised–reduced CCO. However, changes in optical scattering and pathlength could potentially confound the accurate measurement of oxidised–reduced CCO concentration ([oxCCO]) in vivo [2].

The aim of this study is to measure optical scattering, pathlength and changes in chromophore concentration in a cohort of brain-injured patients during normobaric hyperoxia (NBH).

2 Methods

After approval by the institutional Research Ethics Committee and representative consent, recordings were carried out in ten sedated, mechanically ventilated acute brain-injured patients on the neurocritical care unit. These patients were subjected to NBH protocol, which consisted of a 60-min epoch of baseline recording, followed by 60-min epochs where the inspired fraction of oxygen (FiO_2) was increased to 60% and then 100%, followed by a final 30-min epoch where FiO_2 was returned to baseline values. In patients with a baseline FiO_2 of ≥50%, the 60% FiO_2 epoch was omitted.

Systemic physiological monitoring included arterial blood pressure (ABP), pulse oximetry and intermittent measurement of arterial blood gases (ABGs)—including partial pressures of oxygen and carbon dioxide (PaO_2 and $PaCO_2$, respectively). Cerebral monitoring was positioned ipsilaterally to the more injured hemisphere and included transcranial Doppler (TCD) ultrasonography measurement of middle cerebral artery flow velocity (Vmca) (DWL Doppler Box, Compumedics Germany), measurement of the lactate–pyruvate ratio (LPR) by cerebral microdialysis (Dipylon

Medical AB, Solna, Sweden) and continuous measurement of brain tissue oxygen tension ($pbrO_2$) (Licox, Integra Neurosciences, Plainsboro, USA).

The hybrid optical spectrometer has been described in detail elsewhere [3]. Briefly, it comprises two identical broadband spectrometers and a two-channel frequency domain (FD) spectrometer capable of absolute measurements of optical absorption and scattering coefficients (μ_a and μ_s, respectively) at 690, 750, 790 and 850 nm. [HHb], [HbO_2] and [oxCCO] were calculated using the UCLn algorithm [4], fitting to changes in NIR attenuation from 780 to 900 nm. Differential pathlength factor (DPF) was derived from μ_a and μ_s measured at 790 nm during the baseline period of recording, with additional correction applied for the wavelength dependence of pathlength [5]. Optodes were placed in the mid-pupillary line on the forehead, ipsilateral to the TCD and invasive monitoring. Chromophore concentrations derived from the 35 mm source–detector separation are presented.

Cerebral LPR was measured at 15-min intervals and ABGs at 30-min intervals. The remaining systemic, cerebral and NIRS variables were synchronised and monitored continuously, with the mean value from a noise-free window comprising ≥50% of each entire epoch used for subsequent analysis and reporting.

All values are reported as median ± interquartile range (IQR) except where otherwise stated. Probability was calculated using a Wilcoxon signed-rank test comparing the difference between the initial baseline and subsequent study epochs. A Bonferroni correction for repeated measures was applied and $p<0.017$ defined as the level of statistical significance.

3 Results

Patient demographics are summarised in Table 10.1. The NBH protocol induced significant changes in $paO_2$and $pbrO_2$, but not in $paCO_2$ or Vmca (Table 10.2).

The FD spectrometer failed to collect data at 750 nm in two patients and at 790 and 850 nm in one patient. The remaining data revealed no significant change observed in the values of μ_s and DPF at any wavelength (Table 10.3).

Changes in chromophore concentration are shown in Fig. 10.1. There was a statistically significant increases in [HbO_2] and [oxCCO] and decreases in [HHb] during NBH (Δ[HbO_2]: +0.36 and +0.58 μmol l^{-1}; Δ[HHb]: –0.54 and –1.56 μmol l^{-1}; Δ[oxCCO]: +0.15 and +0.28 μmol l^{-1} during 60% FiO_2 and 100% FiO_2 phases, respectively). The baseline median LPR was 24.4 (IQR 22.2–25.7) and showed a statistically significant decrease during NBH (ΔLPR: –2.8 and –3.3 during 100% FiO_2 and return-to-baseline epochs, respectively) and a nonsignificant decrease of –0.8 during the 60% FiO_2 phase ($p<0.019$).

Table 10.1 Data on patient demographics

Patient demographics	
Age (range)	45.5 (23–74)
Sex	7 Females, 3 males
Pathology	4 Traumatic brain injury, 6 subarachnoid haemorrhage
Median Admission Glasgow Coma Score (IQR)	7.5 (4–8)

Table 10.2 Median (IQR) values for monitored physiological variables during four phases of experiment

	Baseline	FiO_2 60%	FiO_2 100%	Return to baseline
ABP (mmHg)	89.9 (81.4–95.2)	94.6 (82.9–99.7)	95.2 (90.7–97.1)	90.7 (84.3–93.2)
paO_2 (kPa)	13.9 (11.7–18.1)	*26.6* (25.8–31.1)*	*52.9* (49.2–57.0)*	11.8 (10.9–14.4)
$paCO_2$ (kPa)	4.8 (4.6–5.0)	5.0 (4.8–5)	4.8 (4.6–4.9)	5.1 (4.7–5.5)
$pbrO_2$ (kPa)	2.8 (1.6–3.4)	*4.0*(2.5–4.7)*	*7.4*(5.6–8.6)*	3.5(1.7–4.7)
Vmca (cm s^{-1})	54.3(51.3–79.6)	56.2(48.7–84.7)	54.3(50.3–83.1)	57.1(51.8–86.1)

Values showing a statistically significant difference from initial baseline are italicised
*$p<0.01$

Table 10.3 Median (IQR) values for measured optical scattering coefficient (μ_s) and differential pathlength factors at four wavelengths

	Baseline	FiO_2 60%	FiO_2 100%	Return to baseline
μ_s 690 nm (cm^{-1})	10.8(9.05–10.8)	10.6(10.1–12.3)	10.5(9.74–12.4)	10.3(9.39–12.5)
μ_s 750 nm (cm^{-1})	9.93(9.02–10.2)	10.0(9.04–10.5)	9.76(9.00–10.6)	9.89(8.99–10.7)
μ_s 790 nm (cm^{-1})	9.48(8.8–9.75)	9.49(8.59–11.5)	9.25(8.25–11.8)	9.10(7.44–11.8)
μ_s 850 nm (cm^{-1})	9.10(9.0–9.70)	9.40(8.76–9.80)	9.36(8.68–9.54)	9.35(8.9–9.4)
DPF 690 nm	8.40(8.33–8.71)	8.634(8.52–9.44)	8.88(8.49–9.57)	8.54(8.03–9.37)
DPF 750 nm	8.15(7.9–8.24)	8.12(7.76–8.44)	8.32(7.96–9.11)	8.15(7.87–9.05)
DPF 790 nm	8.14(8.12–8.28)	8.24(7.91–8.49)	8.21(7.92–8.81)	7.98(7.89–8.63)
DPF 850 nm	8.07(8.06–8.52)	8.00(7.73–8.40)	7.94(7.66–8.37)	8.00(7.83–8.31)

No variable showed statistically significant variation from baseline

4 Discussion

We have used a hybrid optical spectrometer to measure optical scattering, pathlength, changes in oxy- and deoxyhaemoglobin concentrations and changes in CCO oxidation state during NBH in ten critically ill brain-injured patients.

Changes in chromophore concentration, and in particular [oxCCO], have been reported before in a clinical context: the pattern of [oxCCO] change has been related to clinical outcome following cardiopulmonary bypass [6] and our group have previously demonstrated an increase in [oxCCO] with NBH [7] and heterogeneous changes with hypercapnoea [8]. However, changes in optical scattering and pathlength have been proposed as key confounding factors in the measurement of [oxCCO] [2] and the effects of changes in pathlength and scattering on [oxCCO] measurement have been hitherto unknown.

We have adopted a novel approach that combines simultaneous measurement of optical scattering and pathlength in addition to measurement of changes in chromophore concentration and LPR. The concordance between the microdialysis and [oxCCO], combined with the absence of changes in either μ_s or pathlength, suggests that scattering and pathlength changes do not confound the measurement of

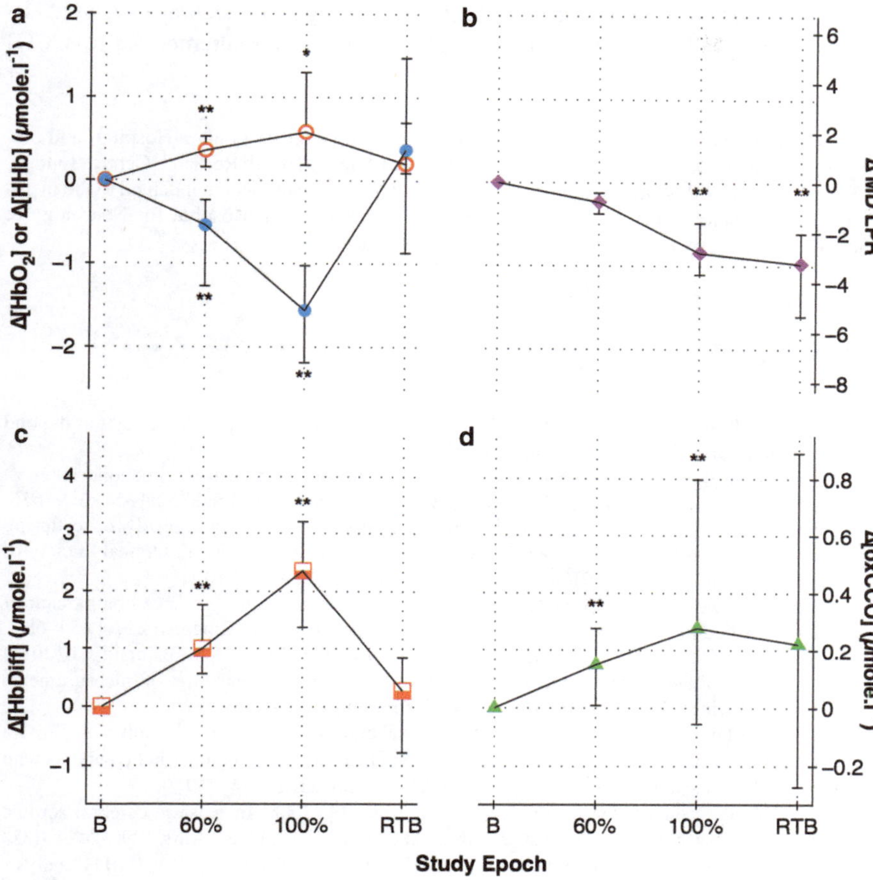

Fig. 10.1 Changes in: (**a**) oxyhaemoglobin (*open circles*) and deoxyhaemoglobin (*solid circles*); (**b**) microdialysate lactate–pyruvate ratio; (**c**) haemoglobin difference; (**d**) oxidised–reduced cytochrome *c* oxidase for each phase of study. *B* baseline, *60%* FiO_2 60%, *100%* FiO_2 100%, *RTB* return to baseline. *$p<0.017$; **$p<0.01$ comparing study epoch to initial baseline

chromophore concentration and that the change in [oxCCO] seen during NBH therefore represents an actual increase in mitochondrial aerobic metabolism.

The pattern of chromophore concentration change that we have observed during NBH is similar to that previously reported by our group using different apparatus in a cohort patients who had a similar median age and admission GCS, but had a primary diagnosis of traumatic brain injury rather than the mixed group of traumatic and vascular brain injuries reported in this study. The magnitude of [HbO_2] increase, however, was smaller in our cohort than that previously reported. The reasons for this are unclear, but may include pathophysiological and instrumentation factors. Interpreting the physiological mechanisms underlying the increase in [oxCCO] in response to NBH is challenging, but may be enhanced by the use of a mathematical model of the cerebral circulation and metabolism [9]. Further work is required to elucidate the relationship between [oxCCO] and changes in cerebral oxygen availability and may aid in the translation of NIRS-derived measurement of [oxCCO] from a research to a clinical tool.

Acknowledgments This work was undertaken at University College London Hospitals and partially funded by the Department of Health's National Institute for Health Research Centres funding scheme, the Medical Research Council and Wellcome Trust. The authors are indebted to the medical and nursing staff of the Neurocritical Care Unit at the National Hospital for Neurology & Neurosurgery, and especially to the study patients and their families.

References

1. Highton D, Elwell C, Smith M (2010) Noninvasive cerebral oximetry: is there light at the end of the tunnel? Curr Opin Anaesthesiol 23(5):576–581
2. Cooper CE, Springett R (1997) Measurement of cytochrome oxidase and mitochondrial energetics by near-infrared spectroscopy. Philos Trans R Soc Lond B Biol Sci 352(1354):669–676
3. Tachtsidis I, Leung T, Ghosh A, Smith M (2010) Multi-wavelength, depth resolved, scattering and pathlength corrected in vivo near-infrared spectroscopy of brain tissue. Optical Society of America Biomedical Optics (BIOMED) BTuB, Miami, FL, 11 Apr 2010
4. Matcher SJ, Elwell CE, Cooper CE, Cope M, Delpy DT (1995) Performance comparison of several published tissue near-infrared spectroscopy algorithms. Anal Biochem 227(1):54–68
5. Gao L, Elwell CE, Kohl-Bareis M, Gramer M, Cooper CE, Leung TS, Tachtsidis I (2011) Effects of assuming constant optical scattering on haemoglobin concentration measurements using NIRS during a Valsalva manoeuvre. Adv Exp Med Biol 701:15–20
6. Kakihana Y, Matsunaga A, Tobo K, Isowaki S, Kawakami M, Tsuneyoshi I, Kanmura Y, Tamura M (2002) Redox behavior of cytochrome oxidase and neurological prognosis in 66 patients who underwent thoracic aortic surgery. Eur J Cardiothorac Surg 21(3):434–439
7. Tisdall MM, Tachtsidis I, Leung TS, Elwell CE, Smith M (2008) Increase in cerebral aerobic metabolism by normobaric hyperoxia after traumatic brain injury. J Neurosurg 109(3):424–432
8. Tachtsidis I, Tisdall MM, Pritchard C, Leung TS, Ghosh A, Elwell CE, Smith M (2011) Analysis of the changes in the oxidation of brain tissue cytochrome-c-oxidase in traumatic brain injury patients during hypercapnoea: a broadband NIRS study. Adv Exp Med Biol 701:9–14
9. Banaji M, Mallet A, Elwell CE, Nicholls P, Cooper CE (2008) A model of brain circulation and metabolism: NIRS signal changes during physiological challenges. PLoS Comput Biol 4(11):e1000212

Chapter 11
Multi-frequency Forced Oscillation Technique Using Impulse Oscillations: Can It Give Mechanical Information about the Lung Periphery?

Hiroshi Hamakawa, Hiroaki Sakai, Ayuko Takahashi, Toru Bando, and Hiroshi Date

Abstract Forced oscillation techniques (FOTs) using sine curved oscillatory waves are used for assessing the lung periphery, but measure only overall respiratory mechanics. Therefore, mathematical models of the respiratory system serve as vehicles to obtain detailed mechanics. Although the simplest model of respiratory mechanics is a simple 3-element series (RIC) model, the constant phase (CP) model is recently used for characterizing respiratory mechanics, which has the advantage of partitioning of respiratory mechanics into airway and tissue components. Meanwhile, FOTs using non-sine curved oscillatory waves are easily applied in patients with severe respiratory diseases because they do not require voluntary apnea. If the latter type of FOTs is as informative as the former, the question arises whether a FOT using non-sine curved oscillatory waves (IOS) could be used to study mechanical properties of the lung periphery. And the CP model should fit the impedance spectra. To answer this, subjects with lymphangioleiomyomatosis (LAM) were recruited as a cohort of patients with lung parenchymal disease. Impedance spectra obtained by the IOS were fitted to the CP and RIC models. Mean values of goodness of fit from the CP and RIC models were 0.978 ± 0.022 and 0.968 ± 0.026, respectively. The extra sum-of-squares F test was used to compare the two mathematical models. The F ratio was 2.37 ± 1.40 and the p-value was 0.29 ± 0.21. Unfortunately, there was no compelling evidence for adopting the CP

H. Hamakawa (✉)
Department of Thoracic Surgery, Kobe City Medical Center General Hospital, 4-6, Minatojimanakamachi, Chuo-ku, Kobe City, Hyogo 650-0046, Japan
e-mail: hamachan@kcho.jp

H. Sakai • T. Bando • H. Date
Department of Thoracic Surgery, Kyoto University, Kyoto, Japan

A. Takahashi
Department of Biomedical Engineering, Boston University, Boston, MA 02215, USA

W.J. Welch et al. (eds.), *Oxygen Transport to Tissue XXXIV*, Advances in Experimental Medicine and Biology 765, DOI 10.1007/978-1-4614-4989-8_11,

model for the evaluation of impedance spectra obtained by IOS. This result might relate to the uncertainty of IOS for detecting mechanical properties of the lung periphery.

Keywords Lung periphery • Multifrequency forced oscillation

1 Introduction

Studying peripheral lung mechanics requires invasive techniques, which are not normally feasible in human patients with severe respiratory disorders. To overcome this obstacle, forced oscillation techniques (FOTs) are accepted as a tool to assess lung mechanics [1]. Spontaneous breathing must be suspended while sine curved oscillatory waves are applied to the airway opening. However, the techniques themselves measure only overall mechanics of the respiratory system. To complement this method and to reduce the need for invasive procedures, mathematical modeling is employed as an alternative approach to obtain mechanical information about the lung periphery. For this purpose, the constant phase (CP) model is used worldwide in the assessment of respiratory mechanics [2]. FOTs using non-sine curved oscillatory waves are also easily applied in patients with respiratory diseases because they do not require voluntary apnea [1, 3]. Nevertheless, there are few reports to date of studies of the mechanical properties of the lung periphery using non-invasive FOTs.

Based on these characteristics, we are interested in whether an FOT using non-sine curved oscillatory waves (impulse oscillometry system: IOS) [4], which are superimposed on tidal breathing, can be used to study mechanical properties of the lung periphery, and if so, which mathematical model would best fit the data. Here, we discuss whether impedance data obtained by IOS, followed by fitting to simple and complicated mathematical models, can be used for the assessment of the lung periphery.

2 Materials and Methods

2.1 Subjects

Respiratory impedance is influenced by gender, age, physique, and disease background, which can affect the conclusions drawn from studies using FOT [3, 5]. Therefore, we chose patients with lymphangioleiomyomatosis (LAM) as a cohort of subjects with diffuse lung disease whose backgrounds were relatively uniform.

The patients with LAM were recruited in our hospital for preoperative assessment and/or evaluation as lung transplant candidates. According to the criteria of the Japanese Society of Lung and Heart-Lung Transplantation, all subjects underwent spirometry and tests to determine static lung volumes, diffusing capacity, and respiratory input impedance. This protocol was approved by the Institutional Review Board at Kyoto University Hospital (E-1067). We conducted a retrospective analysis of prospectively collected data from a cohort of patients with LAM. Subjects with ambiguous results of computation after CP model fitting were excluded from this descriptive study.

The one female without LAM, who was a donor for a living lobar lung transplantation, underwent the same assessments as the patients with LAM and was included as a normal control. Six female subjects were finally enrolled in this study because other subjects less than acceptable lower limit of the CP model fitting were excluded. The mean age was 27.7 ± 4.0 years. The mean height was 160.2 ± 6.6 cm (means ± SD).

2.2 Mathematical Models

The simple 3-element series (RIC) model is the simplest model of the respiratory system, consisting of a single ventilated airspace served by a single conduit [6]. Respiratory input impedance (Z_{rs}) is represented as a function of angular frequency (ω) and consists of the in-phase [$R(\omega)$] and out-of-phase [$X(\omega)$] components. After model fitting, respiratory resistance (R_{rs}) is independent of frequency, where $Z_{rs}(\omega) = R(\omega) + jX(\omega) = R_{rs} + j(\omega I_{rs} - 1/\omega C_{rs})$ and where j is the imaginary unit, I_{rs} is inertia, and C_{rs} is compliance of the respiratory system.

The CP model allows the partitioning of respiratory mechanics into a central airway [airway resistance (R_{aw}), airway inertance (I_{aw})] and a peripheral tissue component [tissue dumping (G), tissue elastance (H)]. Tissue compliance (C_{tis}) is calculated as $1/H$. The CP model is described by the equation:

$$Z(\omega) = R_{aw} + j\omega I_{aw} + \frac{G - -jH}{\omega^{\alpha}},$$

where α is $2/\pi \tan^{-1}(H/G)$.

2.3 Methods

The IOS (Erich Jaeger GmbH, Hoechberg, Germany) was used to obtain respiratory input impedance at discrete frequencies from 3 to 35 Hz. Each observation consisted of at least three tests which did not have artifacts caused by air leakage from the mouth, coughing, breath holding, swallowing, and so on. The raw data of the impedance spectra were fitted to the RIC and CP models (Fig. 11.1). After model fittings, the set of impedance spectra was excluded if the model fitting was out of permissible range. The parameters of each mathematical model with the best value of goodness of fit (R^2) were given as mean and used for statistical evaluations.

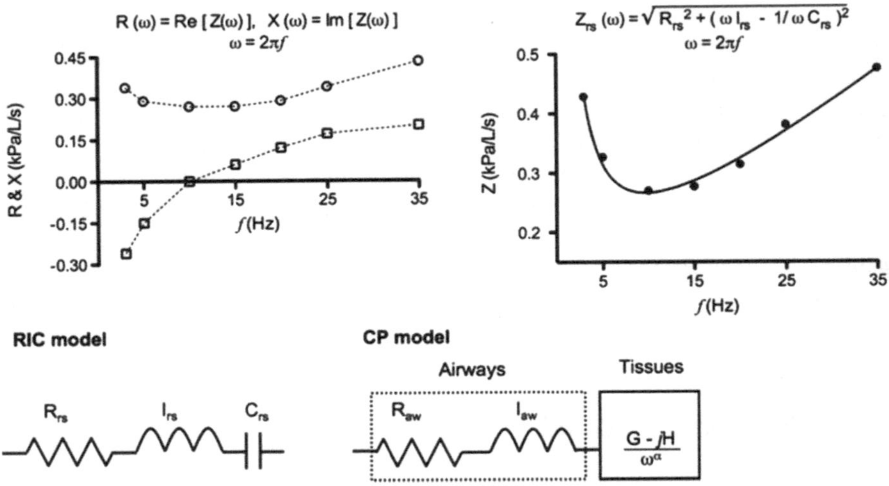

Fig. 11.1 Representative respiratory impedance curve (*upper two graphs*). *Open circle* = real part of respiratory impedance. *Open square* = imaginary part of respiratory impedance. *Closed circle* = respiratory impedance. *Lower figures* are the schematic diagrams of RIC and CP models

The extra sum-of-squares *F* test was used to compare the two mathematical models because they are nested. To evaluate the relationships between lung function parameters and mechanical parameters, Pearson's correlation coefficients (r) were calculated. In this study, forced expiratory volume in 1 s% predicted (%FEV1) and the ratio of diffusing capacity to alveolar volume % predicted (%DLCO/VA) were used. A p-value <0.05 was considered statistically significant.

3 Results

Six female subjects were enrolled in this study. The mean age was 27.7 ± 4.0 years. The mean height was 160.2 ± 6.6 cm. After nonlinear curve fitting, mean values of R^2 from the CP and RIC models were 0.978 ± 0.022 and 0.968 ± 0.026, respectively. The F ratio was 2.37 ± 1.40, and the p-value was 0.29 ± 0.21 (Table 11.1). Both R_{rs} and R_{aw} correlated with %FEV1 ($r^2 = 0.940$ and 0.693, respectively). On the other hand, C_{rs} and C_{tis} did not significantly correlate with %FEV1 or %DLCO/VA (Fig. 11.2).

4 Discussion

Respiratory input impedance using FOTs has been measured since the 1970s [7]. However, the relationship between underlying structure in the lung parenchyma and the imaginary part of respiratory impedance is not well-studied. To overcome

Table 11.1 Mechanical parameters after model fittings

	RIC model fitting			CP model fitting				Significance	
Subject	R_{rs} (kPa/L/s)	I_{rs} (kPa/L)	C_{rs} (L/kPa)	R_{aw} (kPa/L/s)	I_{aw} (kPa/L)	G (kPa/L)	H (kPa/L)	F	p
1	0.27	0.0019	0.14	0.24	0.0020	0.97	3.85	3.38	0.16
2	0.30	0.0012	0.16	0.29	0.0012	0.70	4.11	3.24	0.17
3	0.24	0.0013	0.14	0.21	0.0015	0.99	3.56	2.47	0.21
4	0.32	0.0012	0.17	0.31	0.0013	0.46	4.58	0.19	0.69
5	0.34	0.0011	0.13	0.29	0.0014	1.19	2.59	1.19	0.36
6	0.27	0.0010	0.14	0.25	0.0012	0.93	4.15	3.75	0.15
R^2	0.968 ± 0.026			0.978 ± 0.022					

The mean values of R^2 are expressed as means ± SD

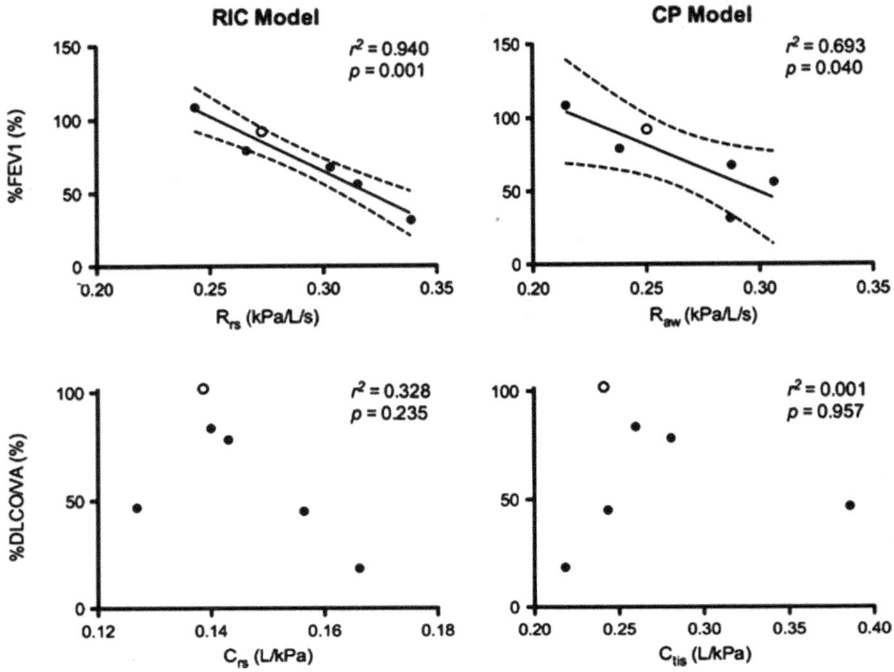

Fig. 11.2 Relationships between mechanical parameters (respiratory and airway resistance, and respiratory and airway compliance) and pulmonary function tests (%FEV1 and %DLCO/VA). *Open circle* = normal control without LAM; *Closed circle* = subject with lymphangioleiomyomatosis; *Solid line* = regression line; *Dotted line* = 95% confidential interval

this obstacle, the CP model is designed as a vehicle for extracting the mechanical properties of peripheral lung tissue. A recent study performed in a mouse model of emphysema demonstrated that the mechanical parameters of CP model obtained from sine curved oscillatory waves described structural changes in the lung parenchyma [8]. Thus, if respiratory impedance is used for studying the mechanics of the lung periphery, the CP model should be applied instead of the RIC model.

%FEV1 is a standard index for assessing and quantifying airflow limitation. Diffusion capacity is an index reflecting the condition of the lung periphery and is correlated with disease severity of LAM [9]. As shown in Fig. 11.2, both the RIC and CP models served as vehicles to obtain mechanical information associated with airflow limitation. However, neither model demonstrated a significant relationship between their mechanical parameters and lung function tests associated with the lung periphery. Furthermore, the results of the CP model fitting in patients with severe pulmonary LAM had a tendency to be ambiguous and could not be included in the comparison of fits.

The most likely explanation is that non-sine curved oscillatory waves in impulses are not able to transmit information about the lung periphery. Furthermore, there was no compelling evidence for adopting the CP model for the evaluation of impedance spectra obtained by IOS. This means that this type of FOT may not provide the mechanical information about the lung periphery. Needless to say, further investigation is necessary to link respiratory mechanics with oxygenating ability for assessing lung diseases.

Acknowledgment This work was supported by Grant-in-Aid for Scientific Research (C) from the Japan Society for the Promotion of Science (20591667).

References

1. Oostveen E, MacLeod D, Lorino H et al (2003) The forced oscillation technique in clinical practice: methodology, recommendations and future developments. Eur Respir J 22(6): 1026–1041
2. Hantos Z, Daroczy B, Suki B, Nagy S, Fredberg JJ (1992) Input impedance and peripheral inhomogeneity of dog lungs. J Appl Physiol 72:168–178
3. Frei J, Jutla J, Kramer G, Hatzakis GE, Ducharme FM, Davis GM (2005) Impulse oscillometry: reference values in children 100 to 150 cm in height and 3 to 10 years of age. Chest 128:1266–1273
4. Smith HJ, Reinhold P, Goldman MD (2005) Forced oscillation technique and impulse oscillometry. Eur Respir Mon 31(5):72–105
5. Guo YF, Herrmann F, Michel JP, Janssens JP (2005) Normal values for respiratory resistance using forced oscillation in subjects >65 years old. Eur Respir J 26:602–608
6. Olson HF Dynamical analogies, 2nd edn. Van Nostarand Reinhold Co., New York, 278 pp
7. Lándsér FJ, Nagles J, Demedts M, Billiet L, van de Woestijne KP (1976) A new method to determine frequency characteristics of the respiratory system. J Appl Physiol 41(1):101–106
8. Wilson AA, Murphy GJ, Hamakawa H et al (2010) Amelioration of emphysema in mice through lentiviral transduction of long-lived pulmonary alveolar macrophages. J Clin Invest 120(1):379–389
9. Taveira-DaSilva AM, Stylianou MP, Hedin CJ et al (2003) Maximal oxygen uptake and severity of disease in lymphangioleiomyomatosis. Am J Respir Crit Care Med 168(12):1427–1431

Chapter 12
NIRS Measurements with Elite Speed Skaters: Comparison Between the Ice Rink and the Laboratory

Catherine Hesford, Marco Cardinale, Stewart Laing, and Chris E. Cooper

Abstract Wearable, wireless near-infrared (NIR) spectrometers were used to compare changes in on-ice short-track skating race simulations over 1,500 m with a 3-min cycle ergometry test at constant power output (400 W). The subjects were six male elite short-track speed skaters. Both protocols elicited a rapid desaturation (ΔTSI%) in the muscle during early stages (initial 20 s); however, asymmetry between right and left legs was seen in ΔTSI% for the skating protocol, but not for cycling. Individual differences between skaters were present in both protocols. Notably, one individual who showed a relatively small TSI% change (−10.7%, group mean = −26.1%) showed a similarly small change during the cycling protocol (−5.8%, group mean = −14.3%). We conclude that NIRS-detected leg asymmetry is due to the specific demands of short-track speed skating. However, heterogeneity between individuals is not specific to the mode of exercise. Whether this is a result of genuine differences in physiology or a reflection of differences in the optical properties of the leg remains to be determined.

Keywords NIRS • Speed skaters

C. Hesford (✉)
School of Biological Sciences, University of Essex, Wivenhoe Park, Colchester CO4 3SQ, UK

British Olympic Medical Institute, University College London Hospital, London, UK
e-mail: chesfo@essex.ac.uk

M. Cardinale • S. Laing
School of Biological Sciences, University of Essex, Wivenhoe Park, Colchester CO4 3SQ, UK

C.E. Cooper
British Olympic Medical Institute, University College London Hospital, London, UK

W.J. Welch et al. (eds.), *Oxygen Transport to Tissue XXXIV*, Advances in Experimental Medicine and Biology 765, DOI 10.1007/978-1-4614-4989-8_12,

1 Introduction

Near-infrared spectroscopy (NIRS) has been used to investigate exercise-specific oxygen metabolism and hemodynamics in the laboratory for a range of sports including: cycling [1, 2]; sailing [3]; and speed skating [4]; for review see [5]. However, the recent development of wearable wireless, portable NIR spectrometers has enabled the investigation of changes in local muscle metabolism and blood flow during exercise in a more realistic sporting setting.

In short-track speed skating, the combination of high intramuscular forces, long duty cycle of the gliding phase in the skating motion, and the low sitting position adopted leads to a restriction of blood flow to the working muscle: the "reduced blood flow hypothesis" [6]. Previous research using NIR techniques has centered on laboratory testing using treadmill skating and has shown that the low sitting position of the skater elicits a lower submaximal VO_2 [7], increased blood lactate concentration [8], and increased quadriceps deoxygenation [4] when compared to upright skating. On-ice tests have shown that skating at higher speeds produces a greater deoxygenation in the quadriceps muscle [8]. The current study utilized a wearable, wireless NIR spectrometer to measure the local muscle oxygenation and blood volume changes during an on-ice race simulation, enabling investigation of the asymmetric effects on right and left legs of skating around a short-track oval. A separate laboratory cycle test with the same subjects was used to confirm whether any asymmetry in NIRS measurements was specific to the technical and biomechanical requirements of the event.

2 Methods

Six elite male short-track speed skaters (mean ± SD age: 23 ± 1.8 years; height: 1.8 ± 0.1 m; mass: 80.1 ± 5.7 kg; mid-thigh skin fold thickness: 6.8 ± 2.2 mm) of Olympic standard took part in this study. All subjects gave their written informed consent prior to participation. The study was approved by the ethics committee of University College London.

Protocol 1 involved each subject completing a 1,500 m on-ice race simulation. Testing took place on a short-track speed skating oval (111.12 m) approved for international competition. Before completing race simulation, subjects were informed to undertake their own warm-up, as they would when preparing for a competitive race. Race simulation TTs were completed individually and subjects were informed to try and achieve the fastest time possible. Protocol 2 was a laboratory-based task which involved 3 min cycling at a constant cadence (90 rpm) and power output (400 W) on an electromagnetically braked cycle ergometer (Excalibur Sport V2, Lode, the Netherlands). This was immediately preceded by a 5-min warm-up at a constant load (90 W) and constant cadence (90 rpm). Protocol 2 took place 4 weeks after protocol 1.

In both protocols, muscle oxygenation in both left and right vastus lateralis was continuously monitored using a recently developed wireless, dual-wavelength Spatially Resolved Spectrometer (Portamon, Artinis Medical Systems, BV, the Netherlands). The unit is self-contained and compact, measuring 83 × 52 × 20 mm and weighing 84 g, including battery. It houses three pairs of light emitting diodes (LEDs), which emit light of wavelengths 760 and 850 nm, and are positioned 30, 35, and 40 mm from the detector. This device has previously been used to investigate muscle oxygenation and hemodynamics in both a controlled laboratory environment [9] and a field setting; thus utilizing the portable, wireless nature of the instrument [10–12]. The devices were positioned on the belly of the vastus lateralis muscle, midway between the greater trochanter of the femur and the lateral epicondyle. To ensure the optodes and detector did not move relative to the subject's skin, the device was fixed into position using surgical tape, and then secured with a bandage, which was wound around the thigh a number of times. The same attachment procedure was undertaken for both protocols. Changes in optical density were converted to changes in oxygenation (ΔHb_{diff}, $\Delta HbO_2 - \Delta HHb$), blood volume ($\Delta tHb$, $\Delta HbO_2 - \Delta HHb$), and an SRS-derived measure of tissue oxygen saturation (%TSI).

3 Results

The 1,500 m race simulation was completed in 151.03 ± 4.04 s. At exercise onset in both on-ice skating and laboratory cycling protocols, there was a rapid decrease in total hemoglobin (tHb), followed by a steady increase throughout the remainder of the exercise. Both protocols also elicited a rapid desaturation (ΔHb_{diff}, ΔTSI%) in the muscle during early stages (initial 20 s). The maximum decrease in saturation (ΔTSI_{max}) for the skating test was Δ26.1% and for the cycling test was Δ14.3%.

By examining the group average data for both protocols (Fig. 12.1), it can be seen that a pattern of asymmetry emerged during the skating race simulation: the right leg TSI remained desaturated, but the left leg TSI gradually increased with time. This asymmetry in skating can also be seen from the observed changes in Hb_{diff} during the race simulation, but there is no apparent asymmetry in the tHb values. However, there is no such pattern in any of the measured parameters during the cycling protocol, in which left and right leg changes are closely matched in magnitude throughout the exercise.

Figure 12.2 presents the ΔtHb values for each individual during both protocols. During the skating protocol, it was noted that the majority of skaters showed an initial reduction in tHb at race start (presumably due to a blood squeezing effect of muscle contraction at exercise onset), followed by a gradual increase in tHb throughout the remainder of the exercise, presumably as a consequence of a blood flow increase. In most cases, this results in a final tHb significantly higher than the baseline value. However, one subject (C) showed very little change in tHb during the

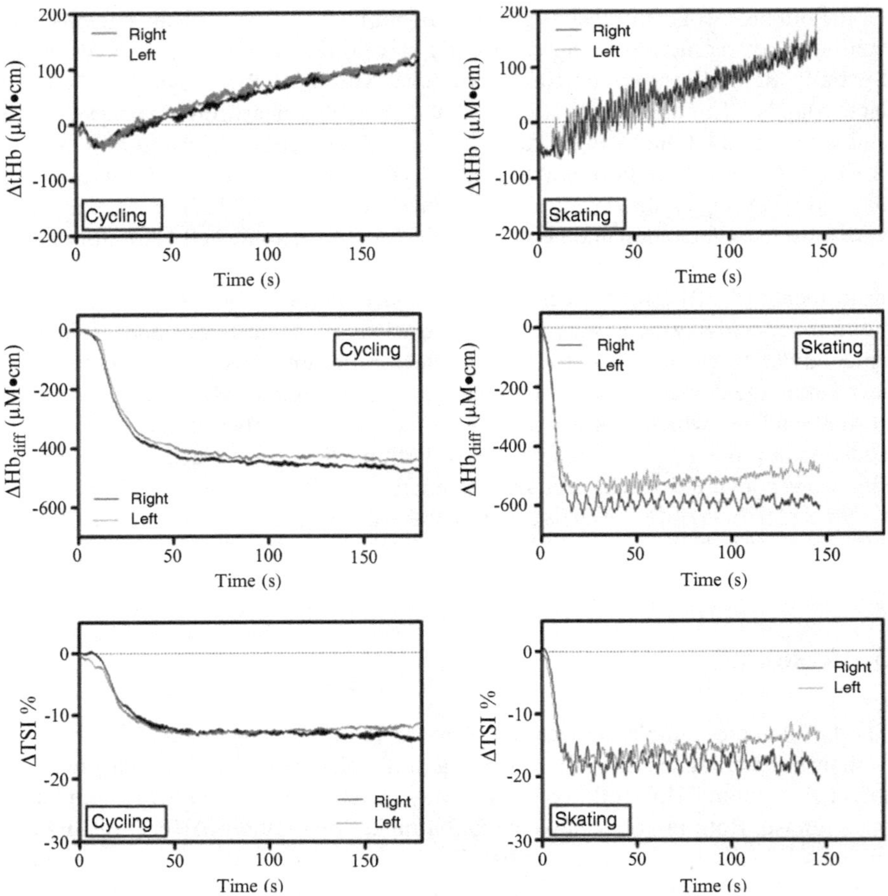

Fig. 12.1 Mean changes from baseline in 3 NIRS-detected parameters (tHb; Hb_{diff}; TSI%) in the right and left VL during both skating and cycling protocols. Following initial desaturation at exercise onset, asymmetry between right and left leg Hb_{diff} and TSI data can be seen during skating, but not during the cycling protocol

race, and his final values in both legs were notably lower than most. The cycling protocol produced a more pronounced difference between the tHb changes of this individual and the rest of the group: Fig. 12.2 shows that during cycling, tHb did not increase in either leg of subject C following the initial decrease at exercise onset. This is in contrast to all other subjects. Subject C also showed the smallest desaturation in both skating and cycling (results not shown).

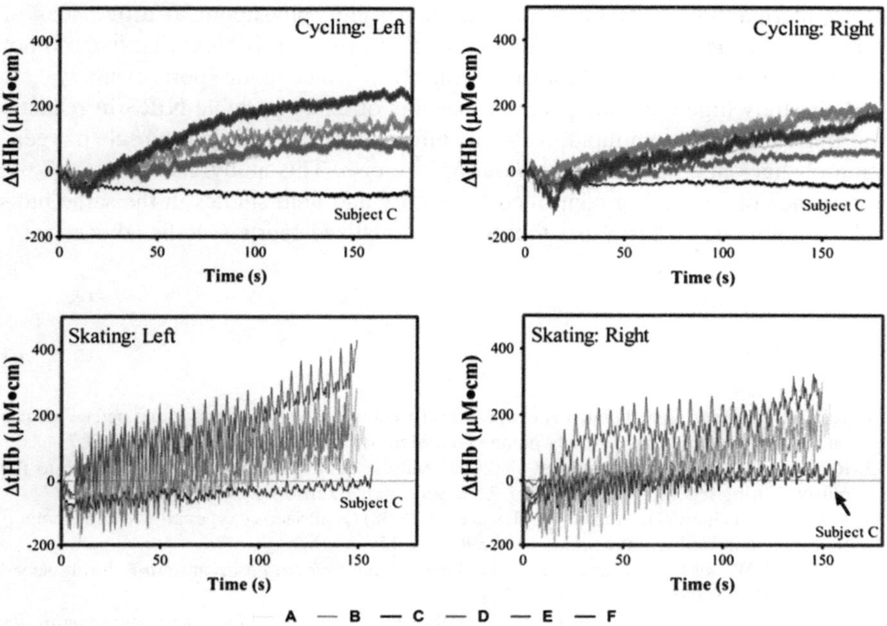

Fig. 12.2 tHb changes in both right and left VL of each subject during both cycling and skating protocols. The high frequency variations seen in skating are due to changes in local blood volume due to different body positions at specific stages of the lap (e.g., straight, corners). This has previously been described [13]

4 Conclusions

The asymmetry observed between the right and left leg during the skating protocol is most likely due to the specific demands of this sport. Skaters travel around the track in an anticlockwise direction, which means the right leg is always the "outside" leg. When travelling around the corners at the highest speeds, the skater travels solely on the right leg, which is in a sustained isometric contraction, whilst the left leg is free of the ice. This technique is called the "hang phase." During this technique, the left leg is able to resaturate, whereas the right leg cannot [13]. We have previously suggested from 500 m studies that the cumulative effect of these "hang" phases could lead to asymmetry in oxygenation and blood volume that persisted throughout the simulation [13]. This study confirmed this effect for the 1,500 m race simulation. The laboratory cycling protocol demonstrated that the asymmetry was due to the specific demands of skating, rather than being due to an intrinsic difference between the right and left quadriceps muscles of this particular subject group.

Individual differences in NIRS-detected parameters were replicated in the sports-specific and laboratory protocols, notably the atypical tHb changes in subject C, which were present in both skating and cycling. It is not possible to tell whether these interindividual variations are due to specific physiological differences or variability in light transport through the leg. However, it is clear that they do not relate to the specifics of the biomechanical performance in the sports event.

This study highlights the potential benefits of using portable NIRS in realistic sporting settings to gain rapid, real-time information pertaining to muscle oxygenation changes and hemodynamics during exercise. This study also illustrates the importance of comparing controlled laboratory and field studies in the same individuals before using NIRS as a tool for individualized sports-specific advice.

References

1. Takaishi T, Ishida K, Katayama K et al (2002) Effect of cycling experience and pedal cadence on the near-infrared spectroscopy parameters. Med Sci Sports Exerc 34(12):2062–2071
2. Racinais S, Bishop D, Denis R et al (2007) Muscle deoxygenation and neural drive to the muscle during repeated sprint cycling. Med Sci Sports Exerc 39(2):268–274
3. Vogiatzis I, Tzineris D, Athanasopoulos D et al (2008) Quadriceps oxygenation during isometric exercise in sailing. Int J Sports Med 29(1):11–15
4. Rundell KW, Nioka S, Chance B (1997) Hemoglobin/myoglobin desaturation during speed skating. Med Sci Sports Exerc 29(2):248–258
5. Hamaoka T, McCully KK, Quaresima V et al (2007) Near-infrared spectroscopy/imaging for monitoring muscle oxygenation and oxidative metabolism in healthy and diseased humans. J Biomed Opt 12(6):16
6. Snyder AC, Foster C (1994) Physiology and nutrition for skating. In: Lamb DR, Murray R (eds) Perspectives in exercise science and sports medicine. Cooper Publishing Group, Carmel, IN, pp 181–219
7. Rundell KW (1996) Compromised oxygen uptake in speed skaters during treadmill in-line skating. Med Sci Sports Exerc 28(1):120–127
8. Foster C, Rundell KW, Snyder AC et al (1999) Evidence for restricted muscle blood flow during speed skating. Med Sci Sports Exerc 31(10):1433–1440
9. Shadgan B, Reid WD, Gharakhanlou R et al (2009) Wireless near-infrared spectroscopy of skeletal muscle oxygenation and hemodynamics during exercise and ischemia. Spectrosc Int J 23(5–6):233–241
10. Buchheit M, Laursen PB, Ahmaidi S (2009) Effect of prior exercise on pulmonary O_2 uptake and estimated muscle capillary blood flow kinetics during moderate-intensity field running in men. J Appl Physiol 107(2):460–470
11. Buchheit M, Bishop D, Haydar B et al (2010) Physiological responses to shuttle repeated-sprint running. Int J Sports Med 31(6):402–409
12. Buchheit M (2010) Performance and physiological responses to repeated-sprint and jump sequences. Eur J Appl Physiol 110(5):1007–1018
13. Hesford C, Laing SJ, Cardinale M, Cooper CE (2012) Asymmetry of quadriceps muscle oxygenation during elite short-track speed skating. Med Sci Sports Exerc 44(3):501–508

Chapter 13
Modelling Cerebrovascular Reactivity: A Novel Near-Infrared Biomarker of Cerebral Autoregulation?

David Highton, Jasmina Panovska-Griffiths, Arnab Ghosh, Ilias Tachtsidis, Murad Banaji, Clare Elwell, and Martin Smith

Abstract Understanding changes in cerebral oxygenation, haemodynamics and metabolism holds the key to individualised, optimised therapy after acute brain injury. Near-infrared spectroscopy (NIRS) offers the potential for non-invasive, continuous bedside measurement of surrogates for these processes. Interest has grown in applying this technique to interpret cerebrovascular pressure reactivity (CVPR), a surrogate of the brain's ability to autoregulate blood flow. We describe a physiological model-based approach to NIRS interpretation which predicts autoregulatory efficiency from a model parameter *k_aut*. Data from three critically brain-injured patients exhibiting a change in CVPR were investigated. An optimal

---

The original version of this chapter was revised. An erratum to this chapter can be found at DOI 10.1007/978-1-4614-4989-8_53

D. Highton (✉)
Neurocritical Care, University College Hospitals, Queen Square, London, UK

National Hospital for Neurology and Neurosurgery, Flat 40 Salisbury Mansions, St Anns Road, London N153TP, UK
e-mail: dave@davidhighton.net

J. Panovska-Griffiths • I. Tachtsidis • C. Elwell
Medical Physics and Bioengineering, University College London, Malet Place, London, UK

A. Ghosh
Neurocritical Care, University College Hospitals, Queen Square, London, UK

Institute of Neurology, University College London, Queen Square, London, UK

M. Banaji
Department of Mathematics, University of Portsmouth, Portsmouth, UK

M. Smith
Neurocritical Care, University College Hospitals, Queen Square, London, UK

Medical Physics and Bioengineering, University College London, Malet Place, London, UK

W.J. Welch et al. (eds.), *Oxygen Transport to Tissue XXXIV*, Advances in Experimental Medicine and Biology 765, DOI 10.1007/978-1-4614-4989-8_13,

value for *k_aut* was determined to minimise the difference between measured and simulated outputs. Optimal values for *k_aut* appropriately tracked changes in CVPR under most circumstances. Further development of this technique could be used to track CVPR providing targets for individualised management of patients with altered vascular reactivity, minimising secondary neurological insults.

Keywords Modelling • Cerebrovascular reactivity

1 Introduction

Cerebral blood flow (CBF) is tightly regulated by cerebral autoregulation (CA), forming a critical link between oxygen supply and demand. Myogenic, metabolic and neurological mechanisms lead to a complex pattern of vascular reactivity over different time scales combining to maintain constant perfusion across a wide range of perfusion pressure. Following brain injury, acute disturbances of CA may lead to hyper or hypo-perfusion and secondary neurological insults; maintaining cerebral perfusion is thus a core goal during the neurointensive care treatment of brain injury. However, delivering this is not straightforward as there is no convenient means of monitoring CBF or CA continuously at the bedside.

Measures of vascular reactivity, derived using surrogates of cerebral blood volume (CBV) or CBF, may be compared with arterial blood pressure (ABP) to investigate efficiency of cerebrovascular pressure reactivity (CVPR) and CA [1]. Recently near-infrared spectroscopy (NIRS) has been investigated in this regard as different NIRS indices reflect aspects of cerebral haemodynamics [2, 3]. Specifically, cerebral tissue oxygen saturation (TOS) and total haemoglobin have been applied as surrogates of CBF and CBV, respectively. When correlated with ABP these indices agree with well-established indices of CVPR [4, 5].

While these modes of analysis are simple and easily performed at the bedside, they do not account for the non-stationary and non-linear complexity within the range of measured signals. A model-based approach might make best use of the available data combining a priori knowledge of complex cerebral physiology with multiple measured variables to establish fully informed physiological predictions. This might account for additional important contributions to our interpretation of NIRS measured signals such as changes in CO_2 or O_2 tension, cerebral metabolic rate ($CMRO_2$) and arterial to venous volume ratio.

We have previously described a physiological model of cerebral haemodynamics, oxygenation and metabolism and used this to aid interpretation of NIRS signals during cerebral physiological challenges in healthy volunteers [6]. The model combines haemodynamic, metabolic and oxygenation components creating simulated outputs of a range of measured signals. Variation in the model parameters from their basal values alters simulated outputs in a way which may mirror changes in underlying physiological processes. The model parameter *k_aut* has been designed to represent changes in the efficiency of CA ranging from 0 with an absence of CA to 1 where it is completely intact. This work translates our model [6] into the pathophysiological context of brain injury. The aim of this work is to use a range of measured signals, including NIRS, to identify a model-derived parameter as a biomarker of CA in individual patients.

2 Methods

Three acutely brain-injured patients showing variation in CVPR were identified from an ongoing multimodal monitoring study in brain-injured patients. This study was approved by the institutional Research Ethics Committee and assent was gained from patient representatives.

For each patient dataset, CVPR was initially characterised using the pressure reactivity index (PRx) and mean velocity index (Mx) [1]. Two 30-min epochs were analysed for each patient, one with reactivity indices <0.3 suggesting intact CA and one >0.3 suggesting loss of CA. NIRS monitoring was performed with the NIRO 100 (Hammamatsu Photonics KK) ipsilateral to intraparenchymal intracranial pressure (ICP) monitoring and transcranial Doppler flow velocity of the middle cerebral artery (Vmca) (DWL Doppler Box, Compumedics, Germany). NIRS measurements included spatially resolved tissue oxygenation index (TOI) and normalised total haemoglobin index (nTHI) representing measures of TOS and total haemoglobin, respectively. Changes in concentration of oxyhaemoglobin ($\Delta[HbO_2]$) and deoxyhaemoglobin ($\Delta[HHb]$) were determined by the modified Beer–Lambert method. Invasive ABP from a radial artery catheter, end tidal CO_2 ($ETCO_2$) and pulse oximetry (SpO_2) were gathered through an Intellivue monitor (Philips, N.V., Amsterdam, The Netherlands). Signals were synchronised, downsampled to 1 Hz and filtered with a lowpass 0.1 Hz fifth-order Butterworth filter to remove high frequency noise and respiratory influences. Of these measured signals ABP, $ETCO_2$ (approximating $PaCO_2$), SpO_2 and ICP were used as model inputs. These produced simulated outputs for CBF, total haemoglobin ([HbT]), $[HbO_2]$, [HHb] and TOS which were compared with their measured counterparts Vmca and NIRS (nTHI, $\Delta[HbO_2]$, $\Delta[HHb]$, TOI).

Optimisation was performed by minimising the difference between measured signals and simulated outputs for Vmca and NIRS finding optimal values for parameter *k_aut* (representing CA) and an additional parameter *u* reflecting cerebral energy demand. Reduction of this additional parameter below its basal level simulating a reduction in cerebral metabolism was required to adequately fit the measured NIRS signals. This seems physiologically plausible because all patients were deeply sedated at the time of study. *k_aut* values produced by these different optimisation strategies were compared to index-based predictions of CVPR for consistency. The difference between simulated outputs and measured signals is expressed as the mean absolute difference between the two. The improvement in the fit following optimisation is given as the percentage difference between measured signals and simulated outputs at basal parameter settings and optimised parameter settings divided by the basal value.

3 Results

A high *k_aut* was associated with intact CVPR and a low value disturbed CVPR in all simulations excluding those optimised on the basis of TOI. An example dataset is shown in Fig. 13.1 demonstrating disturbed CA. It can be seen that simulation with a low value of *k_aut* (0.3), reflecting dramatically impaired autoregulation, simulates Vmca and nTHI most accurately.

Model simulations using different measured signals for optimisation of *k_aut* varied in the relationship between *k_aut* and predicted CVPR. When *k_aut* is optimised by minimising the difference between measured Vmca and simulated CBF alone in all epochs (Table 13.1), there is accurate prediction of Vmca (mean absolute difference 1.98 cm/s). Post-optimisation *k_aut* values are lower in the epochs with reduced CVPR (0.53), suggesting that *k_aut* appropriately reflects the level of CA. When measured NIRS signals are included in this strategy (Table 13.2) it is possible to account for the changes in nTHI by optimising *k_aut* alone. This continues to predict appropriate values of *k_aut* (Table 13.2, column 1). To adequately fit measured and simulated $\Delta[HbO_2]$ and $\Delta[HHb]$, optimisation of *u* reducing cerebral metabolism was required. Again, this approach predicts lower values of *k_aut* (0.47) in those with reduced CVPR. However, to achieve the best fit requires a value for *u* that is unphysiologically low.

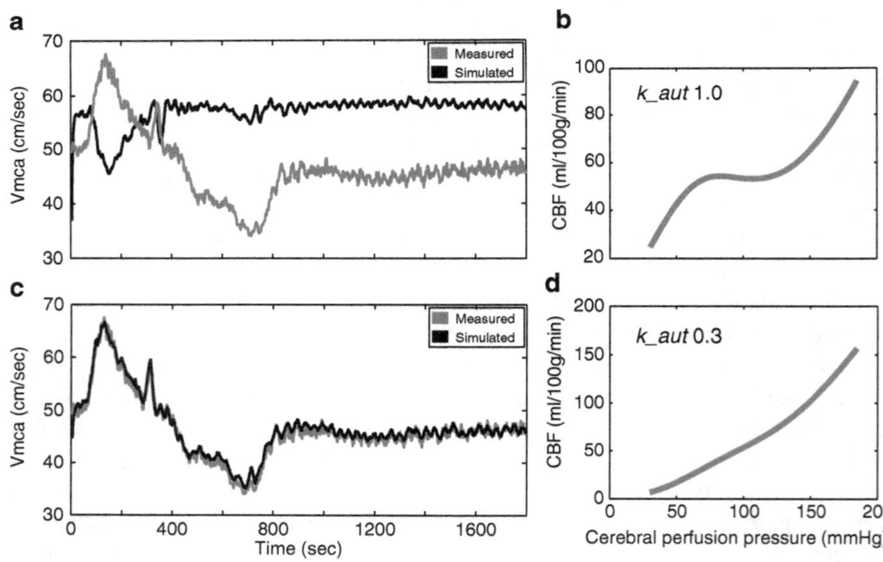

Fig. 13.1 Measured signals and simulated outputs for a patient with low CVPR. (**a**) Measured Vmca and simulated Vmca using the basal *k_aut* value (1.0). (**b**) The steady-state relationship between CPP and CBF using the basal value of *k_aut* reproduces a typical normal static autoregulation curve. (**c**) Measured and simulated Vmca post-optimisation of *k_aut* demonstrate excellent agreement compared with the basal value. This value of *k_aut* (0.3) is low and in the dysautoregulated range suggesting that a loss of CA is required to explain the measured signals. (**d**) The predicted steady state between CPP and CBF using the optimised value for *k_aut* (0.3). This closely resembles a static autoregulation curve with loss of CA

Table 13.1 Optimisation of *k_aut* using simulated CBF against measured Vmca alone

	Improvement (%)	Mean absolute difference	Optimal *k_aut*	
Vmca	59 (35)	1.97 (0.78) cm/s	CVPR intact	0.87 (0.12)
nTHI	28 (25)	0.009 (0.01) au	CVPR lost	0.53 (0.25)
$\Delta[HbO_2]$	−18 (36)	1.86 (2.5) μmol/L		
$\Delta[HHb]$	−52 (65)	1.58 (2.2) μmol/L		
TOI	−70 (75)	17 (2) %		

Mean (SD) improvement between basal *k_aut* and optimised *k_aut* show improved prediction of Vmca and nTHI. Mean (SD) absolute differences between measured signals and simulated outputs are shown demonstrating accurate prediction of Vmca and nTHI. Post-optimisation *k_aut* values appropriately reflect the measured CVPR with a lower mean *k_aut* where CVPR is lost

Table 13.2 Optimisation of *k_aut* and *u* based on different combinations of measured signals

Measured signals used to optimise against	Vmca; nTHI	Vmca; nTHI; $\Delta[HbO_2]$; $\Delta[HHb]$	Vmca; nTHI; $\Delta[HbO_2]$; $\Delta[HHb]$; TOI
Optimal k_aut value			
CVPR intact	0.93 (0.17)	0.83 (0.06)	0.03 (0.52)
CVPR lost	0.37 (0.15)	0.47 (0.55)	0.9 (0.69)
Optimal *u* value	0.50 (0.55)	0.00 (0.00)	0.67 (0.52)
Improvement			
Vmca	70 (28) %	70 (26) %	50 (31) %
nTHI	36 (30) %	36 (31) %	5 (34) %
$\Delta[HbO_2]$	14 (48) %	49 (26) %	−40 (74) %
$\Delta[HHb]$	−6 (82) %	78 (16) %	−26 (106) %
TOI	−536 (894) %	−633 (1,113) %	14 (57) %

For each column different measured signals were compared to model outputs to find optimal values for *k_aut* and *u*. Mean (SD) improvement between basal parameter values and optimised values are shown demonstrating improved post-optimisation prediction of measured signals (excluding TOI). Optimal *k_aut* values for each optimisation strategy are shown and are consistent with levels of measured CVPR except where TOI is included in the optimisation

Inclusion of TOI within the optimisation strategy is problematic and it is not possible to fit TOI well in combination with other measured signals (Table 13.2, column 3). Optimal values for *k_aut* do not reflect the level of predicted CA in this final approach. The behaviour of measured TOI differs significantly from simulated outputs of TOS; large simulated changes in TOS result from large changes in CBF which are not present in the measured TOI (Fig. 13.2).

4 Discussion

We have identified a model parameter *k_aut* which simulates changes in CA and improves prediction of NIRS signals in brain injury. The optimal value of *k_aut* may thus represent a composite biomarker of cerebral autoregulatory function informed from multiple NIRS inputs. This approach aims to form cohesive physiological predictions based on prior knowledge of physiology, maximising the potential of the available data.

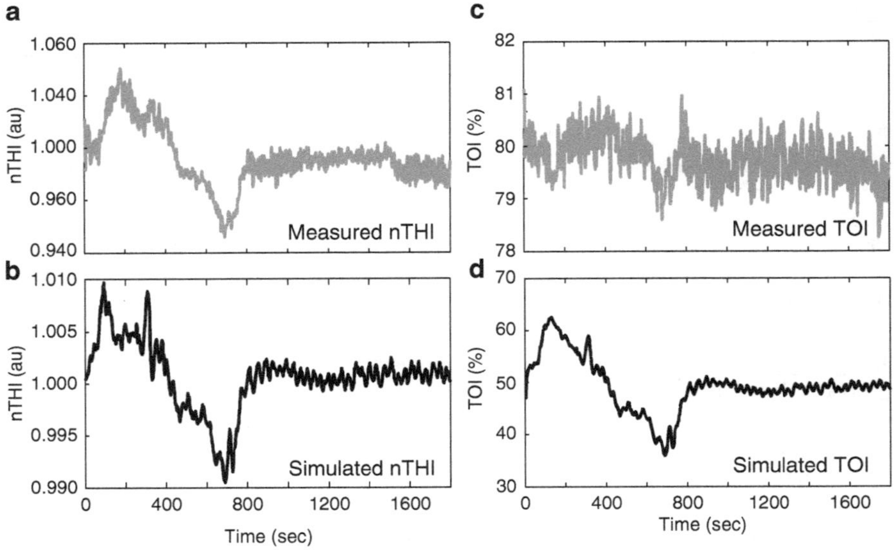

Fig. 13.2 Measured and simulated NIRS outputs from dataset in Fig. 13.1. (**a**) Measured nTHI and (**b**) simulated nTHI demonstrate moderate agreement. (**c**) Measured TOI and (**d**) simulated TOI agree qualitatively only

Fitting nTHI was least problematic probably because fewer physiological processes influence this signal. In comparison $\Delta[HbO_2]$, $\Delta[HHb]$ and TOS encode metabolic components and differential effects of arterial and venous components become more influential. It was impossible to achieve an adequate fit for TOS by varying only the model parameters *k_aut* and *u*. Despite qualitative agreement, the magnitude of variation and baseline saturation showed large discrepancies. It is unlikely that further optimisation within physiological plausibility could explain the lack of variability despite the large changes in CBF observed. However, a differing baseline is more easily explained. Similar observations were made during studies in healthy volunteers [7], but in this case TOS could be explained by adjusting the extracerebral:intracerebral signal weighting to 80:20 or doubling the venous volume, both of which seem unlikely. Studies such as these indicate that accurate prediction and interpretation of TOS might require combined modelling of cerebral physiology and light transport in tissue.

This study was of limited power including only six epochs from three patients. However these datasets demonstrate an extreme of physiological dysfunction with large changes in ABP and CBF, representing an excellent challenge for our model. Further work must include large numbers of patients undergoing a range of physiological challenges to increase the quality of measured signals in patients with lesser degrees of impaired CA. Although this approach has not necessarily been followed for many established indices of CVPR it should be viewed as a prerequisite to translation into the clinic.

With further investigation, model-informed interpretation of NIRS signals might offer enhanced prediction of CA across widely varying physiological and pathophysiological contexts. Prior knowledge of population characteristics and further model simplification should improve computational efficiency and move toward bedside implementation. This form of interpretation progresses beyond simple correlation analyses by combining information from multiple NIRS and systemic measures with a priori knowledge of physiology to provide cohesive predictions of cerebral well-being. Thus, use of *k_aut* as a biomarker of CA efficiency could inform pathophysiology and potentially provide a target for physiological optimisations to improve outcome.

Acknowledgements This work was undertaken at University College London Hospitals and partially funded by the Department of Health's National Institute for Health Research Centres funding scheme. Support has also been provided by the Medical Research Council and Wellcome Trust. The authors are indebted to the medical and nursing staff of the Neurocritical Care Unit at the National Hospital for Neurology & Neurosurgery and to the study patients and their families.

References

1. Czosnyka M, Smielewski P, Kirkpatrick P et al (1998) Continuous monitoring of cerebrovascular pressure-reactivity in head injury. Acta Neurochir Suppl 71:74–77
2. Lee J, Kibler K, Benni P et al (2009) Cerebrovascular reactivity measured by near-infrared spectroscopy. Stroke 40(5):1820–1826
3. Brady K, Lee J, Kibler K et al (2007) Continuous time-domain analysis of cerebrovascular autoregulation using near-infrared spectroscopy. Stroke 38(10):2818–2825
4. Zweifel C, Castellani G, Czosnyka M et al (2010) Noninvasive monitoring of cerebrovascular reactivity with near infrared spectroscopy in head-injured patients. J Neurotrauma 27(11):1951–1958
5. Zweifel C, Castellani G, Czosnyka M et al (2010) Continuous assessment of cerebral autoregulation with near-infrared spectroscopy in adults after subarachnoid hemorrhage. Stroke 41(9):1963–1968
6. Banaji M, Mallet A, Elwell C et al (2008) A model of brain circulation and metabolism: NIRS signal changes during physiological challenges. PLoS Comput Biol 4(11):e1000212
7. Moroz T, Banaji M, Tisdall M et al (2012) Development of a model to aid NIRS data interpretation: results from a hypercapnia study in healthy adults. Adv Exp Med Biol 737:293–300

Chapter 14
Oxygen Delivery Deficit in Exercise with Rapid Ascent to High Altitude

Luke Holdsworth and Christopher Wolff

Abstract This study of high altitude physiology was undertaken during an 11-day expedition to the Himalaya with ascent to Annapurna base camp (4,130 m) reaching it on the sixth day. Fourteen male UK residents (13 aged 16–17 years; 1 adult) measured arterial oxygen saturation (SaO_2) and heart rate (HR) at rest and at 2 min exercise (30 cm step), daily, after arrival at each altitude. Precision was limited by availability of only one oximeter (CMS50-DLP model, Contec Medical Systems, Qinhuangdao, P.R. China). Mean HR correlated (negatively) with SaO_2 both for rest ($HR=-1.7974\times SaO_2\%+236.33$, $r=0.841$, $p=0.001$) and exercise ($HR=-0.8834\times SaO_2\%+226.14$, $r=0.711$ $p<0.02$). Four subjects individually showed significant HR/SaO_2 correlations at rest (nos. 10, 11, 12 and 13) and one, subject 11, in exercise. SaO_2 in exercise was lower than at rest (SaO_2, exercise $=1.5835\times SaO_2$, rest $-\,59.177$, $r=0.987$, $p<0.001$). The product, $HR\times SaO_2$, calculated as a surrogate for oxygen delivery (DO_2, Brierley et al., Adv Exp Med Biol 737:207–212, 2012), from mean values was approximately constant for rest, suggesting good cardiac output (CO) compensation for de-saturation. The $HR\times SaO_2$ for exercise, however, showed a dramatic fall at the highest altitude. Since this deficit occurred at the highest altitude, following 2 days of rapid ascent, there was probably impairment of adequate oxygen delivery (DO_2) at this point. Correlation, HR versus SaO_2 for exercise, was highly significant, with greater significance ($HR=-1.798\times SaO_2+281.83$, $r=0.769$, $p=0.01$) on omission of the values for the highest ascent point (4,130 m), where the reduced $HR\times SaO_2$ occurred. In conclusion, oxygen delivery is sustained well here except where there are the extra stresses of rapid ascent and exercise.

L. Holdsworth
The King's School, Grantham, UK

C. Wolff (✉)
Clinical Pharmacology and Anaesthetics, William Harvey Research Institute, Barts and The London, London, UK

St Bartholomew's Hospital, 52 Victoria Park, Cambridge CB4 3EL, UK
e-mail: chriswolff@doctors.org.uk

W.J. Welch et al. (eds.), *Oxygen Transport to Tissue XXXIV*, Advances in Experimental Medicine and Biology 765, DOI 10.1007/978-1-4614-4989-8_14,

1 Introduction

The ascent to high altitude, with exposure to low inspired oxygen tension (hypobaric hypoxia), results in reduction in alveolar and arterial oxygen tension (PaO_2) and arterial saturation (SaO_2). The level at which SaO_2 settles depends on rate of ascent, metabolic rate and individual response. Montgomery et al. [1] showed that tolerance to ascent varied considerably with the type of ACE gene. The differences were related to SaO_2 of the subjects. After a sudden/rapid change in altitude, and hence barometric pressure, changes in SaO_2 are initially maximal with gradual recovery to a new steady level over the next 3–5 days [2].

An illustration of the wide range of SaO_2 seen with ascent to altitude (two ascents to around 5,000 m) comes from the study of Brierley et al. [3] where mean SaO_2 for each of eight individuals covered a range from 81.9 to 93.2% at rest. The study showed a significant relation between higher AMS scores and lower SaO_2 values. This correlation also applied between AMS scores and the lower SaO_2 values obtained from moderate exercise. $SaO_2 \times$ Heart rate, a surrogate for oxygen delivery, remained near constant suggesting cardiac output compensation for the lower values of SaO_2 also illustrated by highly significant negative correlations between individual heart rates (HR) and SaO_2 values for each of seven of the eight subjects. The subject sustaining the highest SaO_2 showed no such correlation suggesting all compensation in his case had been respiratory.

The ascent profile and altitude reached for the Brierley study are important. This is because all subjects stayed at 3,324 m for the first 5 days, giving time for good acclimatization. After this two ascents were made to a little over 5,000 m with an intervening period at 3,324 m. Since the whole trip took 28 days the ascent profile can be regarded as highly conservative.

For the present study it was planned to, again, measure SaO_2 and HR at rest and in exercise during an ascent of Annapurna. However, the ascent profile consisted of daily ascents over a period of 6 days to reach a maximum of 4,130 m. As a result of practical problems the last two ascents were 575 m and 925 m. Fourteen school boys took part and, with the availability of only one oximeter, single daily measurements were made following each ascent (or later, descent). The whole trip lasted 11 days with missed measurements on day 10. The idea was to explore the range of SaO_2 and the adequacy of compensation in this fast ascent.

2 Methods

During an 11-day expedition to the Himalaya 14 subjects ascended Annapurna to 4,130 m over an initial period of 6 days. All participants (all males) were from the sixth form at the King's school at Grantham in the UK. Resting heart rate (HR) and arterial oxygen saturation (SaO_2) were measured daily (oximeter model, CMS50DLP Contec Medical Systems Ltd., Qinhuangdao, China) following arrival at each new location. Rest measurement followed at least 5 min sitting quietly. Following resting measurements, subjects carried out a 2 min period of exercise (stepping on and off a 30 cm step). SaO_2 and HR were measured at the end of the 2 min period.

3 Results

Average daily SaO_2 values for 13 subjects are shown, both for rest and exercise, in Fig. 14.1; on the left against height. On the right exercise SaO_2 values have been plotted against the higher rest values.

In Fig. 14.2 daily values for SaO_2 and heart rate (HR) show some tendency to change in opposite directions. This raises the possibility of HR, and possibly cardiac output (CO), compensation for low SaO_2.

Individual plots of HR versus SaO_2 for rest and exercise showed a range of significance from $p=0.001$ to non-significant. The plots of three subjects with consistently significant HR versus SaO_2 regressions are shown in Fig. 14.3. Compensatory activity is strongly supported.

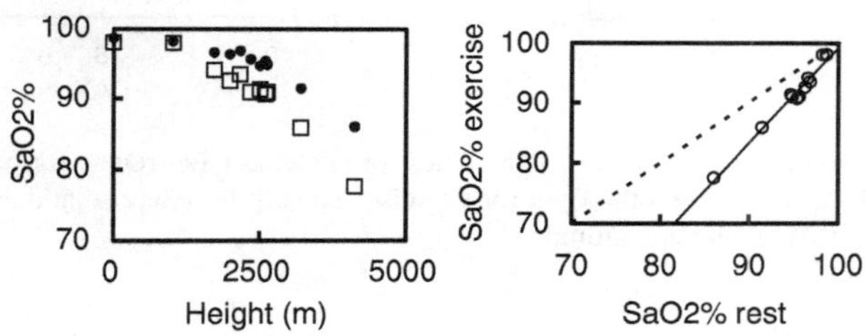

Fig. 14.1 Here SaO_2 average values for 13 subjects are shown for each day, plotted on the *left* against altitude both for rest (*black diamonds*) and exercise (*open squares*). The lowest values are at the highest altitudes. On the *right* the exercise SaO_2 is plotted against the resting value $SaO_2\%Ex = 1.5835 \times SaO_2\%rest - 59.177$) with a line of identity (*dashed line*)

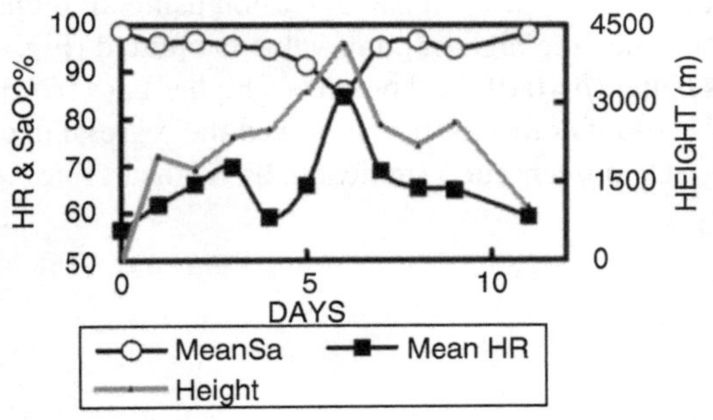

Fig. 14.2 Here we see progress day by day with the altitude (height) shown along with resting $SaO_2\%$ and heart rate (mean of all subject rest values for each day). There is an approximate tendency for the heart rate to be high when SaO_2 is low and vice versa

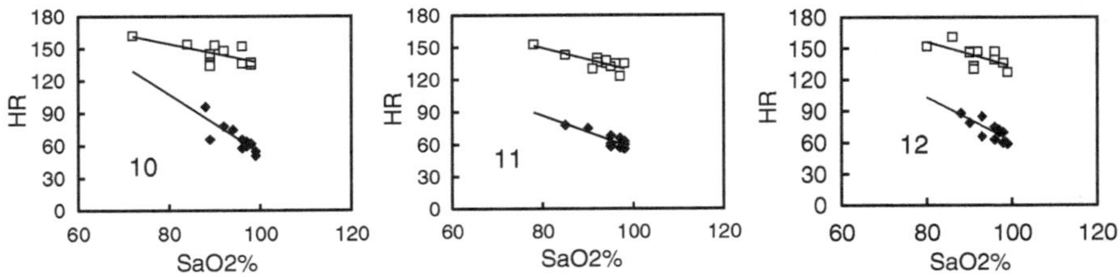

Fig. 14.3 The panels each show the relationship of heart rate to oxygen saturation for three individuals in whom the relationships were significant. For subject 10 at rest Heart rate $= -2.7055 \times SaO_2\% + 323.48$, $p < 0.01$, in exercise $y = -0.8797x + 224.78$, $p < 0.05$, where $y = HR$ and $x = SaO_2$. Similarly, for subject 11, at rest $y = -1.5308x + 208.96$, $p = 0.001$ and in exercise $y = -1.0765x + 235.4$, $p < 0.01$. For subject 12, $y = -2.1032x + 271.35$, $p < 0.01$ at rest and $y = -1.1862x + 251.08$, $p < 0.05$

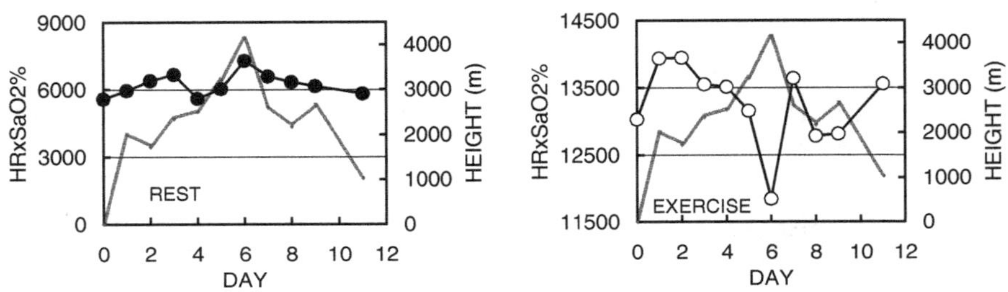

Fig. 14.4 The two panels show daily mean values for $HR \times SaO_2$ (surrogate for oxygen delivery, DO_2); on the left for rest and on the right for exercise. The dip for exercise at day 6 corresponds with maximum rate of ascent and altitude

The possibility that CO increases compensatorily for low SaO_2 is partly examined by the use of the surrogate for oxygen delivery, $HR \times SaO_2$, plotted in Fig. 14.4 (from daily means). For rest (left panel) $HR \times SaO_2$ variation is small. For exercise a sharp fall occurs on day 6, at the highest altitude.

Regression of average HR versus SaO_2 for rest was highly significant ($p = 0.001$, see left panel of Fig. 14.5). For exercise regression using all points was significant at $p < 0.02$ (0.013). When the highest point value is omitted (Fig. 14.5, right panel) the regression fits better ($p = 0.01$) and predicts a higher HR (167 rather than 153) at the SaO_2 value (77.5%) if compensation followed the regression equation.

It appears that CO compensation (indicated by the heart rate) tends to fail at the highest altitude.

4 Discussion

In the study of Brierley et al. [3] initial acclimatization at 3,324 m for 5 days allowed all subjects to maintain good HR compensation for low SaO_2 values throughout their trip. Evidence was clear cut from recordings including maximum and minimum

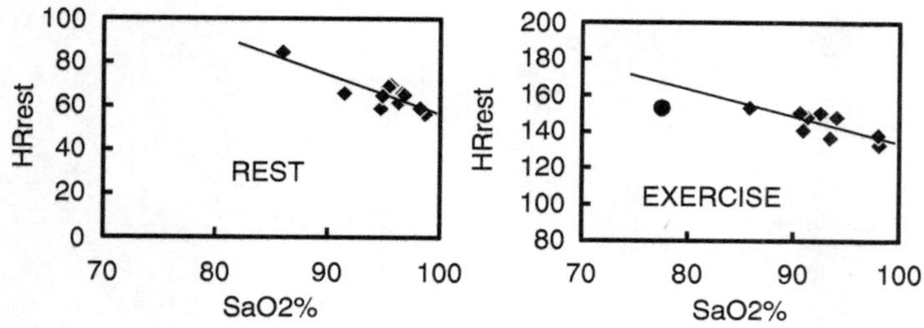

Fig. 14.5 Average values of heart rate (HR) are plotted against SaO_2 for rest (on the *left*) and exercise (*right*). Regression for rest is $HR = -1.7974 \times SaO_2 + 26.33$, $r = 0.841$, $p = 0.001$. For exercise—all points, $HR = -0.8834 \times SaO_2 + 226.14$, $r = 0.712$, $p < 0.02$. The regression line in the figure omits the highest altitude point and the regression is then $HR = -1.798 \times SaO_2 + 281.83$, $r = 0.769$, $p = 0.01$. HR 152.9 (at SaO_2 of 77.54%) is 14.2 beats per minute lower than would be the case (167.1) with full compensation

SaO_2 at each time point (twice daily), individual highly significant HR versus SaO_2 relationships and near constancy of the $HR \times SaO_2$ product (surrogate for DO_2) for both rest and exercise. In the present study measurements were made once a day (soon after arrival at a new altitude) with around twice as many subjects. A single SaO_2 value was recorded for rest and for exercise using only one oximeter. Evidence for HR compensation for low SaO_2 was present for average daily values with several subjects having significant individual HR versus SaO_2 relationships.

The partial breakdown of compensation in exercise at the highest altitude (Figs. 14.4 and 14.5) followed 2 days of rapid ascent. This study supports the idea that cardiac output increases when SaO_2 is low [1, 4] so long as acclimatization is adequate, but breaks down otherwise especially with the stress of exercise.

Acknowledgements To King's School Grantham for their encouragement and help in money raising activities and to Kevin Adams (staff member and mountaineer) and guide on the trek.

References

1. Montgomery HE, Marshall RM, Hemingway H, Myerson S, Clarkson P, Dollery C et al (1998) Human gene for physical performance. Nature (London) 393:221–222
2. Wolff CB, Barry P, Collier DJ (2002) Cardiovascular and respiratory adjustments at altitude sustain cerebral oxygen delivery—Severinghaus revisited. Comp Biochem Physiol A Mol Integr Physiol 132:221–229
3. Brierley G, Parks T, Wolff CB (2012) The relationship of acute mountain sickness to arterial oxygen saturation at altitudes of 3324 to 5176 m. Adv Exp Med Biol 737:207–212
4. Bell M, Thake CD, Wolff CB (2011) Effect of inspiration of 12% O_2 (balance N_2) on cardiac output, respiration, oxygen saturation and oxygen delivery. Adv Exp Med Biol 701:327–332

Chapter 15
Oscillations in Cerebral Haemodynamics in Patients with Falciparum Malaria

Christina Kolyva, Hugh Kingston, Ilias Tachtsidis, Sanjib Mohanty, Saroj Mishra, Rajya Patnaik, Richard J. Maude, Arjen M. Dondorp, and Clare E. Elwell

Abstract Spontaneous oscillations in cerebral haemodynamics studied with near-infrared spectroscopy (NIRS), become impaired in several pathological conditions. We assessed the spectral characteristics of these oscillations in 20 patients with falciparum malaria admitted to Ispat General Hospital, Rourkela, India. Monitoring included continuous frontal lobe NIRS recordings within 24 h of admission (*Day 0*), together with single measurements of a number of clinical and chemical markers recorded on admission. Seven patients returned for follow-up measurements on recovery (*FU*). A 2,048 sampling-point segment of oxygenated haemoglobin concentration ($[\Delta HbO_2]$) data was subjected to Fourier analysis per patient, and power spectral density was derived over the very low frequency (VLF: 0.02–0.04 Hz), low frequency (LF: 0.04–0.15 Hz) and high frequency (HF: 0.15–0.4 Hz) bands. At *Day 0*, VLF spectral power was 21.1 ± 16.4, LF power 7.2 ± 4.6 and HF power 2.6 ± 5.0, with VLF power being statistically significantly higher than LF and HF ($P < 0.005$). VLF power tended to decrease in the severely ill patients and correlated negatively with heart rate ($r = 0.57$, $P < 0.01$), while LF power correlated

---

The original version of this chapter was revised. An erratum to this chapter can be found at DOI 10.1007/978-1-4614-4989-8_53

C. Kolyva • I. Tachtsidis • C.E. Elwell (✉)
Department of Medical Physics and Bioengineering, University College London,
Malet Place Engineering Building, Gower Street, London WC1E 6BT, UK
e-mail: celwell@medphys.ucl.ac.uk

H. Kingston • A.M. Dondorp
Mahidol-Oxford Tropical Medicine Research Unit, Mahidol University, Bangkok, Thailand

S. Mohanty • S. Mishra • R. Patnaik
Department of Internal Medicine, Ispat General Hospital, Rourkela, India

R.J. Maude
Centre for Tropical Medicine, University of Oxford, Oxford, UK

W.J. Welch et al. (eds.), *Oxygen Transport to Tissue XXXIV*, Advances in Experimental Medicine and Biology 765, DOI 10.1007/978-1-4614-4989-8_15,

positively with aural body temperature (r=0.49, P<0.05). In all but one of the patients who returned for *FU* measurements, VLF power increased after recovery. This may be related to autonomic dysfunction in severe malaria, a topic of little research to date. The present study demonstrated that application of NIRS in a resource-poor setting is feasible and has potential as a research tool.

Keywords Cerebral hemodynamics • Falciparum malaria

1 Introduction

Falciparum malaria is a major public health problem in the developing world. It is caused by the protozoan parasite *Plasmodium falciparum* and transmitted to humans when an infected female *Anopheles* mosquito takes a blood meal, injecting sporozoites into the bloodstream as a by-product. Sporozoites multiply inside hepatocytes for an average of 6 days before releasing around 10^5 merozoites into the bloodstream. These mature and multiply inside red blood cells (RBCs) during a 48-h incubation period after which the infected RBCs burst, enabling the off-spring from each burst cell to infect around 8–10 new RBCs. With a total parasite burden of 10^7 to 10^8 the patient becomes febrile with flulike symptoms, while severe disease with involvement of multiple organs can develop when the parasite number exceeds 10^{11} to 10^{12}. At some point, often 1 week after presentation, some of the merozoites divide into male and female gametocytes that are taken up when mosquitoes aspirate blood from an infected subject, closing the life cycle of the parasite [1].

Patients infected by *P. falciparum* experience additional adverse symptoms, compared to other malaria species, due to the unique feature of this particular parasite to cause infected RBCs to adhere to the capillary and venular endothelium of various organs, especially the brain. This sequestration leads to reduced microvascular flow, which is further aggravated by the reduced deformability of an infected subject's RBCs [2].

Near-infrared spectroscopy (NIRS) has been previously used to investigate tissue oxygenation and metabolism in the adult brain [3]. Due to its portable, low-cost and non-invasive nature, it lends itself to environments with limited resources in terms of clinical monitoring. NIRS provides valuable information not only through the magnitudes of the measured parameters but also as a trend measurement. Such measurements in healthy adults have revealed slow oscillations in cerebral NIRS recordings [4], which are known to be impaired by pathological conditions like Alzheimer's disease [5], cerebral microangiopathy [6] and cerebral infarction [7]. Their frequencies are distinctly below the heart rate and respiration frequencies, they are spontaneous and their origin is controversial [4].

The aim of the present study was to assess the spectral characteristics of the spontaneous oscillations in transcranial NIRS recordings in patients with malaria and on recovery and to investigate the mechanism causing them.

2 Methods

Twenty non-sedated patients (14 males; age range 19–70 years) with asexual *P. falciparum* parasitaemia were studied at the Ispat General Hospital, Rourkela, India. Severe malaria was defined according to the modified Hien criteria [8] and cerebral malaria was defined as a Glasgow coma score (GCS) of <11. The study was approved by the Ispat General Hospital Ethics Committee and written informed consent was obtained from all patients or their families. All patients were receiving standard antimalarial treatment at the time of study, according to current guidelines [17].

Physical examination of the patient was performed on admission (*Day 0*) and a venous blood sample was taken to confirm the diagnosis by detection of asexual stage parasites on a peripheral blood film. The term *Day 0* is used throughout this study to indicate the first 24 h of admission. With the patient lying in the supine position and resting, frontal lobe continuous wave NIRS recordings were obtained at 6 Hz during *Day 0* over the course of at least 30 min (NIRO 300, Hamamatsu Photonics). The source–detector separation was 4 cm. Pulse oximetry measurements were simultaneously obtained at 1 Hz (Radical-7, Masimo). Seven of the 20 patients who consented to provide data in *Day 0* returned for follow-up measurements on recovery (*FU*), at a median of 15 days after *Day 0*.

Initial patient evaluation upon admission included single records of clinical parameters such as aural body temperature, GCS, mean blood pressure, heart rate (HR) and arterial oxygen saturation, and chemical markers, such as alanine transaminase, bilirubin, creatine kinase, total haemoglobin, base excess and lactate. From the NIRS measurements and assuming a differential pathlength factor of 6.26 [9], concentration changes of oxygenated [ΔHbO_2] and deoxygenated haemoglobin [ΔHHb] were derived according to the modified Beer–Lambert law, as well as tissue oxygenation index (TOI) by application of the spatially resolved spectroscopy technique. Pulse oximetry provided measures of oxygen saturation, HR, methaemoglobin, total haemoglobin and plethysmography variability index (an index of the variability of the plethysmographic waveform).

All NIRS-derived data were resampled every 0.16 s (6.25 Hz). For each patient, a 327.68 s (=2,048 × 0.16 s) data segment was selected and only this selection was included in the analysis described from this point onwards. The selection was based first upon identifying 2,048-point segments during which the standard deviation of the corresponding TOI data was less than 10% of the mean value (as a way of selecting data with as little noise as possible). Linear regression analysis was then performed on these segments (TOI vs. time) and the segment with the slope closest to zero was selected for further analysis (as a way of selecting data with stable TOI). [ΔHbO_2] and [ΔHHb] data were linearly detrended and averages for TOI and oximetry data were calculated. Because spontaneous oscillations are most prominent in the [ΔHbO_2] signal [4], the [ΔHbO_2] data were run through a fast Fourier transform algorithm, without prior filtering. From the derived power spectral density (PSD), the spectral power of three frequency bands was determined, in line with previous analysis [10]: 0.02–0.04 Hz (very low frequency; VLF), 0.04–0.15 Hz (low frequency; LF) and 0.15–0.4 Hz (high frequency; HF). Spectral power was defined as the area under the PSD curve over a given frequency band, divided by the corresponding frequency range (0.02, 0.11 and 0.25 Hz for the VLF, LF and HF spectral power, respectively), in order to render valid comparisons of spectral power between different frequency bands. For noise reduction purposes, the spectral power of each frequency band was then normalised with the spectral power of the whole PSD curve.

Statistical significance was set to $P<0.05$. Means were compared with Student's *t*-tests and multiple testing was accounted for via the Bonferroni correction. Linear regression analysis was used to investigate relations between different parameters.

3 Results

Table 15.1 summarises the baseline characteristics of the patients. Five patients had cerebral, nine non-cerebral severe and six uncomplicated falciparum malaria.

Day 0: The spectral power of the VLF band was statistically significantly higher than that of the LF and HF bands (Table 15.2). Cases with cerebral malaria tended to have lower spectral power of the VLF band compared to non-cerebral severe cases and uncomplicated cases (Table 15.2). There was no difference in TOI between malaria groups and overall TOI was $58.1 \pm 6.8\%$ (mean ± SD). Figure 15.1 shows examples of NIRS data from two age-matched patients, a case with cerebral malaria who did not survive and an uncomplicated case who made a full recovery, with striking differences in the oscillatory pattern of the signals.

Linear regression analysis was implemented in order to investigate possible relationships between VLF, LF and HF spectral power with all clinical parameters recorded upon admission and during *Day 0*. Oximetry data were available only in

Table 15.1 Patient demographics on admission

	Cerebral	Non-cerebral severe	Uncomplicated
n	5	9	6
Age	25 (24–25)	45 (22–70)	44 (19–65)
Body temperature (°C)	37.6 (35.5–39.4)	37.3 (35.9–39.4)	37.3 (36.5–38.6)
GCS	7 (4–9)	13 (11–15)	15 (15–15)
Systolic blood pressure (mmHg)	133 (125–145)	118 (102–147)	125 (105–139)
Diastolic blood pressure (mmHg)	80 (63–105)	76 (59–96)	74 (65–92)
Heart rate (bpm)	130 (98–149)	101 (75–135)	93 (74–101)
Haemoglobin (g/dl)	9.0 (4.8–13)	9.8 (5.8–13.5)	12.9 (9.3–15.5)
Creatinine (mg/dl)	2.32 (1.9–2.8)	2.09 (1–4.5)	1.07 (0.7–1.6)
Total bilirubin (mg/dl)	8.6 (0.7–27.2)	5.2 (0.4–24.5)	1.3 (0.4–2)

Table entries are mean (range)

Table 15.2 *Day 0* slow oscillation spectral characteristics

	All	Cerebral	Non-cerebral	Uncomplicated
Norm. VLF power	21.1±16.4	8.8±6.8	21.2±13.9	31.3±20.1
Norm. LF power	7.2±4.6*	8.0±6.1	6.8±3.6	7.0±5.5
Norm. HF power	2.6±5.0*,**	2.4±3.3	3.7±7.0*	1.1±0.7*
TOI (%)	58.1±6.8	58.1±5.4	59.5±8.5	55.9±5.4

VLF very low frequency, *LF* low frequency, *HF* high frequency, *TOI* tissue oxygenation index
*$P<0.05$ comparing VLF to LF or HF; **$P<0.05$ comparing LF to HF
Table entries are mean±SD

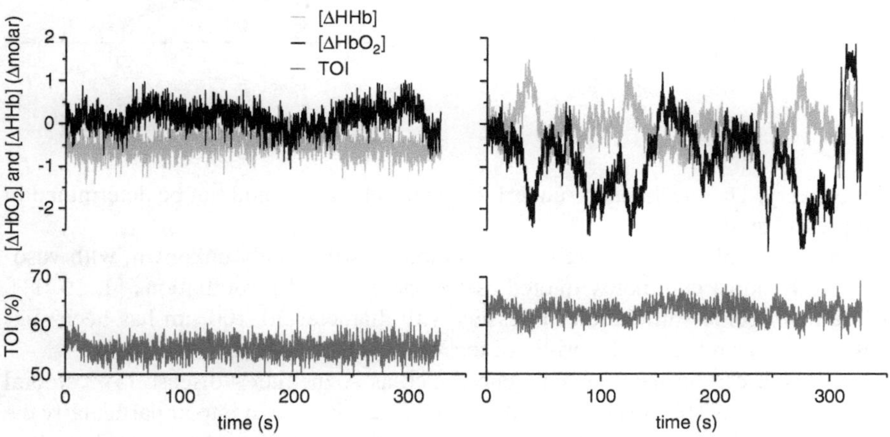

Fig. 15.1 Concentration changes of oxygenated [ΔHbO_2] and deoxygenated haemoglobin [ΔHHb] (*top*) and TOI (*bottom*) in patients with cerebral (*left*) and uncomplicated (*right*) malaria

13 of the 20 patients who participated in the study. The strongest correlation for VLF spectral power was with heart rate on admission ($Y=-0.42X+65.88$, $r=0.572$, $P<0.01$, $n=20$) and the strongest correlation for LF power was with body temperature ($Y=1.97X-66.32$, $r=0.488$, $P<0.05$, $n=20$).

Follow-up: The spectral power of the VLF band increased after recovery in six of the seven patients for whom follow-up measurements were available (Fig. 15.2), on average from 26.2±19.8 to 36.7±23.0 (P=NS). No obvious pattern was observed for the corresponding changes in LF and HF spectral power.

4 Discussion and Conclusions

Our study on the spectral characteristics of the slow spontaneous oscillations observed in transcranial NIRS recordings from patients with falciparum malaria showed that the VLF spectral power was decreased in cerebral malaria and increased

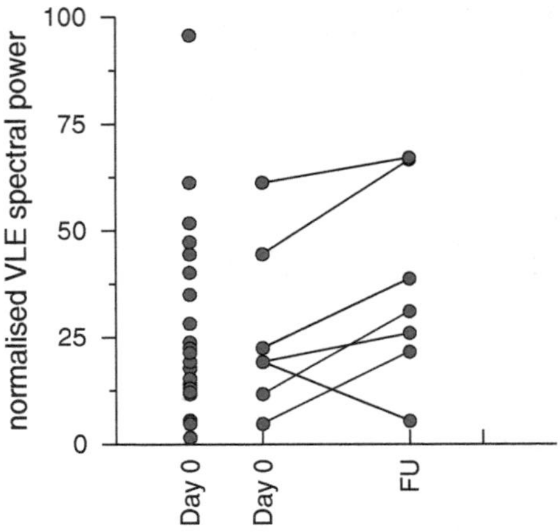

Fig. 15.2 VLF spectral power in patients with falciparum malaria, on the day of admission (*Day 0*) and after recovery (*FU*). The spectral power of the VLF band increased upon recovery in all but one of the patients for whom follow-up measurements were available

on recovery. The mechanisms underlying these changes could not be determined in the present study.

The origin of the slow cerebral oscillations is still largely unknown, with vasomotion the most commonly quoted as a cause for the LF oscillations [4, 10, 11], while neurogenic stimulation in vessels with diameter 50–100 μm has been suggested by researchers as the origin of the VLF oscillations [12].

Although certain pathological states, such as Alzheimer's disease [5], cerebral microangiopathy [6] and cerebral infarction [7], are known to impair particularly the LF oscillations, malaria was found to have an effect exclusively on the VLF oscillations. We therefore assume that the observed changes in severe malaria are mediated through a different mechanism. Low nitric oxide bioavailability [13], hypocapnia [14] and raised concentrations of isoprostanes [8] likely affect the regulation of vascular tone in falciparum malaria. Cerebrovascular resistance has been found to be raised in adults with cerebral malaria displaying normal reactivity to changes in pCO_2, implying preserved autoregulation [15]. It would be interesting to see if the PSD in a different vascular bed was similar to those recorded transcranially. Autonomic control of haemodynamics has received little attention in severe malaria, but could be disturbed in cerebral malaria. Orthostatic hypotension is common in patients with falciparum malaria and is accompanied by a failure of compensatory reflex cardio-acceleration, implying autonomic nervous system dysfunction [16].

We analysed approximately 328 s worth of data from each patient, although significantly longer recordings were available. Due to the restlessness of patients, several recordings were scattered with motion artefacts and in order to analyse the same length of data for all patients we could use only 328 s (2,048 sampling points). We also acknowledge other concerns, such as the limited power of the Fourier analysis in signals with time-varying frequency content, the fact that the recordings were at baseline only without including some kind of intervention and the possibility that the haemoglobin absorption spectrum might be altered in the presence of the parasite product hemozoin.

Despite these shortcomings, the present study constitutes a good first evaluation on changes in slow cerebral oscillations in patients with severe and cerebral malaria. We demonstrated that application of NIRS in a resource-poor setting is possible owing to the non-invasiveness, portability and low cost of this technology and is a promising research tool.

Acknowledgements The authors would like to thank the patients who participated in this study and the doctors and nurses of Ispat General Hospital. Funding from the Central London Research Network and the Wellcome Trust is gratefully acknowledged.

References

1. Cook GC, Zumla AI (2009) Manson's tropical diseases. Saunders Elsevier, Philadelphia, PA
2. Mishra SK, Newton CR (2009) Diagnosis and management of the neurological complications of falciparum malaria. Nat Rev Neurol 5:189–198
3. Highton D, Elwell C, Smith M (2010) Noninvasive cerebral oximetry: is there light at the end of the tunnel? Curr Opin Anaesthesiol 23:576–581
4. Obrig H, Neufang M, Wenzel R et al (2000) Spontaneous low frequency oscillations of cerebral hemodynamics and metabolism in human adults. Neuroimage 12:623–639
5. van Beek AH, Lagro J, Olde-Rikkert MG et al (2010) Oscillations in cerebral blood flow and cortical oxygenation in Alzheimer's disease. Neurobiol Aging 33(2):428.e21–428.e31
6. Schroeter ML, Bucheler MM, Preul C et al (2005) Spontaneous slow hemodynamic oscillations are impaired in cerebral microangiopathy. J Cereb Blood Flow Metab 25:1675–1684
7. Li Z, Wang Y, Li Y et al (2010) Wavelet analysis of cerebral oxygenation signal measured by near infrared spectroscopy in subjects with cerebral infarction. Microvasc Res 80:142–147
8. Charunwatthana P, Abul Faiz M, Ruangveerayut R et al (2009) N-acetylcysteine as adjunctive treatment in severe malaria: a randomized, double-blinded placebo-controlled clinical trial. Crit Care Med 37:516–522
9. Duncan A, Meek JH, Clemence M et al (1995) Optical pathlength measurements on adult head, calf and forearm and the head of the newborn infant using phase resolved optical spectroscopy. Phys Med Biol 40:295–304
10. Tachtsidis I, Elwell CE, Leung TS et al (2004) Investigation of cerebral haemodynamics by near-infrared spectroscopy in young healthy volunteers reveals posture-dependent spontaneous oscillations. Physiol Meas 25:437–445
11. Mayhew JE, Askew S, Zheng Y et al (1996) Cerebral vasomotion: a 0.1-Hz oscillation in reflected light imaging of neural activity. Neuroimage 4:183–193
12. Intaglietta M (1990) Vasomotion and flowmotion: physiological mechanisms and clinical evidence. Vasc Med 1:101–112
13. Yeo TW, Lampah DA, Gitawati R et al (2007) Impaired nitric oxide bioavailability and L-arginine reversible endothelial dysfunction in adults with falciparum malaria. J Exp Med 204:2693–2704
14. Laffey JG, Kavanagh BP (2002) Hypocapnia. N Engl J Med 347:43–53
15. Warrell DA, White NJ, Veall N et al (1988) Cerebral anaerobic glycolysis and reduced cerebral oxygen transport in human cerebral malaria. Lancet 2:534–538
16. Supanaranond W, Davis TM, Pukrittayakamee S et al (1993) Abnormal circulatory control in falciparum malaria: the effects of antimalarial drugs. Eur J Clin Pharmacol 44:325–329
17. WHO (2000) Guidelines for the treatment of malaria, second edition, 2010

Chapter 16
Effect of Spinal Anesthesia for Elective Cesarean Section on Cerebral Blood Oxygenation Changes: Comparison of Hyperbaric and Isobaric Bupivacaine

Yuko Kondo, Kaoru Sakatani, Noriya Hirose, Takeshi Maeda, Jitsu Kato, Setsuro Ogawa, and Yoichi Katayama

Abstract We used near-infrared spectroscopy (NIRS) to evaluate cerebral blood oxygenation changes in subjects undergoing cesarean section under spinal anesthesia (SP) with hyperbaric bupivacaine (group H, 27 subjects) or isobaric bupivacaine (group I, 15 subjects). In group H, total-Hb, oxy-Hb, and mean blood pressure (MBP) within 20 min after SP were significantly lower than the baseline values. In contrast, there was no significant change from baseline in total-Hb, oxy-Hb, or MBP in group I after SP. Total-Hb and MBP in group H were significantly lower than those in group I within 10 min after SP. There was no significant change of deoxy-Hb, tissue oxygen index, or heart rate from baseline in either of the groups. These results suggest that isobaric bupivacaine may be superior to hyperbaric bupivacaine for preventing a decrease of maternal cerebral blood flow after SP for cesarean section.

Keywords Cesarean • Bupivacaine

Y. Kondo (✉) • T. Maeda
Division of Neurosurgery, Department of Neurological Surgery,
Nihon University School of Medicine, 30-1, Oyaguchi-Kamicho, Itabashi-Ku,
Tokyo, 173-8610, Japan

Division of Anesthesiology, Department of Anesthesiology,
Nihon University School of Medicine, Tokyo, Japan
e-mail: yk-mikan@d8.dion.ne.jp

K. Sakatani
Division of Optical Brain Engineering, Department of Neurological Surgery,
Nihon University School of Medicine, Tokyo, Japan

N. Hirose • J. Kato • S. Ogawa
Division of Anesthesiology, Department of Anesthesiology,
Nihon University School of Medicine, Tokyo, Japan

Y. Katayama
Division of Neurosurgery, Department of Neurological Surgery,
Nihon University School of Medicine, 30-1, Oyaguchi-Kamicho, Itabashi-Ku,
Tokyo, 173-8610, Japan

W.J. Welch et al. (eds.), *Oxygen Transport to Tissue XXXIV*, Advances in Experimental Medicine and Biology 765, DOI 10.1007/978-1-4614-4989-8_16,

1 Introduction

Maternal hypotension is a common side effect after spinal anesthesia for cesarean section. Severe hypotension after spinal anesthesia may result in decreased cerebral blood flow and oxygenation, leading to maternal symptoms such as nausea, vomiting, and dizziness [1, 2]. Indeed, the relationship between hypotension and decrease in cerebral blood oxygenation (CBO) after spinal anesthesia has been evaluated by near-infrared spectroscopy (NIRS) [2], and it was found that isobaric bupivacaine has less influence on blood pressure than hyperbaric bupivacaine [3]. However, differences between the effects of hyperbaric and isobaric bupivacaine on cerebral blood flow and oxygenation have not been examined in detail. Therefore, we employed NIRS to evaluate CBO changes after spinal anesthesia with hyperbaric or isobaric bupivacaine for cesarean section.

2 Methods

We studied 47 ASA (American Society of Anesthesiologists) physical status I–II female patients (mean age 33.2 ± 5.7 years) scheduled for elective cesarean section under spinal anesthesia. Patient's profiles are shown in Table 16.1. Of the 47 patients, 32 received spinal anesthesia with hyperbaric bupivacaine (group H) and 15 with isobaric bupivacaine (group I). This study was approved by the Committee for Clinical Trials and Research on Humans of Nihon University School of Medicine, and each subject gave informed consent to participate.

2.1 Monitoring

The patients were monitored with automatic noninvasive blood pressure (NIBP) measurement on the right arm, pulse oximetry, and electrocardiography. We measured CBO changes in the forehead using NIRS (Niro pulse, Hamamatsu Photonics, Japan).

Table 16.1 Patients' characteristics and anesthetic data

	Group H ($n=27$)	Group I ($n=15$)
Age (years)	32.7 ± 5.7	34.4 ± 5.6
Height (cm)	158.3 ± 4.4	158.0 ± 6.1
Weight (kg)	61.2 ± 9.4	64.3 ± 12.1
Gestational age (weeks)	36–39	32–42
Dose of bupivacaine (mg)	11.5 ± 1.0	11.0 ± 0.5
Spinal block level at 20 min after spinal anesthesia (median range)	Th 3/3	Th 3/3
Incidence of hypotension (%)	24 (88.9)	3 (20)
Incidence of nausea and vomiting (%)	5 (18.5)	0 (0)
Total fluid (ml)	747.3 ± 135.9	762.4 ± 148.4

The NIR light from three laser diodes (775, 810, and 850 nm) is directed to the head through a fiber-optic bundle, and the reflected light is transmitted to a multisegment photodiode detector array. The NIRS system measures the concentrations of oxy-Hb, deoxy-Hb, total hemoglobin (total-Hb; oxy-Hb+deoxy-Hb), and the tissue oxygen index (TOI). We started measurements of CBO changes, mean arterial blood pressure (MBP), and heart rate (HR) before the anesthetic procedure and recorded these parameters every 1 min until 20 min after spinal anesthesia.

2.2 *Anesthetic Procedure*

Patients did not receive premedication. In the operating room, each patient was placed in the supine position and received 3 L/min of oxygen through a face mask and colloidal solution at a rate of 20 ml/kg/h. Three minutes later, baseline measurements (oxy-, deoxy-, total-Hb concentrations, TOI, MBP, and HR) were recorded. Then, the patient was placed in the lateral position to receive spinal and epidural anesthesia. An epidural catheter was inserted into L1/2 interspace and the lumbar puncture was performed at the L3/4 interspace. Conditions of anesthesia are summarized in Table 16.1. The patient was returned to the supine position immediately after spinal anesthesia. If hypotension (MBP <80% of baseline value or systolic blood pressure <90 mmHg) was observed, the patient was treated with ephedrine (i.v. bolus) and/or returned to left lateral uterine displacement.

2.3 *Data Analysis*

All results are expressed as mean ± SD or as median and range. MBP and CBO variables for each time interval were compared with the baseline values by analysis of variance (ANOVA) for repeated measurements. If the ANOVA revealed a significant interaction, post hoc analysis was performed using the Tukey–Kramer multiple comparison test when applicable. $P < 0.05$ was considered to represent a statistically significant difference.

3 Results

Data of five patients were excluded from this study due to technical problems during NIRS measurements. Hyperbaric bupivacaine was used in 27 patients (group H), and isobaric bupivacaine in 15 patients (group I). Patient's profiles and anesthesia data are shown in Table 16.1.

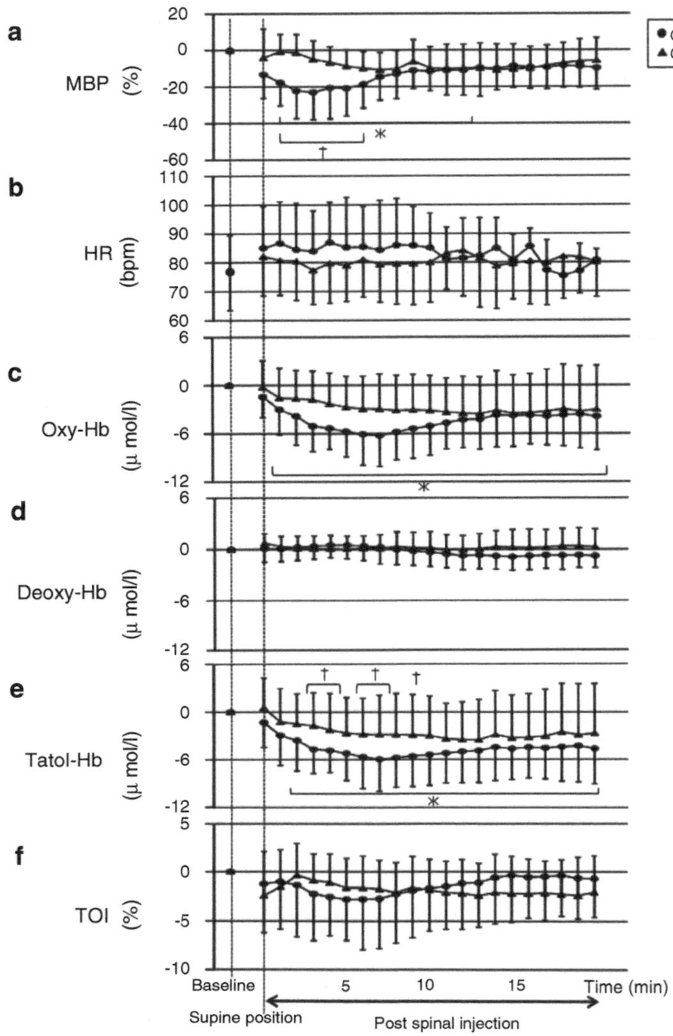

Fig. 16.1 MBP (**a**), HR (**b**), Oxy-Hb (**c**), Deoxy-Hb (**d**), Total-Hb (**e**), and TOI (**f**) versus time (data are mean ± SD). Changes in MBP, Hb concentrations, and TOI from baseline. *Oxy-Hb, total-Hb, and MBP were significantly decreased when compared with baseline values in group H ($P<0.05$). †Total-Hb and MBP in group H were significantly lower than those in group I ($P<0.05$)

Twenty-four patients (88.9%) in group H and three patients (20%) in group I suffered severe hypotension, which was symptomatic (nausea, vomiting, and dizziness) in five (18.5%) in group H; however, the three patients in group I exhibited no symptoms. Figure 16.1 compares mean changes of MBP, HR, oxy-Hb, deoxy-Hb, total-Hb concentrations, and TOI between group I and group H. In group H, oxy-Hb, total-Hb, and MBP within 20 min after spinal anesthesia were significantly lower than the baseline values ($P<0.05$), while HR, deoxy-Hb, and TOI did not change significantly. In contrast, there was no significant change in any parameter in group I. Total-Hb and MBP in group H were significantly lower than those in group I within 10 min after spinal anesthesia ($P<0.05$).

4 Discussion

Our results show that spinal anesthesia with hyperbaric bupivacaine is associated with a decrease of oxy-Hb and total-Hb, which indicates a decrease of cerebral blood flow (CBF) and cerebral blood volume (CBV) [4, 5]. In contrast, spinal anesthesia with isobaric bupivacaine did not cause hypotension or CBO changes. These results suggest that maternal CBF may be decreased after spinal anesthesia with hyperbaric bupivacaine, but not with isobaric bupivacaine. These findings are consistent with a report indicating that hyperbaric bupivacaine caused hypotension and reduction of CBF [2]. Interestingly, hyperbaric bupivacaine did not change deoxy-Hb in the present study. This suggests that the reduction of CBF caused by hyperbaric bupivacaine might be moderate since changes of deoxy-Hb during ischemia are dependent on the degree of reduction of CBF [6]. It should be noted, however, that patients anesthetized with hyperbaric bupivacaine complained of symptoms such as nausea, suggesting that the reduction of CBF might have been sufficient to influence brain function. It is interesting that, although the degree of hypotension was within the functional range of cerebral autoregulation, spinal anesthesia with hyperbaric bupivacaine caused a reduction of CBF. It has been reported that an abrupt decrease of systemic blood pressure within the threshold of autoregulation can decrease CBF [7]. In addition, spinal anesthesia tends to rapidly decrease systemic blood pressure, for a period of several minutes [8]. These results suggest that the reduction of CBF under spinal anesthesia with hyperbaric bupivacaine was caused by a mild but rapid decrease in systemic blood pressure.

In general, hyperbaric bupivacaine tends to increase the anesthetic level more rapidly than isobaric bupivacaine and, therefore, hemodynamic changes are more severe with hyperbaric bupivacaine than with isobaric bupivacaine [3]. In addition, even a small dose of local anesthetic may induce hypotension, since the amount of cerebrospinal fluid decreases in the lumbosacral area in late pregnancy [9, 10]. These observations suggest that isobaric bupivacaine is more suitable for spinal anesthesia for cesarean section.

5 Conclusions

Measurements of CBO changes indicate that isobaric bupivacaine is more suitable than hyperbaric bupivacaine to prevent a reduction of CBF after spinal anesthesia for cesarean section.

Acknowledgments This research was partly supported by Japan Science and Technology Agency, under the Strategic Promotion of Innovative Research and Development Program, and a Grant-in-Aid from the Ministry of Education, Culture, Sports, Sciences and Technology of Japan (B23300247).

References

1. Datta S et al (1982) Method of ephedrine administration and nausea and hypotension during spinal anesthesia for cesarean section. Anesthesiology 56:68–70
2. Berlac PA, Rasmussen YH (2005) Per-operative cerebral near-infrared spectroscopy (NIRS) predicts maternal hypotension during elective caesarean delivery in spinal anesthesia. Int J Obstet Anesth 14:26–31
3. Gessel EF et al (1991) Comparison of hypobaric, hyperbaric, and isobaric solutions of bupivacaine during continuous spinal anesthesia. Anesth Analg 72:779–784
4. Ferrari M et al (1992) Effects of graded hypotension on cerebral blood flow, blood volume, and mean transit time in dogs. Am J Physiol 262:1908–1914
5. Pryds O et al (1990) Carbon dioxide-related changes in cerebral blood volume and cerebral blood flow in mechanically ventilated preterm neonates: comparison of near infrared spectrophotometry and 133 Xenon clearance. Pediatric Res 27:445–449
6. Baikoussis NG et al (2010) Baseline cerebral oximetry values in cardiac and vascular surgery patients: a prospective observational study. J Cardiothorac Surg 5:41
7. Zhang R et al (2002) Autonomic neural control of dynamic cerebral autoregulation in humans. Circulation 106:1814–1820
8. Langesaeter E et al (2008) Continuous invasive blood pressure and cardiac output monitoring during cesarean delivery. Anesthesiology 109:856–863
9. Higuchi H et al (2004) Influence of lumbosacral cerebrospinal fluid density, velocity, and volume on extent and duration of plain bupivacaine spinal anesthesia. Anesthesiology 100:106–114
10. Jawan B et al (1993) Spread of spinal anaesthesia for caesarean section in singleton and twin pregnancies. Br J Anaesth 70:639–641

Chapter 17
DCX-Expressing Neurons Decrease in the Retrosplenial Cortex after Global Brain Ischemia

Nobuo Kutsuna, Yoshihiro Murata, Takashi Eriguchi, Yoshiyuki Takada, Hideki Oshima, Kaoru Sakatani, and Yoichi Katayama

Abstract Many studies have demonstrated cognitive function disorders including space learning disorders after global brain ischemia (GBI). Previous research on space perception and learning has indicated that the retrosplenial cortex (RS) is strongly involved. We performed immunostaining with doublecortin (DCX) for neurons with plasticity potential in the RS and investigated the neuronal numbers to assess the changes of plasticity in the RS following GBI. We employed male Sprague–Dawley rats and carried out bilateral carotid arterial occlusion for 10 min as a GBI model (control, $n=5$; GBI model, $n=5$). We counted the right and left hemispheres separately on two serial sections, for a total of four regions per animal to examine the differences in expression related to GBI. Additionally, we performed Fluoro-Jade B (FJB) staining to investigate the cause of any DCX-expressing neuron decrease. The total number of DCX-expressing neurons was 1,652 and 912 in the controls and GBI model, respectively. The mean number of DCX-expressing neurons per unit area was significantly lower in the GBI model than in the controls. FJB positive neurons were not found in the RS, while many were present in the hippocampus CA1 after GBI. The decrease of DCX-expressing neurons in the RS indicated a plasticity decrease following GBI. The lack of FJB positive neurons in the RS after GBI suggested that the decrease of DCX-expressing neurons in the RS was not due to neuronal cell death in contrast to the hippocampus CA1, while the FJB positive neurons in the hippocampus indicated a delayed neuronal cell death as observed in many previous studies.

Keyword Brain ischemia

N. Kutsuna (✉) • Y. Murata • T. Eriguchi • Y. Takada • H. Oshima • Y. Katayama
Division of Neurosurgery, Department of Neurological Surgery, Nihon University School of Medicine, 30-1 Oyaguchikami-cho, Itabashi-ku, Tokyo 173-8610, Japan
e-mail: nkutsuna@yahoo.co.jp

K. Sakatani
Division of Optical Brain Engineering, Department of Neurological Surgery, Nihon University School of Medicine, Tokyo, Japan

W.J. Welch et al. (eds.), *Oxygen Transport to Tissue XXXIV*, Advances in Experimental Medicine and Biology 765, DOI 10.1007/978-1-4614-4989-8_17,

1 Introduction

The causes of global brain ischemia (GBI) include cardiac arrest, low blood pressure shock, and vascular disorders [1–3]. Many studies have demonstrated cognitive function disorders including space learning disorders after GBI. Most experimental investigations have focused on the hippocampus and revealed that delayed neuronal cell death in the CA1 and CA3 increased following GBI [4–6], but the association between cognitive impairment and cell death was unclear [7–10]. Research on space perception and learning has indicated that the retrosplenial cortex (RS) which forms the caudal cingulate cortex is strongly involved [11–14]. However, little work has been done on the RS after GBI, whereas the hippocampus has been well investigated.

In the developing brain, doublecortin (DCX) contributes to the radial migration of immature neurons from the paraventricle to cortex, nuclear translocation, and the extension of axons and dendrites [15]. Based on these functions, DCX-expressing neurons are considered immature and thought to contribute strongly to structural plasticity. It has been reported that neurons having DCX are present in the cerebral neocortex and allocortex of the mature individual, and it has been suggested that they are associated with the structural plasticity of the brain [16–18].

The RS has a synaptic plasticity potential [19]. We performed immunostaining with DCX for neurons with structural plasticity potential in the RS and investigated the neuronal numbers to assess the alterations of the neuronal circuits and plasticity in the RS following GBI.

2 Materials and Methods

2.1 GBI Model

We employed male Sprague–Dawley rats (body weight 250–300 g). The procedure to make transient global cerebral ischemia model has been reported previously [25]. Anesthesia was performed with pentobarbital (Somnopentyl, Kyoritsu Seiyaku; 15 mg/kg body weight) intraperitoneally, N_2O, and isoflurane. Obliteration time of bilateral internal carotid artery was 10 min (control, $n=5$; GBI model, $n=5$). The animals were purchased from Charles River Laboratories and bred at the animal housing facility of Nihon University. The colony was maintained at 22–23 °C on a 12-h light/dark cycle (lights on at 08:00). After 7 days, the rats were transcardially perfusion-fixed with lactated Ringer's solution, followed by perfusion of 4 % paraformaldehyde. Coronal serial brain sections (50 μm) were cut in the frontal plane on a vibratome and stored at 4 °C in PBS (pH 7.4) before being analyzed.

All experimental procedures were conducted in accordance with the Guidelines for the Care and Use of Mammals in Neuroscience and Behavioral Research (National Research Council, National Academy Press, Washington, DC, 2003) and approved by the Animal Care and Use Committee of Nihon University.

Table 17.1 Primary antibodies

Antigen	Class of antibody	Dilution	Manufacturer
Doublecortin (DCX)	Polyclonal goat	1:2,000	Santa Cruz Biotechnology
PSA-NCAM	Monoclonal mouse IgG	1:500	Chemicon

2.2 *Immunohistochemistry*

We performed immunostaining and fluorescence immunostaining with DCX and polysialic acid-neural cell adhesion molecule (PSA-NCAM) to observe the neurons with plasticity. The primary antibodies used in this study are shown in Table 17.1. We counted the right and left hemispheres separately on two serial sections, for a total of four regions per animal to examine the differences in expression related to GBI.

2.3 *Fluoro-Jade B Staining*

Additionally we performed Fluoro-Jade B (FJB) staining to investigate the cause of any DCX-expressing neuron decrease. The FJB staining procedure was as reported previously [20].

2.4 *Measurement and Analysis*

We used Biozero (BZ-8000; Keyence, Japan) and a BZ-Analyzer (Keyence, Japan) to prepare the microphotographs, and a VH-Analyzer (Keyence, Japan) which had unbiased three-dimensional stereological software to count the numbers of immunostain-positive cells. We used Neurolucida (Version 3; MicroBrightField, USA) to analyze the area and cell counts without double counting of positive cells in the ACC and RS, and then estimated the cell counts per unit area in the control and GBI model. We employed the statistical software SPSS Statistics 17.0 for data analysis. We performed the Mann–Whitney test to compare unrelated groups.

3 Results

3.1 *Number of DCX-Expressing Neurons and Colocalization with PSA-NCAM*

We identified the RS by referring to the rat atlas of Paxinos and Watson (1982) [21] and measured the neuronal DCX expression in immunostained coronal sections subjected to DCX immunohistochemistry (Fig. 17.1).

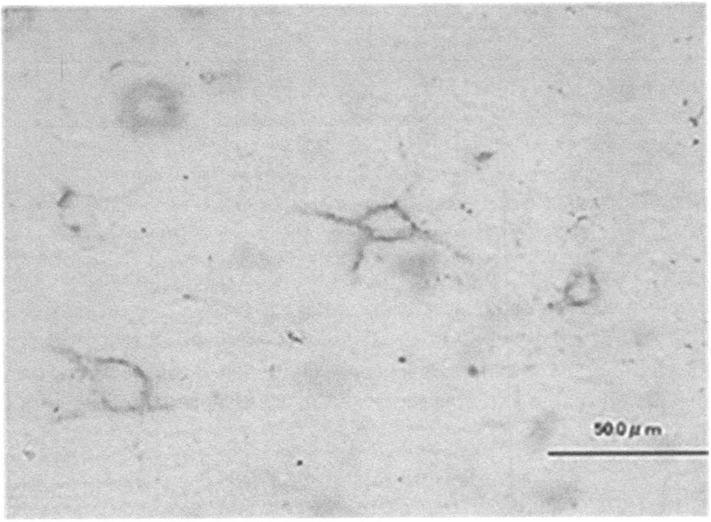

Fig. 17.1 Doublecortin (DCX)-expressing neurons in the retrosplenial cortex (RS) (diaminobenzidine reaction; scale bar, 50 μm; bregma posterior, 3.30 mm). This image was obtained from the RS of a global brain ischemia model (layer: III/IV). DCX-expressing neurons were similarly present in the RS of the control

The total number of DCX-expressing neurons was 1,652 in the controls and 912 in the GBI model. The mean number of DCX-expressing neurons per unit area (mean ± SD) was 67 ± 14.78 neurons/mm^2 in the controls and 34 ± 8.63 neurons/mm^2 in the GBI model. The mean number of DCX-expressing neurons was significantly lower in the GBI model than in the controls (Mann–Whitney test, $p < 0.001$) (Fig. 17.2).

PSA-NCAM represents one of the immature neuronal markers. We investigated the colocalization of DCX with PSA-NCAM in the controls and GBI model. The results showed that few DCX-expressing neurons expressed PSA-NCAM in the RS, while many DCX-expressing neurons were colocalized with PSA-NCAM within the hippocampus subgranular zone in which neurogenesis had occurred.

3.2 *FJB Staining*

FJB staining is used for the identification of neuronal cell death. As regards the RS, there were no positive cells in the controls and GBI model, respectively, while many positive neurons were present in the hippocampus CA1 of the GBI model. This number for the RS was not appropriate for the decreased number of DCX-expressing neurons.

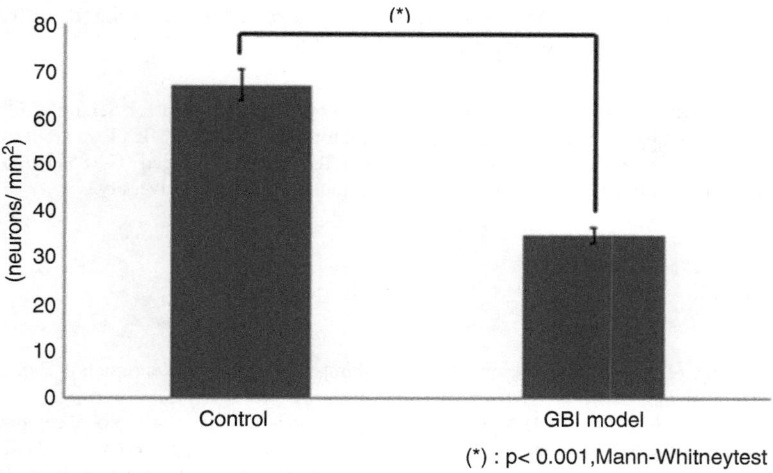

Fig. 17.2 In the global brain ischemia model ($n=5$), significantly less doublecortin (DCX)-expressing neurons were observed in the retrosplenial cortex (RS) than in the control ($n=5$) per unit area ($n=20$ samples/subregion; right and left hemispheres of two sections from each rat)

4 Discussion and Conclusion

Transient GBI could lead to neuronal differentiation of immature neurons in the RS followed by a decreased structural plasticity in the mature individual brain. The present data showed that the DCX-expressing neurons of the RS were significantly decreased after transient GBI. Following lethal ischemia, neuronal cell death occurs in the cortex [22]. The results for FJB staining revealed no FJB positive cells after GBI in the RS, indicating that neurons which fall into cell death are rare after GBI in the RS. Since cell death does not occur as the number of DCX-expressing neurons decreases, it is considered that DCX disappears from the neurons. In relation to the function of DCX, the neuron advances to a more mature form during the stages of neuronal development, and it is inferred that the neuron has a reduced function to grow dendrites and axons to form new synapses without neuronal cell death [15].

As described above, the RS plays an important role in space learning and memory. It has been suggested that information based on movement from the occipito-parietal cortex spreads in the space reference frame within the RS [23]. There are cells with a function to make cephalic presentation signals [23]. Also, the neurons make signals responding immediately to the animal body position and body movement as well as cephalic presentation [24]. The information associated with space navigation is unified within the RS.

We consider that many DCX-expressing neurons could respond to the stimulation of space perception and form local neuronal circuits according to new space information in the developed brain. A decrease of these neurons after GBI may lead to a decrease in the RS plastic potential and, as a result, be associated with space learning failure and memory dysfunction.

Acknowledgments This work was supported in part by a Grant-in-Aid for Scientific Research (C-20591725) and by the Strategic Research Program for Brain Science (MEXT), a grant from the Ministry of Education, Culture, Sports, Science and Technology of Japan (C-18591614) and a grant for the promotion of industry–university collaboration at Nihon University.

References

1. Garcia JH (1992) The evolution of brain infarcts: a review. J Neuropathol Exp Neurol 51:387–393
2. Mossakowski MJ et al (1994) Abnormalities of the blood–brain barrier in global cerebral ischemia in rats due to experimental cardiac arrest. Acta Neurochir Suppl (Wien) 60:274–276
3. Teschendorf P et al (2008) Time course of caspase activation in selectively vulnerable brain areas following global cerebral ischemia due to cardiac arrest in rats. Neurosci Lett 448:194–199
4. Cummings JL et al (1984) Amnesia with hippocampal lesions after cardiopulmonary arrest. Neurology 34:679–681
5. Auer RN et al (1989) Neurobehavioral deficit due to ischemic brain damage limited to half of the CA1 sector of the hippocampus. J Neurosci 9:1641–1647
6. Gionet TX et al (1991) Forebrain ischemia induces selective behavioral impairments associated with hippocampal injury in rats. Stroke 22:1040–1047
7. Rod MR et al (1990) The relationship of structural ischemic brain damage to neurobehavioural deficit: the effect of postischemic MK-801. Can J Psychol 44:196–209
8. Kiyota Y et al (1991) Relationship between brain damage and memory impairment in rats exposed to transient forebrain ischemia. Brain Res 538:295–302
9. Green EJ et al (1992) Protective effects of brain hypothermia on behavior and histopathology following global cerebral ischemia in rats. Brain Res 580:197–204
10. Jaspers RM et al (1990) Spatial learning is affected by transient occlusion of common carotid arteries (2VO): comparison of behavioural and histopathological changes after '2VO' and 'four-vessel-occlusion' in rats. Neurosci Lett 117:149–153
11. Devinsky O et al (1995) Contributions of anterior cingulate cortex to behaviour. Brain 118:279–306
12. Desgranges B et al (1998) The functional neuroanatomy of episodic memory: the role of the frontal lobes, the hippocampal formation, and other areas. Neuroimage 8:198–213
13. Maddock RJ (1999) The retrosplenial cortex and emotion: new insights from functional neuroimaging of the human brain. Trends Neurosci 22:310–316
14. Maguire EA (2001) The retrosplenial contribution to human navigation: a review of lesion and neuroimaging findings. Scand J Psychol 42:225–238
15. Friocourt G et al (2007) Both doublecortin and doublecortin-like kinase play a role in cortical interneuron migration. J Neurosci 27:3875–3883
16. Xiong K et al (2008) Doublecortin-expressing cells are present in layer II across the adult guinea pig cerebral cortex: partial colocalization with mature interneuron markers. Exp Neurol 211:271–282
17. Luzzati F et al (2008) DCX and PSA-NCAM expression identifies a population of neurons preferentially distributed in associative areas of different pallial derivatives and vertebrate species. Cereb Cortex 19:1028–1041
18. Cai Y et al (2009) Doublecortin expression in adult cat and primate cerebral cortex relates to immature neurons that develop into GABAergic subgroups. Exp Neurol 216:342–356

19. Garden DL et al (2009) Anterior thalamic lesions stop synaptic plasticity in retrosplenial cortex slices: expanding the pathology of diencephalic amnesia. Brain 132:1847–1857
20. Schmued LC et al (2000) Fluoro-Jade B: a high affinity fluorescent marker for the localization of neuronal degeneration. Brain Res 874:123–130
21. Paxinos G et al (2007) The rat brain in stereotaxic coordinates, 6th edn. Academic, San Diego, CA
22. Burda J et al (2006) Delayed postconditioning initiates additive mechanism necessary for survival of selectively vulnerable neurons after transient ischemia in rat brain. Cell Mol Neurobiol 26:1141–1151
23. Chen LL et al (1994) Head-direction cells in the rat posterior cortex. I. Anatomical distribution and behavioral modulation. Exp Brain Res 101:8–23
24. Cho J et al (2001) Head direction, place, and movement correlates for cells in the rat retrosplenial cortex. Behav Neurosci 115:3–25
25. Yagita Y et al (2001) Neurogenesis by progenitor cells in the ischemic adult rat hippocampus. Stroke 32:1890–1896

Chapter 18
Calibration and Validation Scheme for In Vivo Spectroscopic Imaging of Tissue Oxygenation

Maritoni Litorja, Robert Chang, Jeeseong Hwang, David W. Allen, Karel Zuzak, Eleanor Wehner, Sara Best, Edward Livingston, and Jeffrey Cadeddu

Abstract The determination of the level of oxygenation in optically accessible tissues using multispectral or hyperspectral imaging (HSI) of oxy- and deoxyhemoglobin has special appeal in clinical work due to its noninvasiveness, ease of use, and capability of providing molecular and anatomical information at near video rates during surgery. In this paper we refer to an example of the use of HSI in monitoring oxygenation of kidneys during partial nephrectomy. In a study using porcine models, it was found that artery-only clamping left the kidney better oxygenated, as opposed to simultaneously clamping the artery and the vein. A subsequent study correlates gradations in blood flow by partial clamping during the surgical procedure with postoperative renal function via assessment of creatinine level. We discuss the various contributions to the uncertainty of the oxygen saturation measured by this remote-sensing imaging technique in medical application.

Keywords Spectroscopic imaging • Tissue oxygenation

1 Introduction

There are certain biomedical applications where a method of measuring tissue oxygenation through remote sensing is needed. For example, during surgery, a clinician may want to monitor the extent of oxygenation over a large spatial area

M. Litorja (✉) • R. Chang • J. Hwang • D.W. Allen
National Institute of Standards and Technology,
100 Bureau Dr. Stop 8443, Gaithersburg, MD 20899, USA
e-mail: litorja@nist.gov

K. Zuzak
Digital Light Innovations, Dallas, TX, USA

E. Wehner • S. Best • E. Livingston • J. Cadeddu
University of Texas Southwestern Medical Center, Dallas, TX, USA

W.J. Welch et al. (eds.), *Oxygen Transport to Tissue XXXIV*, Advances in Experimental Medicine and Biology 765, DOI 10.1007/978-1-4614-4989-8_18,

to aid in decision-making. Hyperspectral imaging (HSI), also known as chemical imaging spectroscopy, has proven to be useful for clinical oxygenation monitoring.

Hyperspectral images are two-dimensional spatial maps (x, y) with a third-dimensional component of spectral bands (λ), such that each spatial point has a spectral profile. The spectral bands are chosen to coincide with wavelength ranges where the component of interest has spectral features that stand out above the background. The bandwidth is chosen according to the specific precision tolerance of the application. The user determines the spatial resolution needed for the application with the use of the appropriate optics, e.g., camera lens, endoscope, or microscope. Hyperspectral image datacubes (x, y, λ) can be acquired in different ways. Spatial or spectral scanning systems may be employed to construct the hyperspectral datacube. Bearman and Levenson summarize the various technological variants that exist and are currently used in biological imaging [1].

2 Description of Hyperspectral Imaging Oximetry

This spectroscopic imaging of oxygenation uses hemoglobin saturation as a tissue oxygenation metric. This is based on Beer's law and similar to that used in cuvette oximetry where oxyhemoglobin in drawn blood is extracted and analyzed spectrophotometrically [2]. The clinical equipment used in this study is described in detail elsewhere [3]. Briefly, it consists of a digital micromirror device (DMD)-based spectrally tunable light source illuminating the surgical site via collimating optics at near normal incidence. The source scrolls through every nanometer from 520 to 645 nm, synchronous with the acquisition of the image by a charge-coupled device (CCD) camera placed directly normal to the sample. The acquired reflectance spectrum is converted to apparent absorbance and regressed using linear least squares method against a multivariate calibration model of oxy- and deoxygenated hemoglobin reference absorbance spectra [4, 5]. The output is a tissue oxygenation map of the scene where for each pixel a value of oxygenation expressed as % HbO_2 is presented.

$$\%\,HbO_2 = 100 \times \frac{[HbO_2]}{([HbO_2] + [Hb])}. \tag{18.1}$$

This is similar to the expression for oxygen saturation, % SO_2 with the exception that the [] here which denotes concentration, is uncorrected for scattering. The data acquisition, analysis, and display of results take a few seconds, adequate for a temporally sensitive application like surgery. This can be shortened to 0.3 s by sending two shaped spectral illumination corresponding to oxy- and deoxyhemoglobin instead of the 126 sequential bands [3]. Ambient room lights must be turned off during the acquisition or the camera optics must be baffled to minimize stray light that could affect the signal.

3 Clinical Research Application

The clinical hyperspectral imager was used to monitor oxygenation on the surface tissues of porcine kidneys during partial nephrectomy, where only the diseased part of the kidney is removed. The object of the surgical study is to find improvements to the procedure that can lead to better outcomes. Since the kidney is highly perfused, the renal artery is temporarily clamped while the surgeon removes part of the kidney and closes the blood vessels at the site of the resection. During this temporary occlusion, the kidney is starved of oxygen, which can lead to damage to the nephrons and possible subsequent renal dysfunction.

In a study of the relative merits of clamping the artery-only or both the artery and the vein during partial nephrectomy, it was found through the use of the HSI oximeter that the artery-only clamping left the kidney better oxygenated. This lent an explanation to a decades-long empirical observation that artery-only clamping left the kidney with a higher tolerance to ischemia [6]. Figure 18.1 shows the kidney oxygenation map expressed as % HbO_2 before, during, and after a clamp is applied on the renal artery during partial nephrectomy. A follow-up study on pigs was conducted where the blood flow was reduced to 0%, 10%, and 25% of the baseline blood flow. The group with 25% blood flow showed better post-operative kidney function relative to the 0% blood flow group, as shown by creatinine levels from a blood test taken a week later [7, 8].

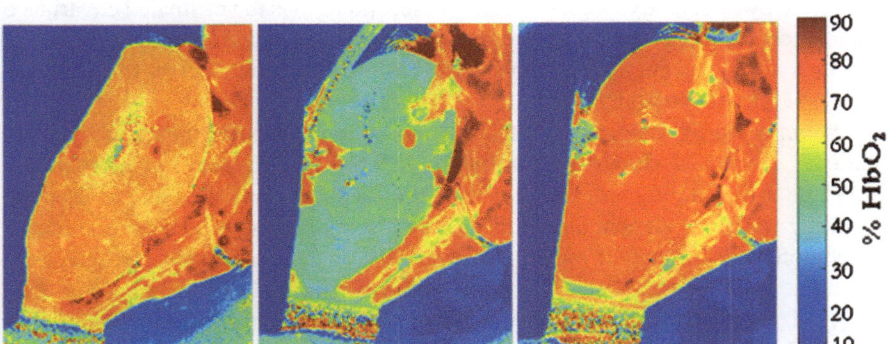

Fig. 18.1 % HbO_2 in the kidney (*left*) before the renal artery is clamped, (*middle*) during the period of clamping when the oxygenation dropped to ~50% and stayed at that level for the 60-min ischemic interval, and (*right*) recovered to baseline level during reperfusion

4 Determining Uncertainty of Measurement

The light has very limited penetration into the kidney tissue in the 520–645 nm range, getting absorbed by the hemoglobins in the surface tissues and partly reflected back. The reflected light is directly measured by the imager, with the assumption that what is not absorbed is reflected. The raw data are not useful without the accompanying measurement of light reflected off of a diffuse sheet of known reflectance factor—a reflectance blank. The reflected light from the sample is ratioed to that from the blank for each spatial point on the imager. This procedure serves to correct for the light source spectral distribution and the detector's responsivity curve. It also compensates the illumination field nonuniformities to some degree. It should be noted that the illuminator–tissue distance and angle is highly variable in the clinical setting and the blank reflectance needs to be taken at the same distance and angle as the tissues.

Analysis of the measurement uncertainty starts with specifying the measurand. In this case, it is the reflected light signal at each wavelength (S_λ) ratioed to the reflected light signal from the reflectance blank ($S_{b,\lambda}$) for a given spatial point.

$$R_i = \frac{S_\lambda}{S_{b,\lambda}} \cdot R_{b,\lambda}. \tag{18.2}$$

The value $R_{b,\lambda}$ is the absolute reflectance of the blank, which is often a standard calibrated diffuse reflector [9].

Total uncertainty of any measurement is given by [10]

$$u_c^2(y) = \sum_{i=1}^{n} \left(\frac{\partial f}{\partial x_i} \right)^2 u^2(x_i) + 2 \sum_{i=1}^{n-1} \sum_{j=i+1}^{n} \left(\frac{\partial f}{\partial x_j} \right) \cdot r(x_i x_j) \cdot u(x_i) u(x_j). \tag{18.3}$$

It is the root-sum-of-squares of each independent uncertainty contribution $u^2(x_i)$ multiplied by its sensitivity coefficient. The second term is the estimated covariance between contributors. In HSI, where data are correlated in both spectral and spatial domains, this is not negligible. Table 18.1 is a list of the contributors to the uncertainty in HSI oximetry.

The reflected light signal uncertainty contributions are instrumental and scene-related. Figure 18.2 shows the bandwidth differences between two instruments, A and B. If the calibration model was acquired with one and the other is used for acquiring unknown sample data, this will be a source of error. Ideally, one uses a calibration model taken with the same instrument with the same optical configuration, but that is not always practical. However, if the measuring instrument used is characterized, then instrumental shortcomings can be corrected for. The ratioing to a standard blank eliminates some of the instrumental factors. Scene-related factors are due to illumination and detection geometry and sample topography. For this particular application, because the measurement is a ratio of oxyhemoglobin to the total, it is not sensitive to irradiance changes due to topography. Otherwise, reflectance from spatial points off-normal to the detector has to be cosine-corrected.

Table 18.1 Sources of uncertainty to oxygen saturation measurement by hyperspectral imaging

Sources of uncertainty	Description
Reflectance signal	
Instrument-related	Stray light, wavelength accuracy, bandwidth, sensor nonlinearities and nonuniformities, drift, source and detector noise
Scene-related	Inhomogeneity of light field, sample topology leading to shading, glare, geometric factors
Calibration and validation model	
Sample-specific	Other absorbing constituents, light scattering, motion of sample
Statistical analysis	Robustness and selectivity of statistical model

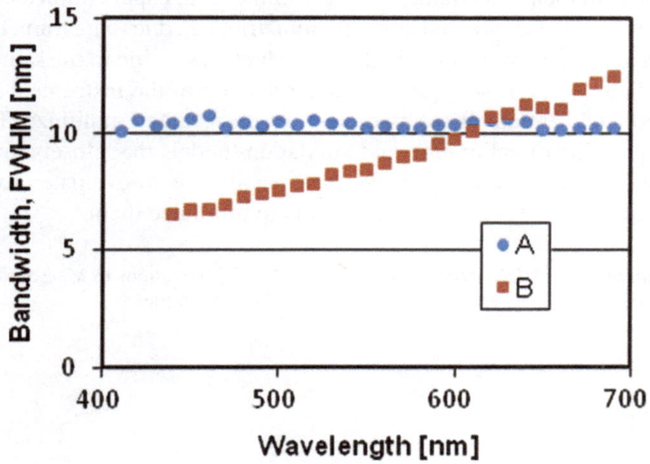

Fig. 18.2 Two instruments with different bandwidths. Stray light differences, if uncorrected, is a source of uncertainty if one is used for calibration and the other for routine acquisition

Quantitative prediction of a component in complex samples requires a calibration model [11]. Reference spectra, often of the pure isolated compound at known concentrations, are used. Validation mixtures are made in the laboratory to closely represent tissue samples. These are then tested using the calibration model to check the prediction accuracy of the discriminant algorithm used [12]. For hemoglobin, this often involves heparinized whole blood immersed in liquid with controllable scattering properties. This allows computation of the scattering contributions [13] since tissue light scattering confounds true absorbance. Validation models based on these tissue phantoms often do not capture the complexity of real tissues leading to inaccuracy of quantitative prediction. Hyperspectral image data collection from the clinical study allows us to gather a large number of spectra that could be used to estimate other interfering components' spectra. Unsupervised statistical methods for component classification are used for the task of finding the most relevant components [14]. We can computationally build more robust validation models based on the reduced clinical hyperspectral data to complement data from the tissue phantoms.

5 Summary and Conclusion

HSI is used to visualize oxygenation of tissues during surgery. The methodology is a spectrophotometric detection of oxy- and deoxyhemoglobin over a large spatial area encompassing a whole kidney during partial nephrectomy. Various contributors to the measurement uncertainty for this technique and application are identified. These can be broadly categorized as (1) radiometric, i.e., due to instrumentation and its operation and (2) analytical methodology, which is specific to the sample system being interrogated. The first requires characterization of the instrument in the way that it is used, and is relatively easy to quantify, correct, and minimize. The second requires preparation of calibration and validation models that closely approximate the real tissues. The objective of uncertainty analysis is to evaluate these relative contributions and apply commensurate efforts to minimize them.

Acknowledgments The NIST work is funded by the NIST Innovations in Measurement Science Award. The clinical research is funded by the UTSW Dept. of Urology.

References

1. Bearman G, Levenson R (2003) Biological imaging spectroscopy. In: Vo-Dinh T (ed) Biomedical photonics handbook. CRC, Boca Raton, FL
2. Flewelling R (2000) Non-invasive optical monitoring. In: Bronzino JD (ed) The biomedical engineering handbook, 2nd edn. CRC, Boca Raton, FL
3. Zuzak KJ, Francis RP, Wehner E et al (2011) Active DLP® hyperspectral illumination: a non-invasive, in vivo system characterization visualizing tissue oxygenation at near video rates. Anal Chem 83(19):7424–7430
4. Zuzak KJ, Schaeberle MD, Gladwin MT et al (2001) Noninvasive determination of spatially resolved and time-resolved tissue perfusion in humans during nitric oxide inhibition and inhalation by use of a visible-reflectance hyperspectral imaging technique. Circulation 104: 2905–2910
5. Zuzak KJ, Schaeberle MD, Lewis EN et al (2002) Visible reflectance hyperspectral imaging: characterization of a noninvasive, in vivo system for determining tissue perfusion. Anal Chem 74:2021–2028
6. Tracy CR, Terrell JD, Francis RP et al (2010) Characterization of renal ischemia using DLP hyperspectral imaging: a pilot study comparing artery-only occlusion versus artery and vein occlusion. J Endourol 24:321–325
7. Best SL, Thapa A, Holzer MS et al (2011) Assessment of renal oxygenation during partial nephrectomy using DLP® hyperspectral imaging. Proc SPIE 7932:793202-1-8

8. Best SL, Thapa A, Holzer MJ et al (2011) Minimal arterial in-flow protects renal oxygenation and function during porcine partial nephrectomy: confirmation by hyperspectral imaging. Urology 185(4):e610–e611
9. Barnes PY, Early EA, Parr AC (1998) Spectral reflectance. Natl Inst Stand Technol. Special Publication 250-48
10. Joint Committee for Guides in Metrology (2008) Guide to the expression of uncertainty in measurement. JCGM100:2008. http://www.bipm.org/utils/common/documents/jcgm/JCGM_100_2008_E.pdf. Accessed 15 Jun 2011
11. Martens HA, Naes T (1989) Multivariate calibration. Wiley, New York, NY
12. ASTM E1655-05 (2005) Standard practices for infrared multivariate quantitative analysis
13. Pogue BW, Patterson MS (2006) Review of tissue simulating phantoms for optical spectroscopy, imaging and dosimetry. J Biomed Opt 11(4):041102-1-16
14. Allen DA, Rice JP, Wehner E et al (2011) An examination of spectral diversity of medical scenes for hyperspectral projection. Proc SPIE 7932:793205-1-6

Chapter 19
Considering the Vascular Hypothesis of Alzheimer's Disease: Effect of Copper Associated Amyloid on Red Blood Cells

Heather R. Lucas and Joseph M. Rifkind

Abstract The vascular hypothesis of Alzheimer's disease (AD) considers cerebral hypoperfusion as a primary trigger for neuronal dysfunction. We have previously reported that red blood cells (RBCs) bind amyloid, which are the characteristic deposits found in AD brains, and interact with amyloid on the vasculature [1–3]. Oxidative stress triggered by these RBC/amyloid interactions could impair oxygen delivery. Recent literature has implicated copper bound amyloid-β peptide (CuAβ) and the associated production of reactive oxygen species (ROS) as one of the primary factors contributing to AD pathology. In this work, we have investigated CuAβ generated RBC oxidative stress. $A\beta_{1-40}$ peptide with a stoichiometric amount of copper bound was produced and compared to the metal-free form of the peptide. Different aggregation states of the peptides were isolated and incubated with RBCs for 15 h. Interestingly, CuAβ stimulated a pronounced increase in red cell oxidative stress as indicated by increased hemoglobin (Hb) oxidation, increased formation of fluorescent heme degradation products, and a decrease in RBC deformability. These findings demonstrate a potential role for CuAβ in promoting vascular oxidative stress leading to impaired cerebral oxygen delivery, which may contribute to neurodegeneration associated with AD.

Keywords Copper–amyloid • Deformability • Heme degradation • Hemoglobin • Oxidative stress

H.R. Lucas • J.M. Rifkind (✉)
Molecular Dynamics Section, National Institute on Aging, National Institutes of Health, 251 Bayview Blvd. 05B129, Baltimore, MD 21224, USA
e-mail: rifkindj@mail.nih.gov

W.J. Welch et al. (eds.), *Oxygen Transport to Tissue XXXIV*, Advances in Experimental Medicine and Biology 765, DOI 10.1007/978-1-4614-4989-8_19,

1 Introduction

Alzheimer's disease (AD) is a progressive degenerative disorder of the central nervous system that ultimately results in the loss of cognitive function. At present, 5.1 million Americans have been diagnosed with AD. One of the hallmarks of AD is the existence of amyloid plaques, which are composed primarily of Aβ peptide in a fibrillar conformation, within the postmortem brain of disease victims. Even though these plaques reflect neuronal loss, it has not been established that these plaques cause AD pathology. In fact, the decline of cognitive functions actually precedes plaque formation. Therefore, the relationship between amyloid plaques and AD pathology is not established and needs to be further studied.

Several competing hypotheses have been proposed to explain AD pathology. For example, toxicity has been ascribed to the insoluble amyloid plaques [4]. Conversely, the lethal form of Aβ has been suggested as an intermediate species referred to as protofibrils (or oligomers) that form along the aggregation pathway [5, 6]. Additionally, copper dyshomeostasis leading to ROS generation via Fenton-like chemistry has been suggested as the primary cause of AD [7, 8]. Finally, there is a suggestion that disease onset occurs within the vasculature, eventually leading to a breach of the blood–brain barrier [9, 10]. A combination of these hypotheses inspired the study described herein.

Vascular damage due to Aβ or CuAβ toxicity would lead to inefficient oxygen delivery resulting in brain hypoperfusion, followed by mild cognitive impairment and eventually neurodegeneration. In support of this, AD subjects commonly suffer from vascular inflammation and 95% have cerebral amyloid angiopathy (CAA), which is described as a vascular lesion made up of Aβ deposits on the blood vessels [11]. These deposits promote degeneration of the vessel wall, eventually causing cerebral microbleeds, which could be a source of metal dyshomeostasis [9]. Interestingly, it has recently been shown that vascular lesions from subjects with both AD and CAA had elevated levels of copper and Aβ [11].

Since blood flowing through the capillaries of AD subjects constantly encounters Aβ deposits found on the vasculature, we are interested in determining whether or not the red cell contributes to AD pathology. In this study, RBC oxidative stress following incubation with Aβ and CuAβ was measured by (a) determining Hb oxidation or metHb within the hemolysate, (b) measuring the level of fluorescent heme degradation products, and (c) evaluating the ability of the cell to deform. Hb oxidation and the presence of elevated levels of heme degradation products reflect a pool of un-neutralized ROS within the red cell. Impaired deformability suggests oxidative damage to the cytoskeleton.

2 Methods

2.1 Sample Preparation

A 250 μM solution of $A\beta_{1-40}$ peptide (BioSource) was prepared by dissolution of the lyophilized powder in Dulbecco's phosphate-buffered saline (DPBS) without Ca^{2+} and Mg^{2+}. $A\beta_{1-40}$ was used for this study because it is the dominant peptide fragment found in cerebrovascular plaques. The metallated form of $A\beta_{1-40}$ was prepared by adding a stoichiometric amount of $Cu^{II}SO_4$ (Sigma Aldrich). All samples were then incubated at 37°C while rotating for 0, 6, 12, 24, or 72 h. Harvesting different peptide conformers at various time points allows different intermediate aggregation states of Aβ and CuAβ to be examined. Peptide conformers were stored at −80°C.

RBCs were washed twice with 20-fold DPBS and centrifuged (~1,125 × *g*, 4°C, 10 min) for removal of plasma and buffy coat. This was repeated two times before final dilution to 50% HCT in DPBS. Each amyloid conformer (20 μM) was added to a freshly washed sample of RBCs. The samples were then gently vortexed to ensure proper mixing and incubated at 37°C for 15 h before final sample analysis. For all experiments, the control sample consists of washed RBCs without added amyloid, incubated at 37°C for 15 h. In addition, RBCs incubated with 20 μM $Cu^{II}SO_4$ were examined to confirm that the observed oxidative stress was not from free metal. Since the results following addition of copper were very similar to the control, this data was averaged with the RBC control data. Each measurement was repeated two to three times and the mean value was used for sample analysis. The results are an average of six experiments, $n = 6$.

2.2 Sample Analysis

The percentage of oxyHb (λ_{max}, 577 nm; ε, 15.4 M^{-1} cm^{-1}) and metHb (λ_{max}, 630 nm; ε, 3.7 M^{-1} cm^{-1}) within the hemolysate was determined using a Perkin Elmer Lambda 35 UV–visible spectrometer. The amount of heme degradation was determined by measuring the fluorescence intensity of a 50 μM sample of total Hb (oxyHb + metHb) from hemolyzed RBCs using a Perkin Elmer LS 50B spectrofluorometer (λ_{ex}, 321 nm; slit widths, 10 nm). The fluorescence intensity at 465 nm was used to determine the extent of heme degradation.

RBC deformability was measured using a microfluidic RheoScan-D slit-flow ektacytometer (Rheo Meditech) [12]. RBCs (6 μl) were suspended in 4% polyvinylpyrrolidone 360 solution (600 μL), supplied by Rheo Meditech within a microfluidic chip. The sample flows through a micro-channel under a range of shear stresses and the diffraction pattern from the laser that is directed through the deformed cells is analyzed by a microcomputer. Based upon the geometry of the elliptical diffraction pattern, an elongation index (EI) is calculated at each shear stress, $EI = (L - W)/(L + W)$, where L and W are the length and width of the diffraction pattern.

2.3 Statistical Analysis

Origin 8.1 (Microcal Software, Northhampton, MA) was used for sample analysis of the data and plotted as mean values ± standard deviation. Analysis of variance (ANOVA) was used to test the significance of the results regarding the effect of Aβ and CuAβ conformers on RBCs.

3 Results and Discussions

3.1 Evidence of CuAβ Initiated Hb Oxidation and Membrane Associated Heme Degradation Products

Hemolysate from various samples were analyzed for metHb in order to determine the degree of oxidative stress within the red cell. As shown in Fig. 19.1, Aβ conformers that lack coordinated copper result in only a minor increase in the amount of metHb in comparison to CuAβ. This increase in Hb oxidation for all aggregates is significant, $p<0.001$; however, the dramatic increase for CuAβ is significantly higher than for Aβ, $p<0.0001$. Interestingly, the 6 h CuAβ conformer, which has not yet formed large fibrils, exhibits the most oxidized Hb within the hemolysate, with the percentage of metHb increasing from 7 to 48%. In fact, this amount of metHb for the 6 h CuAβ conformer is significantly higher than the other CuAβ aggregates, $p<0.0001$. This may suggest that an intermediate state of CuAβ is most toxic to the red cell.

Previously, it was shown by our group that fluorescent heme degradation products are formed during Hb oxidation and slowly accumulate on the cell membrane [13, 14].

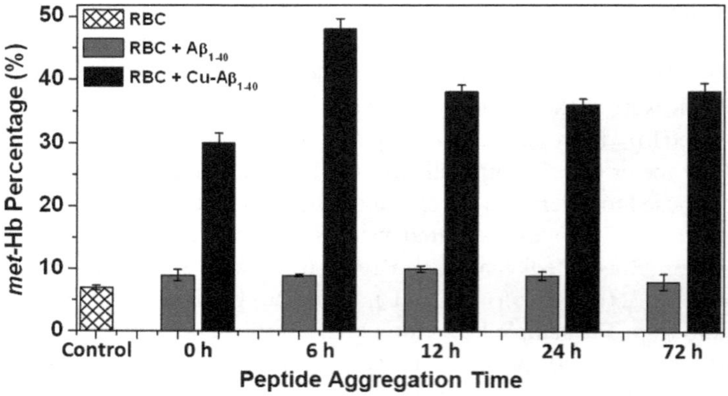

Fig. 19.1 Percentage of metHb within hemolysate from red cells incubated with Aβ and CuAβ conformers as an indicator of RBC oxidative stress

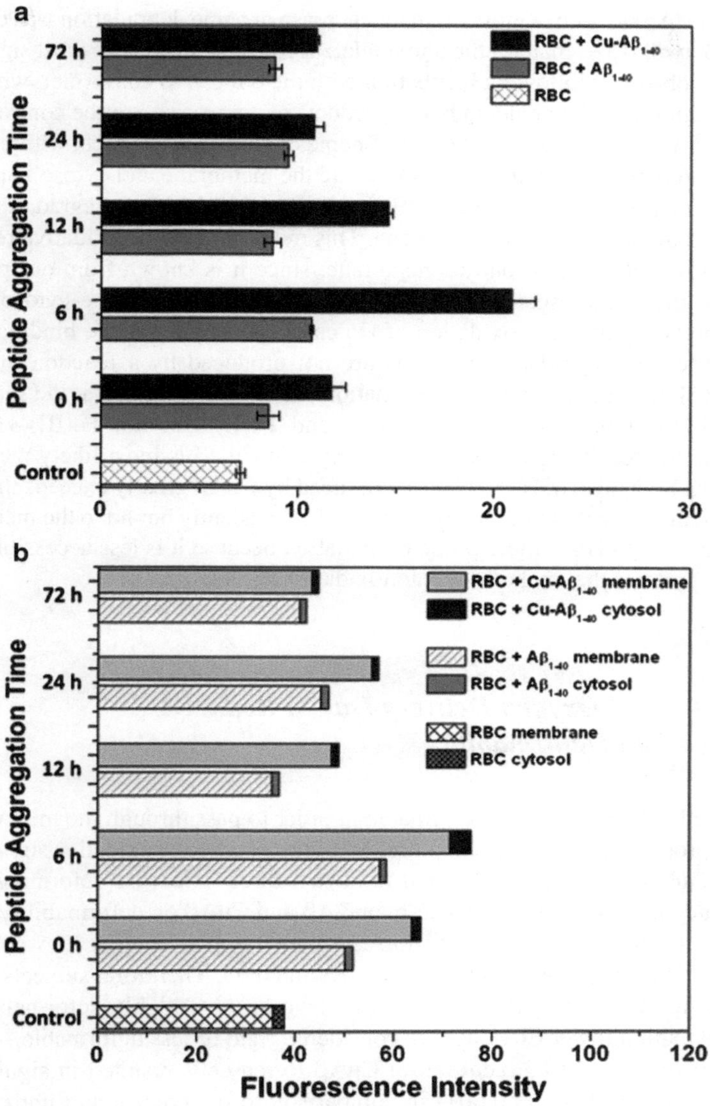

Fig. 19.2 Fluorescence data corresponding to the amount of heme degradation products within hemolysate from RBC samples incubated with Aβ and CuAβ conformers. Data plotted are from before (**a**) and after (**b**) fractionation of the membrane and cytosolic portions

Higher levels of these fluorescent species provide an integrated measure of the oxidative stress experienced by the red cell. In the experiments described herein, an increase in heme degradation products was detected with addition of Aβ that is statistically significant, $p<0.001$ (Fig. 19.2a). Consistent with the aforementioned results for Hb oxidation, a more dramatic increase in heme degradation was detected for CuAβ, $p < 0.0001$, than for the unmetallated Aβ conformers. The most significant effect was observed following incubation with the 6 h CuAβ conformer, which had triple the amount of heme degradation products in comparison to the control.

In order to analyze the distribution of heme degradation products within the red cell, the samples were centrifuged to separate the membrane and cytosolic portions of the hemolysate. As shown in Fig. 19.2b, almost all of the heme degradation products were associated with the membrane. This indicates that the oxidative reactions promoted by CuAβ occur on the membrane, since it is known from our previous work that these fluorescent products are not transferred from the cytosol to the membrane [13]. Although oxidation of Hb can increase membrane binding, membrane associated degradation products are not produced by a reaction involving metHb [15]. Instead, these heme degradation products are formed at the same time as metHb by a reaction involving H_2O_2 and Fe(II)–Hb, not Fe(III)–Hb [16]. Therefore, the results described herein imply that CuAβ is most likely membrane bound. The high level of Hb oxidation produced by CuAβ greatly exceeds the membrane binding sites of Hb but likely involves Hb transiently bound to the membrane. Membrane-bound Hb is more prone to oxidation because it is less accessible to the cytosolic proteins that make up the antioxidant network.

3.2 Impaired Oxygen Delivery Due to Reduced Red Cell Deformability

RBCs need to be sufficiently deformable in order to pass through the microcirculation and properly deliver oxygen to tissues. Since a source of oxidative stress bound to the membrane would be expected to have a direct effect on deformability, we investigated the effect of membrane-bound Aβ and CuAβ on deformability. To test this, the EI value was determined for each cell sample. Typical EI values from human subjects commonly vary between 0.3 and 0.35. Therefore, subjects with EI values within this range are considered to have healthy, highly deformable cells. Cells that exhibit lower EI values are considered rigid or less deformable.

As shown in Fig. 19.3, addition of CuAβ to red cells resulted in significantly reduced deformability ($p < 0.001$) in comparison to the control and unmetallated Aβ conformers. Again, incubation of RBCs with the 6 h CuAβ conformer resulted in a more dramatic change with the means significantly different than for the other CuAβ conformers, $p < 0.0001$. Interestingly, incubation of red cells with the unmetallated 6 h Aβ conformer also resulted in a significantly reduced EI value ($p < 0.001$). Yet, all other RBC samples incubated with unmetallated Aβ aggregates remained highly deformable and the overall means were not statistically significant. This result could have been caused by a trace amount of copper within the DPBS buffer that preferentially coordinates the 6 h Aβ conformer. More work is necessary to examine this possibility.

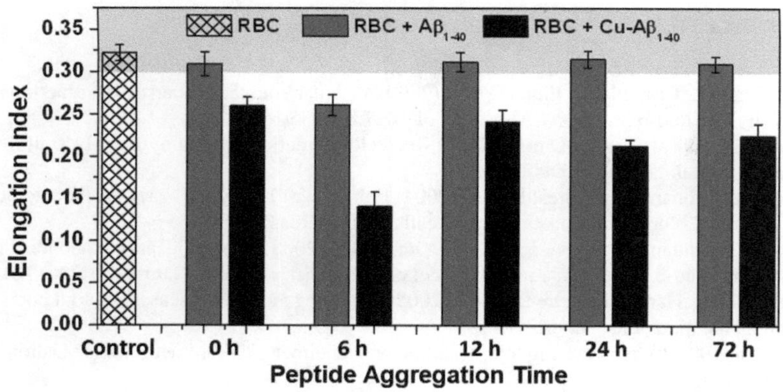

Fig. 19.3 Deformability data for RBC samples incubated with Aβ and CuAβ conformers as an indicator of damage to the RBC membrane; typical range for healthy RBCs is *shaded*, EI ≥ 0.3

4 Conclusions

Oxidative stress within the red cell that results from high ROS content will oxidize Hb to metHb, reducing the oxygen binding capacity of the RBC. In addition, a reduction in the deformation ability of the RBC will result in impaired delivery of oxygen. Following incubation of RBCs with different aggregate forms of Aβ peptide and CuAβ, we find that the copper-bound form overall results in increased RBC oxidative stress. This effect is most pronounced with the 6 h conformer, supportive of increased toxicity of the intermediate protofibril. This enhanced red cell oxidative stress results in Hb oxidation and subsequent membrane bound heme degradation products as well as impaired RBC deformability. The suggested mechanism involves CuAβ associating with the membrane producing ROS. This could result from an electron transfer reaction that occurs when CuAβ is near deoxyHb yielding metHb and a reduced cuprous species that can readily react with O_2 to form ROS. For these ROS to produce heme degradation and reduced deformability they need to form in close proximity to the membrane in order to avoid scavenging by cytosolic proteins. Overall, this attack on the membrane would lead to altered red cell function and impaired oxygen transport. Also, this same pool of ROS could leak out of the red cell and damage nearby cells or tissues.

Acknowledgment This research was supported by the Intramural Research Program of the NIH, National Institute on Aging.

References

1. Nagababu E, Usatyuk PV, Enika D et al (2009) Vascular endothelial barrier dysfunction mediated by amyloid-b proteins. J Alzheimers Dis 17:845–854
2. Ravi LB, Poosala S, Ahn D et al (2005) Red cell interactions with amyloid-b1-40 fibrils in a murine model. Neurobiol Dis 19:28–37
3. Ravi LB, Mohanty JG, Chrest FJ et al (2004) Influence of b-amyloid fibrils on the interactions between red blood cells and endothelial cells. Neurol Res 26:579–585
4. Meyer-Luehmann M, Spires-Jones TL, Prada C et al (2008) Rapid appearance and local toxicity of amyloid-b plaques in a mouse model of Alzheimer's disease. Nature 451:720–724
5. Lashuel HA, Hartley D, Petre BM et al (2002) Neurodegenerative disease: amyloid pores from pathogenic mutations. Nature 418:291
6. Rauk A (2008) Why is the amyloid b peptide of Alzheimer's disease neurotoxic? Dalton Trans 14(10):1273–1282
7. Zatta P, Drago D, Bolognin S et al (2009) Alzheimer's disease, metal ions and metal homeostatic therapy. Trends Pharmacol Sci 30:346–355
8. Gaggelli E, Kozlowski H, Valensin D et al (2006) Copper homeostasis and neurodegenerative disorders (Alzheimer's, prion, and Parkinson's diseases and amyotrophic lateral sclerosis). Chem Rev 106:1995–2044
9. Stone J (2008) What initiates the formation of senile plaques? The origin of Alzheimer-like dementias in capillary haemorrhages. Med Hypotheses 71:347–359
10. de la Torre JC (2002) Alzheimer disease as a vascular disorder: nosological evidence. Stroke 33:1152–1162
11. Schrag M, Crofton A, Zabel M et al (2011) Effect of cerebral amyloid angiopathy on brain iron, copper, and zinc in Alzheimer's disease. J Alzheimers Dis 24:137–149
12. Shin S, Hou JX, Suh JS et al (2007) Validation and application of a microfluidic ektacytometer (RheoScan-D) in measuring erythrocyte deformability. Clin Hemorheol Micro 37:319–328
13. Nagababu E, Mohanty JG, Bhamidipaty S et al (2010) Role of the membrane in the formation of heme degradation products in red blood cells. Life Sci 86:133–138
14. Rifkind JM, Ramasamy S, Manoharan PT et al (2004) Redox reactions of hemoglobin. Antioxid Redox Signal 6:657–666
15. Nagababu E, Rifkind JM (1998) Formation of fluorescent heme degradation products during the oxidation of hemoglobin by hydrogen peroxide. Biochem Biophys Res Commun 247:592–596
16. Nagababu E, Rifkind JM (2000) Reaction of hydrogen peroxide with ferrylhemoglobin: superoxide production and heme degradation. Biochemistry 39:12503–12511

Chapter 20
The Role of Mitochondrial Proteomic Analysis in Radiological Accidents and Terrorism

David Maguire, Bingrong Zhang, Amy Zhang, Lurong Zhang, and Paul Okunieff

Abstract In the wake of the 9/11 terrorist attacks and the recent Level 7 nuclear event at the Fukushima Daiichi plant, there has been heightened awareness of the possibility of radiological terrorism and accidents and the need for techniques to estimate radiation levels after such events. A number of approaches to monitoring radiation using biological markers have been published, including physical techniques, cytogenetic approaches, and direct, DNA-analysis approaches. Each approach has the potential to provide information that may be applied to the triage of an exposed population, but problems with development and application of devices or lengthy analyses limit their potential for widespread application. We present a post-irradiation observation with the potential for development into a rapid point-of-care device. Using simple mitochondrial proteomic analysis, we investigated irradiated and nonirradiated murine mitochondria and identified a protein mobility shift occurring at 2–3 Gy. We discuss the implications of this finding both in terms of possible mechanisms and potential applications in bio-radiation monitoring.

Keywords Mitochondrial proteomic analysis • Radiation

1 Introduction

Humans have evolved with an array of mechanisms, mostly involving DNA repair, to ameliorate the effects of low-dose radiation. Since the detonation of the first atomic bomb in the New Mexico desert on July 16, 1945, and two nuclear devices subsequently used against Japan, some 2,000 additional nuclear weapons have been detonated in nuclear tests by eight nations, contributing to background radiation

D. Maguire (✉) • B. Zhang • A. Zhang • L. Zhang • P. Okunieff
Department of Radiation Oncology, University of Florida,
P.O. Box 103633, Gainesville, FL 32610, USA
e-mail: slh@ufl.edu

W.J. Welch et al. (eds.), *Oxygen Transport to Tissue XXXIV*, Advances in Experimental Medicine and Biology 765, DOI 10.1007/978-1-4614-4989-8_20,

levels for some populations. Of the tens of thousands of medical and industrial devices using radioactive isotopes, there have been two recorded significant releases from medical imaging/treatment devices involving 137-cesium and cobalt-60 in Brazil and Mexico, respectively. Other releases of radioactivity are generally associated with military activity, including the loss of eight submarines from approximately 200 naval nuclear power plants.

Additional radioactive exposure of human populations has resulted from the two recorded Level 7 nuclear power plant incidents at Chernobyl in the former USSR and at Fukushima Daiichi in Japan. Although 31 deaths were directly attributed to the Chernobyl meltdown, it is believed to have subsequently led to the deaths of many more people directly exposed to high radiation levels. An unknown number of citizens of the USSR and surrounding countries suffered injuries when exposed to lower levels of irradiation. Radiation worker exposure, which is permitted up to tenfold the background level, is routinely monitored by various programs involving film badges or thermoluminescent disks, but no such physical devices are routinely applied to monitoring the general public's radiation exposure. In fact, no population-wide programs to routinely monitor public levels of irradiation currently exist. Since the recent Level 7 release of radioactivity in Japan and in light of the increased potential for radiological terrorism since the events of 9/11, a pressing need exists for effective technologies that can be used to monitor the general public's radiation exposure.

Approaches to monitoring radiation in vivo were reviewed in 2008 [1], and since then a number of other biodosimeters have been championed, including direct DNA analyses [2–5], teeth and nail dosimetry [6–10], and whole cell approaches. In addition to these single-parameter analyses, a range of other multiparametric approaches have been advocated, such as proteomic approaches focusing on urine [11], macrophages [12], lymphocytes [13], serum [14, 15], and cytokine cascades [16].

Mitochondrial proteomics have burgeoned since publication of a method to electrophoretically separate membrane complexes of mitochondria while retaining enzymatic activities [17]. Gas chromatography–mass spectrometry (GC–MS) and second dimension separation techniques, involving protein subunit denaturation, provide additional information about subunit molecular weight.

In this study, we investigated the mitochondrial proteome of two murine strains following irradiation at various levels. Using simple mitochondrial proteomic analysis, we examined mitochondria before and after irradiation and identified a protein mobility shift occurring over 2–3 Gy in one murine strain. We discuss the implications of this finding in terms of possible mechanisms and their potential application in bioradiation monitoring and eventual development into a rapid point-of-care device.

2 Methods

Male and female, 7 to 8-week-old, BALB/c (radiosensitive) and C57BL/6 (radioresistant) mice (National Cancer Institute's Mouse Repository, Frederick, MD) were housed in a microisolator or a colony management room at the University of Rochester (UR, Rochester, NY) on a 12-h light/dark cycle and fed a standard diet.

Murine colony management followed the UR Institute Animal Protocol, and care and use procedures were approved by the UR Institute Animal Care and Use Committee. Except for breeding mice, mice were immobilized in plastic boxes and sham-irradiated (as a control) or exposed to 3 or 5 Gy total body irradiation (TBI) 5 days after arrival. Radiation was delivered with a 137-cesium gamma source at a dose rate of 1.75 Gy/min. After irradiation, mice were observed for survival and monitored daily for 1 month. At different time points, mice were euthanized and hepatic and renal tissues were collected. Collected tissues were either used immediately or frozen at −70°C until use.

Tissues were placed into ice-cold buffer *A* (250 mM sucrose, 1 mM ethylene glycol tetraacetic acid (EGTA), and 10 mM 4-(2-hydroxyethyl)-1-piperazineethanesulfonic acid (HEPES), pH 7.4). Tissues were chopped finely, and buffer *A* was changed 3–5 times during the procedure to remove blood cells and glycogen. The chopped tissues were homogenized with a glass-on-teflon potter homogenizer for up to 10 min. The homogenized tissues were centrifuged for 10 min at 500 × *g* at 4°C. Supernatants were transferred into new centrifugation tubes and centrifuged for 10 min at 12,000 × *g* at 4°C. Sediments were resuspended with buffer *B* (250 mM sucrose and 10 mM HEPES, pH 7.4) and centrifuged again for 10 min at 12,000 × *g* at 4°C. The pellets were resuspended in 0.7 ml buffer *B* and transferred into 1.5 ml eppendorf tubes. Isolated mitochondria were kept on ice and used for experiments within 4 h of completion of the isolation procedure. For protein assays, isolated mitochondria were diluted in cell lysis buffer (100 mM NaCl, 10 mM Tris-CI (pH 8.0), 25 mM EDTA (pH 8.0), 0.5% sodium dodecyl sulfate). A BCA Protein Assay (Thermo Scientific, Rockford, IL) was used to calculate sample protein concentration by reference to a protein standard curve using bovine serum albumin as standard. Protein estimations were used to normalize sample loading of gels.

Blue Native gels were used to separate and detect mitochondrial membrane complexes. Purified mitochondria from the liver and kidney were resolved on Native polyacrylamide gel electrophoresis (PAGE) Bis-Tris gels according to the manufacturer's instructions (MitoSciences, Eugene, Oregon). Briefly, an amount of purified mitochondria (equal to 0.4 mg total protein) was used. Membrane proteins were extracted by the addition of 40 μL of 750 mM aminocaproic acid, 40 μL of I mM Bis-Tris/HCI (pH 7.0), and 5 μL 10% lauryl maltoside. Samples were incubated on ice for 30 min and centrifuged at 100,000 × *g* for 30 min at 4°C. Supernatants were transferred to new tubes, followed by addition of 2.5 μL of 5% solution/suspension of Coomassie Blue G in 500 mM aminocaproic acid. Samples (approximately 75 μg) were resolved by Native PAGE Bis-Tris gel electrophoresis. After electrophoresis, gels were recorded using an image capture system.

3 Results

Figure 20.1 shows the results of Blue Native PAGE gel examinations of extracted mitochondria. Before irradiation, there was no difference between the first dimension (gradient) electrophoretic patterns of the mitochondrial membrane complexes

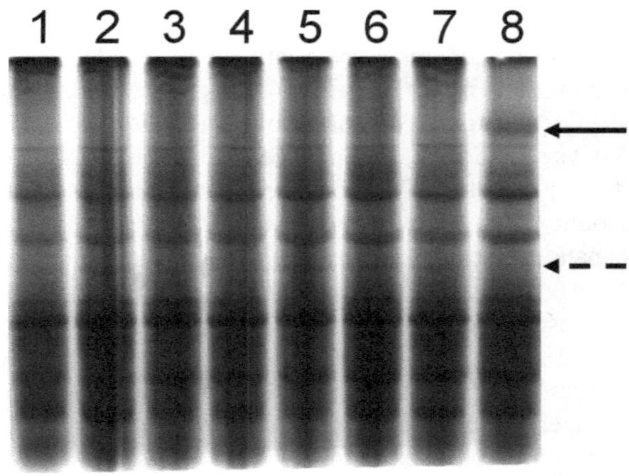

Fig. 20.1 First dimension BN PAGE gel electrophoresis of murine hepatic mitochondrial membrane proteins. *Lanes 1–4*: BALB/c, nonirradiated; *Lane 5*: C57BL/6, nonirradiated; *Lanes 6–7*: C57BL/6, 3 Gy; *Lane 8*: C57BL/6, 5 Gy. The *solid arrow* indicates position of band appearing in *Lane 8* after irradiation. The *dashed arrow* indicates position of band present before irradiation

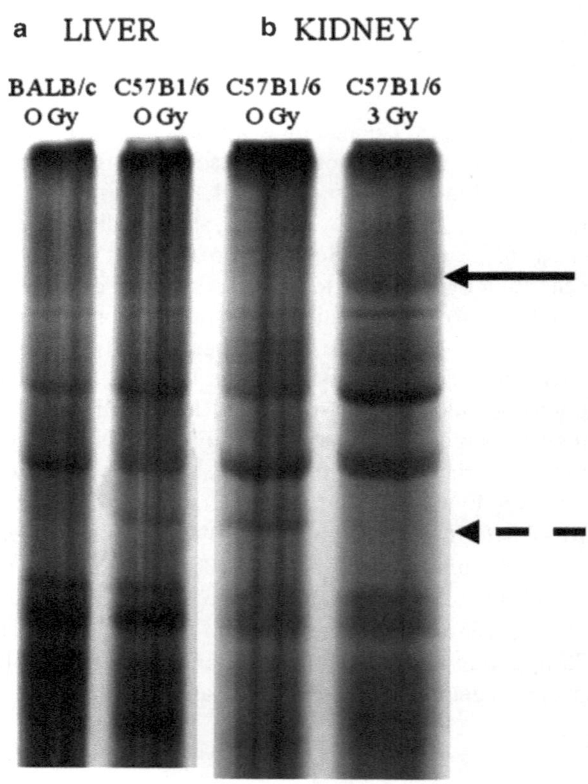

Fig. 20.2 One dimensional BN PAGE of (**a**) hepatic mitochondrial proteins from two murine strains and (**b**) C57BL/6 renal mitochondrial proteins before and after irradiation (3 Gy). *Arrows* indicate position of band lost (*dashed line*) or formed (*solid line*) after irradiation

of the two pure-bred strains (Fig. 20.2). As expected, there were minor differences between the patterns of hepatic and renal mitochondria. After TBI at a relatively low dose, there were dramatic changes in the patterns for the radioresistant C57BL/6 strain.

Table 20.1 The effect of total body irradiation on mitochondrial protein mobility

Ionizing radiation (Gy)	BALB/c Liver	BALB/c Kidney	C57BL/6 Liver	C57BL/6 Kidney
0	0 (3)	0 (3)	0 (3)	0 (3)
3	1 (4)	2 (5)	0 (4)	2 (5)
5	3 (4)	3 (4)	2 (4)	3 (4)

Note. Total number of mice in parentheses

No discernible change was exhibited up to 3 Gy; however, at 5 Gy, one low MW complex was lost and a new high MW band appeared. This change was reflected in the renal mitochondria of the same strain but at a slightly lower dose (Fig. 20.2). Table 20.1 shows the overall effect of radiation on mitochondrial protein mobility.

4 Discussion

A number of potential biodosimeters exist [2–15] for application in different settings following irradiation after accidental exposure or nuclear terrorism events. Biodosimeters have an advantage over physical devices, such as Geiger–Muller (GM) tubes, insofar as they reflect actual damage to exposed tissue. They may detect changes that have developed in the period following irradiation, thus giving an insight into an otherwise unrecorded radiation history. Widespread use of GM tubes to monitor exposure of the general public to 131-iodine is only valuable when the isotope has been absorbed from the air. There is currently no public monitoring of radiation exposure, although a limited program using physical dosimeters was introduced for children in Japan following the Fukushima Daiichi Level 7 event.

Despite its labor-intensive nature, dicentric analysis was considered the gold standard of biodosimetry [16]. Tooth dosimetry [6] has the potential to provide information about inhaled airborne radioactive material and external radiation. Circulating DNA is also a short-term biodosimeter (up to about 5 days after irradiation) for total body or partial body exposure [2]. The latter two methods are being developed as field-deployable techniques following irradiation of civilian populations or military units. Mitochondrial/nuclear DNA ratios [3, 4] are potentially good tissue biodosimeters; however, because mtDNA is rapidly cleared from blood, they do not offer any prospect as general biodosimeters. A range of multiparametric biodosimeters have also been suggested, including proteomic analyses of urine [11], serum [14, 15], or lymphocytes [13]. Such approaches require highly sophisticated instrumentation but might be of use in retrospective, population-wide investigations after specific irradiation events. Results from some of the proteomic surveys may be applied to develop simpler, antibody-based, field dosimeters. Chen et al. [12] detected 16 spot changes after C57BL/6 mice were irradiated and identified the following proteins: calreticulin, lactoylglutathione lyase, regulator of G-protein signaling 16, and peroxiredoxin-5, a mitochondrial precursor. Our approach has been to investigate the mitochondrial membrane proteome as a potential field for discovery of new biodosimetric targets.

The results presented here indicate the existence of subtle tissue-specific and strain-specific changes in the electrophoretic mobility of proteins extracted in mitochondrial membranes of irradiated mice. These changes occur in the renal tissue of radioresistant C57BL/6 mice after 3 Gy TBI but in the hepatic tissue at 5 Gy. In C57BL/6 mice, a single TBI of approximately 10 Gy is lethal. A dose of 7–8 Gy is lethal in the radiosensitive BALB/c strain. The changes observed in the radioresistant mice are consistent with a change in the association of complex subunits that may involve new associations with non-membrane protein subunits, possibly by protein–protein cross-linking following irradiation. This phenomenon might also occur via reorganization of membrane subunit–subunit interactions. Mitochondria are relatively easy to access as components of cells, such as lymphocytes or platelets, by finger-prick techniques. Development of an antibody-based discriminator for the two forms of mitochondrial membrane complexes described in this work could lead to a simple, individual, point-of-care device to test for critical levels of irradiation either from a single dose or perhaps from accumulated radiation over a period of time.

If confirmed in human tissue, these results may provide a discriminator to identify radioresistant individuals within the workforce and the general population.

Acknowledgments We thank Kate Casey-Sawicki for editing and preparing this manuscript for publication.

References

1. Okunieff P, Chen Y, Maguire DJ et al (2008) Molecular markers of radiation-related normal tissue toxicity. Cancer Metastasis Rev 27:363–374
2. Zhang H, Zhang SB, Sun W et al (2009) B1 sequence-based real-time quantitative PCR: a sensitive method for direct measurement of mouse plasma DNA levels after gamma irradiation. Int J Radiat Oncol Biol Phys 74:1592–1599
3. Zhang H, Maguire D, Swarts S et al (2009) Replication of murine mitochondrial DNA following irradiation. Adv Exp Med Biol 645:43–48
4. Zhang H, Maguire DJ, Zhang M et al (2011) Elevated mitochondrial DNA copy number and POL-gamma expression but decreased expression of TFAM in murine intestine following therapeutic dose irradiation. Adv Exp Med Biol 701:201–206
5. Zhang L, Zhang M, Yang S et al (2010) A new biodosimetric method: branched DNA-based quantitative detection of B1 DNA in mouse plasma. Br J Radiol 83:694–701
6. Brady JM, Aarestad NO, Swartz HM (1968) In vivo dosimetry by electron spin resonance spectroscopy. Health Phys 15:43–47
7. Suzuki H, Tamukai K, Yoshida N et al (2010) Development of a compact electron spin resonance system for measuring ESR signals of irradiated fingernails. Health Phys 98:318–321
8. Wilcox DE, He X, Gui J et al (2010) Dosimetry based on EPR spectral analysis of fingernail clippings. Health Phys 98:309–317
9. Black PJ, Swarts SG (2010) Ex vivo analysis of irradiated fingernails: chemical yields and properties of radiation-induced and mechanically-induced radicals. Health Phys 98:301–308

10. Romanyukha A, Reyes RA, Trompier F et al (2010) Fingernail dosimetry: current status and perspectives. Health Phys 98:296–300
11. Sharma M, Halligan BD, Wakim BT et al (2008) The urine proteome as a biomarker of radiation injury: submitted to proteomics – clinical applications special issue: "renal and urinary proteomics (Thongboonkerd)". Proteomics Clin Appl 2:1065–1086
12. Chen C, Boylan MT, Evans CA et al (2005) Application of two-dimensional difference gel electrophoresis to studying bone marrow macrophages and their in vivo responses to ionizing radiation. J Proteome Res 4:1371–1380
13. Szkanderova S, Vavrova J, Hernychova L et al (2005) Proteome alterations in gamma-irradiated human T-lymphocyte leukemia cells. Radiat Res 163:307–315
14. Guipaud O, Holler V, Buard V et al (2007) Time-course analysis of mouse serum proteome changes following exposure of the skin to ionizing radiation. Proteomics 7:3992–4002
15. Menard C, Johann D, Lowenthal M et al (2006) Discovering clinical biomarkers of ionizing radiation exposure with serum proteomic analysis. Cancer Res 66:1844–1850
16. Romm H, Wilkins RC, Coleman CN et al (2011) Biological dosimetry by the triage dicentric chromosome assay: potential implications for treatment of acute radiation syndrome in radiological mass casualties. Radiat Res 175:397–404
17. Schagger H, von Jagow G (1991) Blue native electrophoresis for isolation of membrane protein complexes in enzymatically active form. Anal Biochem 199:223–231

Chapter 21
Alteration of Plasma Galactose/ *N*-acetylgalactosamine Level After Irradiation

Jun Ma, Deping Han, Mei Zhang, Chun Chen, Bingrong Zhang, Zhenhuan Zhang, Xiaohui Wang, Shanmin Yang, Yansong Guo, Paul Okunieff, and Lurong Zhang

Abstract Although glycoproteins possess a variety of functional and structural roles in intracellular and intercellular activities, the effect of ionizing radiation (IR) on glycosylation is largely unknown. To explore this effect, we established a sandwich assay in which PHA-L, a phytohaemagglutinin that agglutinates leukocytes, was used as a coating layer to capture glycoproteins containing complex oligosaccharides; the bound glycoproteins were then measured. C57BL/6 mice were exposed to 0, 3, 6, or 10 Gy, and the plasma was collected at 6, 12, 18, 24, 48, 72, or 168 h and then analyzed for galactose/*N*-acetylgalactosamine (Gal/GalNAc) containing proteins. We found that (1) the sandwich assay accurately measured the level of glycoproteins, (2) 6–12 h after IR, the amount of glycoproteins containing GalNAc increased, and (3) at 72 and 168 h, 10 Gy was associated with a decrease in Gal/GalNAc. These IR-induced alterations might relate to the release of glycoproteins into the blood and the damage of the proteins and genes that are related to the glycosylation process.

Keywords Galactose/*N*-acetylgalactosamine • Glycosylation • Lectins • PHA-L • Radiation

J. Ma
Institute of Digestive Diseases, Zhengzhou University, Henan, China

D. Han • M. Zhang • C. Chen • B. Zhang • Z. Zhang • X. Wang
S. Yang • Y. Guo • P. Okunieff • L. Zhang (✉)
Department of Radiation Oncology, UF Shands Cancer Center, University of Florida, P.O. Box 100385, Gainesville, FL 32610, USA
e-mail: lurongzhang@ufl.edu

W.J. Welch et al. (eds.), *Oxygen Transport to Tissue XXXIV*, Advances in Experimental Medicine and Biology 765, DOI 10.1007/978-1-4614-4989-8_21,

1 Introduction

Ionizing radiation (IR) damages cells by directly breaking the bonds between molecules and by generating free radicals that alter the structure of macromolecules, such as DNA, proteins, and lipids [1–5]. However, little is known about the effect of IR on glycosylated molecules. Glycosylation, a major posttranslational modification of proteins, is an energy-consuming enzymatic process [6]. Once the protein is synthesized in the rough endoplasmic reticula (ER), it transports into smooth ER where the carbohydrates (glycans) are added in an enzyme-directed and site-specific fashion [7]. While *N*-linked glycans attach to a nitrogen of asparagine or arginine in the Golgi apparatus [8], the *O*-linked glycans attach to the hydroxy oxygen of serine, threonine, tyrosine, hydroxylysine, or hydroxyproline or to oxygens on lipids, such as ceramide [4]. The attached glycans can make up to 60% of the apparent weight of glycoproteins [9], which can exist in both membrane and secretory forms and possess a variety of functional and structural roles in intracellular and intercellular activities [10]. Glycans in proteins and lipids serve as "antennae" to sensor (on receptors of cell membranes) or deliver (on ligands or secretion molecules) the signal carried by glycoproteins [5, 10]. This study examines the effect of IR on the glycosylation process.

The lack of an effective tool is one of the major obstacles to the study of glycoproteins. The molecular biological approach can upregulate or downregulate one enzyme that affects one type of glycosylation in many proteins and lipids; thus, it cannot easily address a specific question. Also, glycans' lack of antigenicity renders the immunological approach ineffective. Lectins, produced naturally as "antibodies" to recognize glycans, are the best tools that can be used to study glycosylation. Thus, to explore the effect of IR on glycosylation, we established a sandwich assay in which PHA-L, a phytohaemagglutinin that agglutinates leukocytes and recognizes complex oligosaccharides [11], was used as a coating layer to capture glycoproteins; thereafter, the amount of bound glycoproteins was measured by biotinylated *Vicia villosa* lectin (VVL/VVA) and *Erythrina cristagalli* lectin (ECL) to recognize galactose/*N*-acetylgalactosamine/mannose on the glycoproteins [12]. These glycoprotein levels were studied as a function of dose and time.

2 Methods

PHA-L, which was used to recognize the complex structure of glycans, was dissolved in 0.005M NaHCO buffer (pH 9.6) at a concentration of 4 μg/ml; 100 μl of this coating solution was added to each well in a 96-well, high-bound, enzyme-linked immunosorbent assay (ELISA) plate. After overnight incubation at room temperature, the unbinding sites were blocked with 5% of Tween-20 at room temperature for 1 h and washed twice with phosphate buffered saline (PBS). Murine plasma (100 μl of 1:100 diluted in PBS) was added to the well in duplicate and incubated overnight at 4°C. After being washed three times with 0.01% Tween

20-PBS, 100 μl of 1 μg/ml of biotinylated VVL/VVA (for recognition of GalNAc) or ECL (for recognition of Gal) was added and incubated at room temperature for 1 h, followed by 100 μl of 2 μg/ml of horseradish peroxidase (HRP), then 100 μl of 3,3′,5,5′-tetramethylbenzidine (TMB) substrate of HRP and 2M H_2SO_4 stopper, and read at A_{450} with a SpectraMax M2 (Molecular Devices, LLC, Sunnyvale, CA).

Male, 8-week-old, C57BL/6 mice were divided into groups of five and subjected to 0, 3, 6, or 10 Gy total body irradiation (TBI). Mice were euthanized at 3, 6, 9, 12, 15, 18, 24, 48, 72, or 168 h after TBI, and plasma was collected and stored at −80°C until analysis.

3 Results

To determine if the lectin ELISA assay could measure the levels of glycoproteins, plasma was diluted 1:200, 1:400, 1:800, or 1:1,600, and then measured for the concentration of glycoproteins. Figure 21.1 shows that as the dilution increased, the bound-glycoprotein concentration decreased. The quadratic relation had a correlation coefficient value of $R = 0.979$.

To determine the radiation-induced alteration as function of time, we examined the alteration patterns of plasma glycoprotein levels at 3, 6, and 10 Gy. As Fig. 21.2a shows, upon exposure to 3 Gy, the levels of glycoproteins containing Gal and/or GalNAc and/or mannose, as captured by PHA-L and detected by VVA/VVL, increased at 6–12 h and then dropped to baseline at 24 h. At 6 Gy, the levels increased significantly at 3 h, dropped at 6–9 h, and then increased again at 12 h. The levels

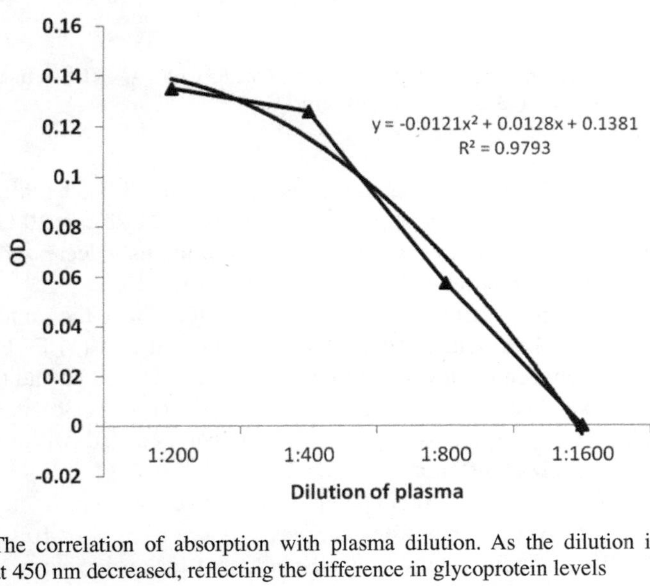

Fig. 21.1 The correlation of absorption with plasma dilution. As the dilution increased, the absorption at 450 nm decreased, reflecting the difference in glycoprotein levels

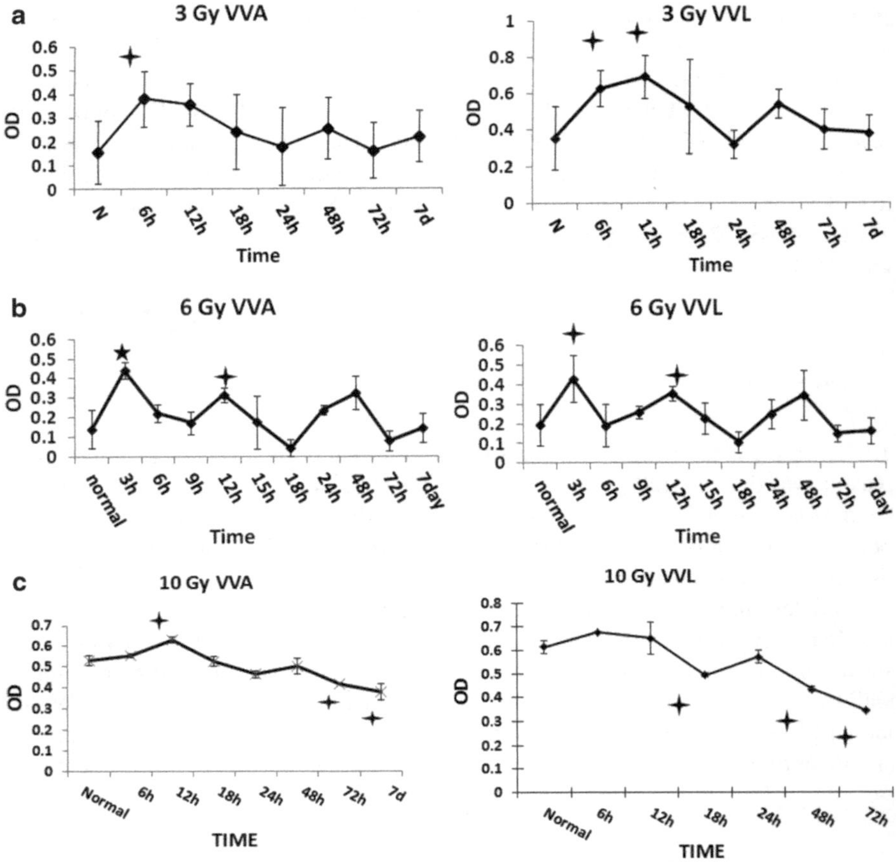

Fig. 21.2 Alteration patterns of plasma glycoproteins detected by VVA/VVL as a function of time after mice are exposed to (**a**) 3 Gy, (**b**) 6 Gy, or (**c**) 10 Gy

decreased again at 18 h and had a tendency to increase at 48 h. The third decrease was at 72 h. At 168 h, the level dropped to baseline (Fig. 21.2b). At 10 Gy, the glycoproteins detected by VVA increased at 12 h but continuously decreased after that; they reached their lowest point at 72 h (Fig. 21.2c).

To explore if different lectins detecting different sugars have the same pattern as VVA/VVL, we used biotinylated ECL. The data showed that (1) ECL yielded a good correlation between the level of glycoproteins containing Gal and the reading value (Fig. 21.3a) and (2) at 3 Gy (Fig. 21.3b) and 6 Gy (Fig. 21.3c), the alteration pattern of plasma glycoproteins was similar to that detected with VVA/VVL, although they recognized different sugars.

To examine if plasma glycoprotein levels are effective for biodosimetry, we measured the levels of glycoproteins from samples exposed to 3, 6, or 10 Gy at 6, 12, 72, or 168 h. Biodosimeters help medical personnel make decisions concerning

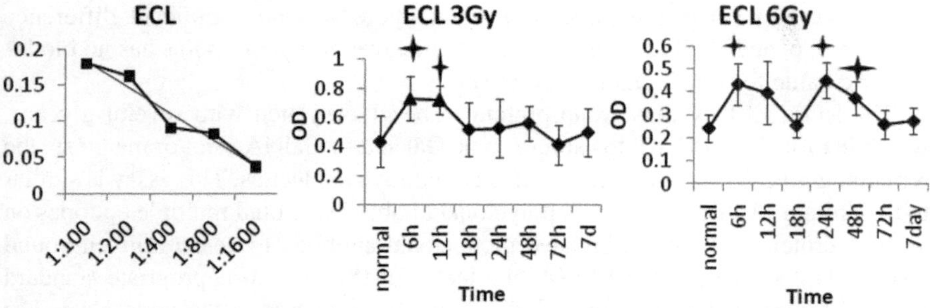

Fig. 21.3 (**a**) The correlation between the level of glycoproteins containing Gal and the reading value. Alteration patterns of plasma glycoproteins detected by ECL as a function of time after mice are exposed to (**b**) 3 Gy or (**c**) 6 Gy

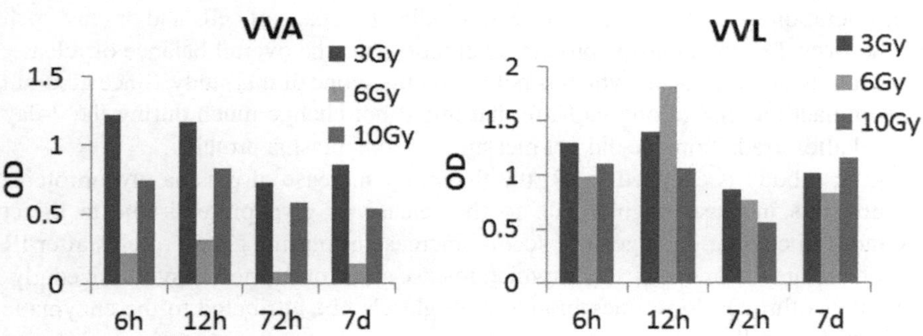

Fig. 21.4 Plasma glycoprotein level is not dose dependent

triage and care for exposed individuals, including (1) antibiotics and supportive care for those exposed to approximately 3 Gy; (2) bone marrow transplantation for those exposed to approximately 4–6 Gy; and (3) palliative care for those exposed to approximately 10 Gy. An effective biodosimeter should exhibit a difference in levels at these different exposure doses; however, the data (Fig. 21.4) showed no difference between glycoprotein levels, indicating that the plasma glycoprotein level is not valuable for biodosimetry.

4 Discussion

This study demonstrated that (1) the lectin ELISA measured plasma levels of glycoproteins containing different oligosaccharides, (2) the alteration patterns of plasma glycoproteins, as detected by VVA/VVL and ECL, were time dependent and dose dependent, and (3) at the same time point, there was no significant difference between 3, 6, and 10 Gy; therefore, the plasma level of glycoproteins has no biodosimetric value for triage in a nuclear event.

The lectin ELISA allows one protein to be glycosylated with several glycans, which in turn allow PHA-L to capture some Gal and/or GalNAc and/or mannose; the remaining glycans are specifically detected biotinylated lectins. This assay is similar to the protein ELISA, which uses polyclonal antibodies to bind multiple epitopes on a given protein and the biotinylated monoclonal antibody to detect a given bound protein. Unlike the protein ELISA, the lectin ELISA has no appropriate standard material to use; hence, different samples must be compared in the same plate. The result is expressed by an absorption value that is proportional to the amount of glycoproteins. Because the reading time for each plate may differ, the OD value can also differ from plate to plate. However, the alteration tendency can be determined by the difference between the controls and the testing groups measured in the same plate. This study focuses on the alteration of plasma glycoproteins following irradiation. The alteration might be related to direct or indirect damage of cells and organs, such as the liver. The total plasma protein level represents the overall balance of release, production, and clearance, which is not within the scope of this study. Since albumin is dominant and has a long half-life that might not change much during the 7-day period after irradiation, we did not measure the total plasma protein.

After a body is exposed to IR, the first early increase of plasma glycoproteins occurs; this increase might relate to the release of glycoproteins due to direct IR-induced cellular damage. The second increase of plasma glycoproteins after IR may be related to apoptotic cells, which release some of the newly synthesized glycoproteins through loose membranes. It might also be attributed to the enzymatic lysis of dead cells.

The early drop (within hours) of plasma glycoproteins may be due to the strong clearance capability of the endoreticulum system overwhelming the release from IR-damaged cells that engraft the free glycoproteins with macrophages and digest them so that glycoprotein fragments cannot be captured or recognized by lectins. The small glycoproteins (< 46,000 Da) can be discharged by the kidneys. The later decrease of plasma glycoproteins at 72 and 168 h after 10 Gy might be due to the ability of IR and its free radicals to damage genes cording the enzymes and proteins related to the glycosylation process.

5 Conclusion

Exploring how IR affects glycosylation and glycoproteins is essential for radiation biology. IR can cause alterations in plasma glycoproteins in a time-dependent and dose-dependent manner. At least 2 waves of alteration of glycoproteins occurred at 6 Gy and 1 wave at 3 Gy or 10 Gy. Ten Gy could significantly damage the cellular machinery for glycosylation and reduce the production of glycoproteins. More research is needed to reveal the complicated effect of radiation on plasma glycoproteins.

Acknowledgments This project is supported in part by U19 AI067733, RC1AI078519, RC2-AI-087580, RC1-AI081274 (NIAID/NIH), and Shands Cancer Center startup funds (University of Florida). We thank Dr. Chihray Liu and his team at UF for ensuring dosimetric accuracy and Kate Casey-Sawicki for editing this manuscript.

References

1. Rothkamm K, Lobrich M (2002) Misrepair of radiation-induced DNA double-strand breaks and its relevance for tumorigenesis and cancer treatment (review). Int J Oncol 21:433–440
2. Rodriguez-Rocha H, Garcia-Garcia A, Panayiotidis MI et al (2011) DNA damage and autophagy. Mutat Res 711:158–166
3. Jagetia GC (2007) Radioprotective potential of plants and herbs against the effects of ionizing radiation. J Clin Biochem Nutr 40:74–81
4. Larkin A, Imperiali B (2011) The expanding horizons of asparagine-linked glycosylation. Biochemistry (Mosc) 50:4411–4426
5. Rambaruth ND, Dwek MV (2011) Cell surface glycan-lectin interactions in tumor metastasis. Acta Histochem 113:591–600
6. Zhou SM, Cheng L, Guo SJ et al (2011) Lectin microarray: a powerful tool for glycan related biomarker discovery. Comb Chem High Throughput Screen 14(8):711–719
7. Varki A, Cummings R, Esko J, Freeze H, Hart G, Marth J (eds) (1999) Essentials of glycobiology. http://www.ncbi.nlm.nih.gov/books/NBK20709/. 19 Jul 2011
8. Larkin A, Imperiali B (2011) The expanding horizons of asparagine-linked glycosylation. Biochemistry 50(21):4411–4426
9. Hammarstrom S (1999) The carcinoembryonic antigen (CEA) family: structures, suggested functions and expression in normal and malignant tissues. Semin Cancer Biol 9:67–81
10. Neelamegham S, Liu G (2011) Systems glycobiology: biochemical reaction networks regulating glycan structure and function. Glycobiology 21(12):1541–1553
11. Medicago (Absolute Lectins) (2011) http://www.medicago.se/sites/default/files/pdf/productsheets/PHA-E_v.01.pdf. 19 Jul 2011
12. Sullivan K (2011) The lectin report. http://www.krispin.com/lectin.html. 19 Jul 2011

Chapter 22
Fibroblast Growth Factor-Peptide Promotes Bone Marrow Recovery After Irradiation

Jun Ma*, Yanqian Hou*, Deping Han, Mei Zhang, Chun Chen, Bingrong Zhang, Zhenhuan Zhang, Xiaohui Wang, Shanmin Yang, Yansong Guo, Paul Okunieff, and Lurong Zhang

Abstract Various members of the fibroblast growth factor (FGF) family mitigate radiation-induced damage. We designed and synthesized the binding domain peptide of FGF-2 (FGF-P) with a dimer form resistant to peptidase and examined its mitigatory effect on murine bone marrow cells. NIH Swiss mice were exposed to different doses of total body irradiation (TBI) and treated with ten doses of 5 mg/kg FGF-P. We achieved the following results: (1) FGF-P stimulated the growth of bone marrow cells harvested from mice exposed to 3 Gy; (2) on day 25 after 6 Gy TBI, the number of leukocytes and granulocytes was higher in the FGF-P group than in the vehicle-alone group; (3) FGF-P significantly increased the number of pro-B and pre-B cells; and (4) FGF-P treatment in vivo increased the long-term hematopoietic stem cells (LT-HSC) in bone marrow. These data reveal the underlying mechanism by which FGF-P rescued a significant percentage of the exposed mice. The increase of LT-HSC in bone marrow leads to a concomitant increase of pro-B and pre-B cells followed by leukocytes and granulocytes, which in turn enhance immunity against infection.

Keywords Bone marrow recovery • Fibroblast growth factor-peptide • Irradiation

---

*Jun Ma and Yanqian Hou contributed equally.

J. Ma
Institute of Digestive Diseases, Zhengzhou University, Henan, China

Y. Hou
Department of Laboratory Medicine, Kongjiang Hospital, Shanghai, China

D. Han • M. Zhang • C. Chen • B. Zhang • Z. Zhang
X. Wang • S. Yang • Y. Guo • P. Okunieff • L. Zhang (✉)
Department of Radiation Oncology, UF Shands Cancer Center, University of Florida, P.O. Box 100385, Gainesville, FL 32610, USA
e-mail: lurongzhang@ufl.edu

W.J. Welch et al. (eds.), *Oxygen Transport to Tissue XXXIV*, Advances in Experimental Medicine and Biology 765, DOI 10.1007/978-1-4614-4989-8_22,

1 Introduction

Ionizing radiation (IR) damages sensitive bone marrow cells, leading to bleeding and lethal infection [1]. After breaks in DNA strands of bone marrow stem cells or damage by free radicals are induced by radiation, cells immediately enter the repair process, and residual cells start to proliferate. Growth factors, which stimulate the renewal of stem cells and the differentiation of their daughter progenitors into functional blood cells, are crucial to the preservation of life after irradiation [2]. A granulocyte colony stimulating factor (G-CSF) that enhances differentiation of myeloid-lineage progenitors and releases granulocytes into peripheral blood is the only Food and Drug Administration (FDA)-approved drug for the mitigation of acute radiation syndromes [3]. Currently, no agent that stimulates stem cell renewal, which is the key to retaining the stem cell pool before cells differentiate, exists to treat victims of IR.

In this study, we utilized a fibroblast growth factor peptide (FGF-P) derived from the binding domain of FGF-2 [4] with modifications for dimer and peptidase resistance to examine its effect on bone marrow recovery and stem cell renewal after irradiation. Since FGF receptors are expressed in all types of stem cells and all states of cells and FGF-2 is the only growth factor used in serum-free media for stem cell culture, FGF clearly plays an important role in maintaining the stem cell pool and promoting differentiated cells. We chose to use the peptide form because (1) lacking a secondary structure, it is much more stable than the parental full-length protein that can be denatured and lose activity at extreme temperatures and in different pH conditions, (2) peptide production is much cheaper than that of protein production, (3) the functional small peptide is normally the binding domain of its corresponding receptor, and (4) it lacks antigenicity.

2 Methods

Eight-week-old, male, NIH Swiss mice (National Cancer Institute, Frederic, MD) were divided into groups of five and exposed to total body irradiation (TBI) with a cesium-137 source (Best Theratronics Ltd., Ottawa, Canada) at a dose rate of 1 Gy/min. Groups exposed to 3 Gy TBI were used for in vitro culture assay for proliferation, while groups exposed to 6 Gy were used for dynamic observation of leukocyte alterations and bone marrow cells with and without FGF-P treatment in vivo. FGF-P was chemically synthesized with 98% purity as assessed by high-performance liquid chromatography (Genemed Synthesis, Inc., San Antonio, TX).

We performed a ^{3}H-Thymidine (^{3}H-TdR, Perkin Elmer, Inc., Waltham, MA) incorporation assay to test the proliferation of irradiated bone marrow cells. Three hours after exposure to 3 Gy TBI, mice were euthanized; thereafter, bone marrow cells were harvested from the femurs and cultured for 3 days in round-bottom, 96-well plates in triplicate with UltraDOMA serum-free medium, with or without 450 ng/ml FGF-P treatment. FGF-2 (20 ng/ml, PepTech Corporation, Burlington, MA) was used as a positive control. On day 2.5, 0.3 μCi ^{3}H-TdR was added to each well and incubated overnight. The incorporated ^{3}H-TdR and bone marrow cells were harvested with a 96-well, auto-harvester and counted by a liquid scintillation 96-well plate reader.

We then monitored the effects of radiation and FGF-P on leukocytes. Forty-eight hours after exposure to 6 Gy TBI, mice were subcutaneously injected with either vehicle alone (saline) or FGF-P (5 mg/kg/day for 5 days and then 5 mg/kg/every other day for 10 days). On days 10, 15, 25, 35, and 50 after irradiation, the number of leukocytes and granulocytes were measured with a HemaTrue analyzer (Heska, Loveland, CO).

Finally, we determined the percentage of pro-B, pre-B, and long-term hematopoietic stem cells (LT-HSC) cells in the bone marrow. Mice were exposed to 6 Gy TBI and treated with ten doses of saline or FGF-P and euthanized on days 25 or 50. Bone marrow cells were harvested from femurs and subjected to antibodies staining (R&D Systems, Minneapolis, MN and BioLegend, Inc., San Diego, CA) for pro-B ($IgM^{-}/CD19^{+}/c\text{-}kit^{+}$) and pre-B ($IgM^{-}/CD19^{+}/CD25^{+}$) cells. Similarly, the LT-HSC were stained ($lin^{-}/Sca\text{-}1^{+}/c\text{-}kit^{+}/CD38^{+}$) and assessed with a C6 flow cytometer (FCM, Accuri Cytometers, Inc., Ann Arbor, MI).

For pro-B and pre-B cell analysis, IgM negative cells in lymphocytes of bone marrow cells were regated by CD19 and c-Kit/CD25. For LT-HSC, lineage (B220, Gr-1, Ter119, CD3, and CD11b) negative cells were further gated using c-kit and Sca-antibodies, then the percentage of CD38 positive cell numbers in this gate were calculated.

3 Results

To determine if FGF-P has a biological effect on irradiated bone marrow cells, we harvested them from mice exposed to 3 Gy TBI, a dose that causes significant damage to bone marrow cells. They were cultured in an UltraDOMA serum-free medium to not only avoid effects caused by other mitogens but also supply sufficient nutrition. To mimic the cell–cell interactions within bone marrow, a cell mixture was used. The result (Fig. 22.1) showed that FGF-P stimulated the proliferation of bone marrow cells ($P<0.04$). In this case, FGF-P had a better effect than the native FGF-2 protein ($P<0.02$). Similar results were obtained from separate experiments with the ^{3}H-TdR incorporation proliferation assay.

To determine when FGF-P exerted its effect on white blood cells, they were counted every day for 50 days. Figure 22.2a shows that the first 15 days were a nadir phase for bone marrow cells in which FGF-P did not exert an effect on the leukocyte count; however, on day 25, the leukocyte count of the FGF-P group was significantly higher than that of the vehicle-control group. On days 35 and 50, no difference was observed. Similar results were obtained with granulocytes (Fig. 22.2b), a major part of the innate immune system that fights invading pathogens.

Bone marrow is a "home" hosting different lineages for regeneration of billions of different blood cells daily. To study the effect of FGF-P on the production of

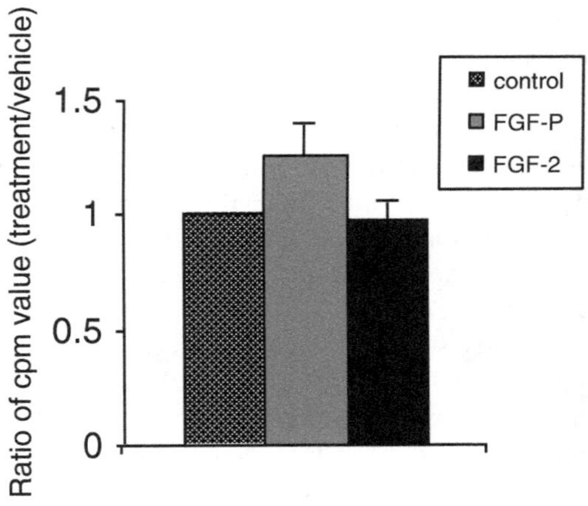

Fig. 22.1 ^{3}H-TdR intake of bone marrow cells from 3 Gy TBI mice. Bone marrow cells were flushed from the femora of irradiated mice and cultured in 96-well plates treated with FGF-P, FGF-2, or vehicle only. ^{3}H-TdR was added simultaneously at the beginning of cell culture

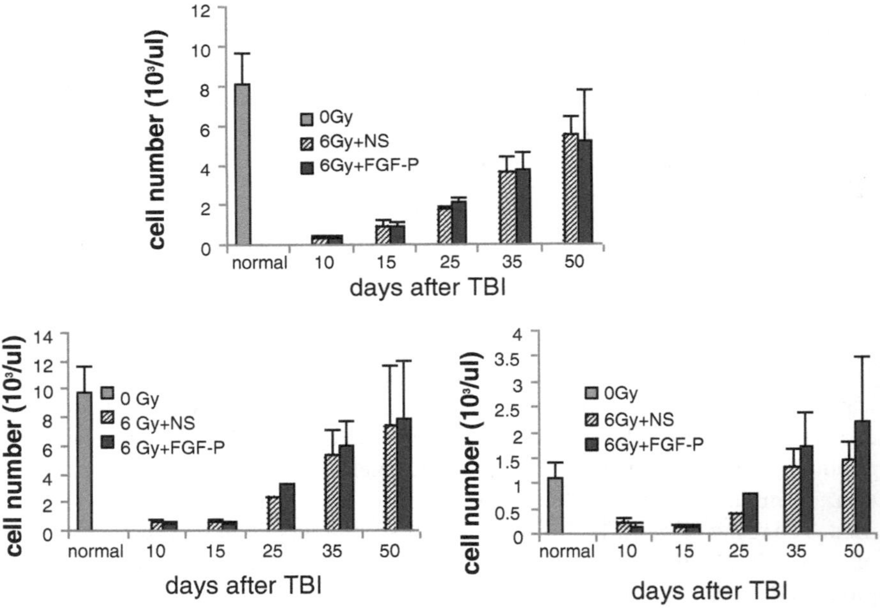

Fig. 22.2 PBC test after treatment in TBI mice. On days 10, 15, 25, 35, and 50 after irradiation, PBC numbers were measured with a HemaTrue analyzer. Between NS control and FGF-P on day 25 (**a**) leukocyte $P<0.01$; (**b**) granulocyte $P<0.01$; and (**c**) lymphocyte $P<0.05$

these lineages, we used pro-B (IgM^{-}/CD19^{+}/c-kit^{+}) and pre-B (IgM^{-}/CD19^{+}/CD25^{+}) cells as test targets because they had distinct markers that allowed the assay to yield clear-cut results. In addition, we investigated the radiation-induced immune response to damaged/mutant cells. The results showed that on day 25 the percentages of pro-B (Fig. 22.3a) and pre-B (Fig. 22.3b) cells in bone marrow were significantly

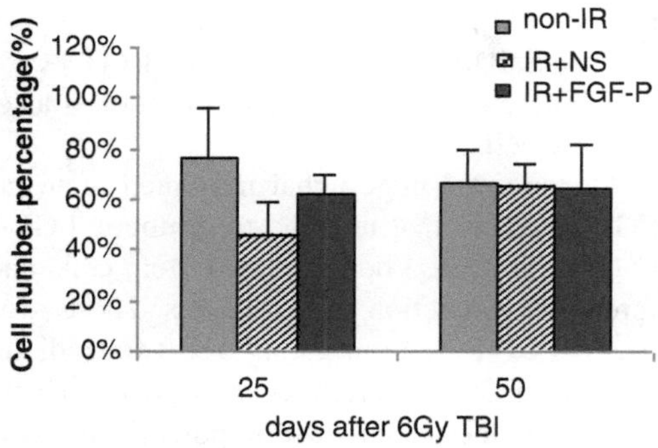

Fig. 22.3 Effect of FGF-P on pro-B and pre-B cells in bone marrow of TBI mice. In FGF-P group on day 25 (**a**) pro-B and (**b**) pre-B were higher than those in NS group ($P<0.05$)

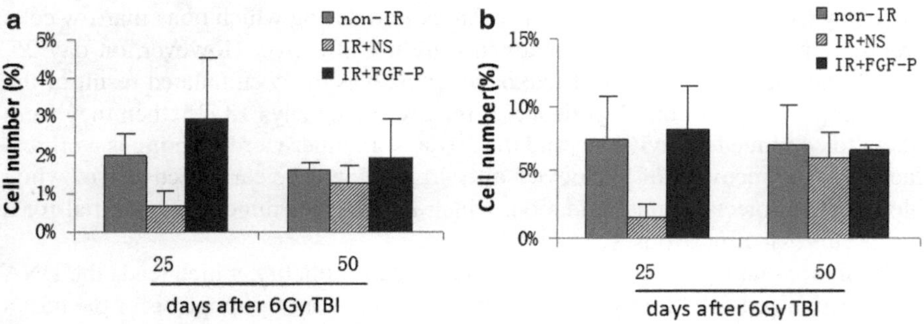

Fig. 22.4 LT-HSC in bone marrow after TBI. LT-HSC in FGF-P group was higher than that in NS group on day 25 after TBI ($P<0.05$)

increased by FGF-P as compared to the control, suggesting that FGF-P might increase the development of certain progenitor cells. The difference disappeared on day 50 when the bone marrow crisis had subsided.

Since it is the source of all leukocytes, the stem cell pool was our focus; thus, we studied the effect of FGF-P on LT-HSC percentages (lin$^-$/Sca-1$^+$/c-kit$^+$/CD38$^+$). Although 6 Gy TBI reduced the percentage of LT-HSC, on day 25 this percentage showed an increase due to FGF-P (Fig. 22.4) that was masked by recovered bone marrow cells on day 50.

4 Discussion

This study demonstrated that (1) in vitro, FGF-P stimulated the proliferation of bone marrow cells, (2) in vivo, FGF-P (5 mg/kg for ten doses) increased the number of leukocytes and granulocytes on day 25, thereby enhancing the host's defense system, (3) FGF-P significantly increased the percentages of pro-B (IgM$^-$/CD19$^+$/c-kit$^+$) and pre-B ($IgM^-/CD19^+/CD25^+$) cells in the bone marrow, which meet the needs of the special immune response, and (4) FGF-P increased LT-HSC percentages, which help to retain the stem cell pool as a source for the differentiation of bone marrow cells.

Our results suggest that irradiated bone marrow cells still produce functional FGF receptors that respond to synthetic FGF-P. Indeed, both in vitro and in vivo, FGF-P stimulated bone marrow stem cells and progenitors and enhanced the differentiation function of leukocytes. These results lay down the foundation for the use of FGF-P as a mitigatory agent for radiation-induced bone marrow syndrome. Notably, FGF-P exerted its effect even when administered 48 h after 6 Gy TBI, making it an especially useful potential agent for nuclear disasters.

The pattern of FGF-P's effect on leukocytes fits well with the patterns of nadir and survival timing in irradiated mice. After irradiation, leukocyte counts drop continuously for a 2-week period. In this latent period during which bone marrow cells "reprogram" themselves, most treatments are not effective. However, on day 25, FGF-P increased the number of leukocytes; this was the accumulated result of the preceding 24 days. In our experience, if mice survived days 18–25, then they were likely to continue to survive beyond the 50 days. In other words, timing is a critical factor for the recovery of leukocytes after irradiation. The early recovery of white blood cells protects against infection, which in turn determines the potential for a host's survival after exposure to radiation.

Radiation-damaged DNA can experience repair infidelity, which leads the DNA to be recognized as a "foreigner" that must be cleared out by apoptosis or the host's immune system. The increased pro-B and pre-B cells fit the needs of the immune response to remove radiation-damaged or mutant cells, including those that might have triggered oncogenes, and thereby ensure homeostasis.

An effective mitigatory agent must not exhaust the stem cell pool by stimulating bone marrow stem cells into differentiation too early. The agent must renew the stem cell pool, while the cells differentiate into progenitors and various lineages. Without an increase in the source of stem cells, the downstream differentiation eventually "dries out" and the recovery power decreases with time. FGF-P increases LT-HSC; through this important mechanism, radiation-induced morbidity may be decreased. Our years of experience with FGF-P indicate that the dosage and timing are critical factors for the success of this treatment.

5 Conclusion

FGF-P, a peptide derived from the binding domain of FGF-2 with modifications for dimer and peptidase resistance, has demonstrated its ability to stimulate the recovery of bone marrow cells after irradiation. Due to its mitigation properties, including increasing the percentages of stem cells, progenitors, and differentiated leukocytes after irradiation, FGF-P could be developed into an effective agent to treat acute bone marrow syndrome due to a nuclear event.

Acknowledgments This project is supported in part by U19 AI067733, RC1AI078519, RC2-AI-087580, RC1-AI081274 (NIAID/NIH), and Shands Cancer Center startup funds (University of Florida). We thank Dr. Chihray Liu and the medical physics faculty at UF for ensuring dosimetric accuracy in these experiments and Kate Casey-Sawicki for editing this manuscript.

References

1. Herodin F, Grenier N, Drouet M (2007) Revisiting therapeutic strategies in radiation casualties. Exp Hematol 35:28–33
2. TMT handbook. http://www.tmthandbook.org/. 18 Jul 2011
3. Layton JE, Hall NE, Connell F et al (2001) Identification of ligand-binding site III on the immunoglobulin-like domain of the granulocyte colony-stimulating factor receptor. J Biol Chem 276:36779–36787
4. Lin X, Takahashi K, Campion SL et al (2006) Synthetic peptide F2A4-K-NS mimics fibroblast growth factor-2 in vitro and is angiogenic in vivo. Int J Mol Med 17:833–839

Chapter 23
Dynamic Two-Photon Imaging of Cerebral Microcirculation Using Fluorescently Labeled Red Blood Cells and Plasma

Kazuto Masamoto, Hiroshi Kawaguchi, Hiroshi Ito, and Iwao Kanno

Abstract To explore the spatiotemporal dynamics of red blood cells (RBCs) and plasma flow in three-dimensional (3D) microvascular networks of the cerebral cortex, we performed two-photon microscopic imaging of the cortical microvasculature in genetically engineered rats in which the RBCs endogenously express green fluorescent protein (GFP). Water-soluble quantum dots (Qdots) were injected intravenously into the animals to label the plasma, and concurrent imaging was performed for GFP-RBCs and Qdot plasma. The RBC and plasma distributions were compared between resting state and forepaw stimulation-induced neural activation. The RBC and plasma images showed detectable signals up to a depth of 0.4 and 0.6 mm from the cortical surface, respectively. A thicker plasma layer (2–5 μm) was seen in venous vessels relative to the arterial vessels. In response to neural activation, the RBCs were redistributed among the parenchymal capillary networks. In addition, individual capillaries showed a variable ratio of RBC and plasma distributions before and after activation, indicative of dynamic changes of hematocrit in single capillaries. These results demonstrate that this transgenic animal model may be useful in further investigating the mechanism that controls dynamic RBC flow in single capillaries and among multiple capillary networks of the cerebral microcirculation.

Keywords Brain microcirculation • Functional imaging • Oxygen demand and supply • Somatosensory cortex

---

K. Masamoto, Ph.D. (✉)
Center for Frontier Science and Engineering, University of Electro-Communications, 1-5-1 Chofugaoka, Chofu, Tokyo 182-8585, Japan

Molecular Imaging Center, National Institute of Radiological Sciences, Chiba, Japan
e-mail: masamoto@mce.uec.ac.jp

H. Kawaguchi • H. Ito • I. Kanno
Molecular Imaging Center, National Institute of Radiological Sciences, Chiba, Japan

W.J. Welch et al. (eds.), *Oxygen Transport to Tissue XXXIV*, Advances in Experimental Medicine and Biology 765, DOI 10.1007/978-1-4614-4989-8_23,

1 Introduction

Two-photon microscopy allows for three-dimensional (3D) microscopic imaging of animal brains in vivo with deep penetration of near-infrared excitation light [1, 2]. In previous studies, we have shown that 3D images of cortical microvasculature can be obtained up to a depth of 0.6 mm in rats with thinned skulls [3] and 0.8 mm in mice with a closed cranial window [4]. However, these studies are limited to only structural imaging and are not investigated for measuring blood flow, which is of particular importance for understanding oxygen transport to tissues in living brains.

With two-photon microscopy, Kleinfeld et al. introduced a line-scanning method for measuring red blood cell (RBC) speed in single capillaries [5]. Later studies have successfully employed a variety of methods for the quantification of capillary RBC speed [6–11]. With a high-speed frame rate (500 frames/s), Tomita et al. created a 2D RBC speed map in rat and mouse cerebral cortex [12–14]. In their studies, fluorescently labeled RBCs were injected to create contrast within the circulation, which allows for the tracking of individual RBCs in multiple vessels simultaneously using a confocal laser-scanning microscope. However, the labeled RBCs represent only a portion of the RBCs resident in the circulation, and thus, vessel occupancy of the RBCs relative to plasma levels, i.e., hematocrit, cannot be resolved using this technique.

In the present study, we performed dynamic two-photon imaging of cortical microvasculature using genetically engineered rats in which whole RBCs were endogenously labeled with green fluorescent protein (GFP). In addition, we injected the animals with water-soluble quantum dots (Qdots) that act as a plasma marker [15]. The selected Qdots had a peak emission in the red spectrum at a wavelength of 605 nm, which allowed for simultaneous imaging of GFP-RBCs and Qdot plasma. We compared RBC and plasma distributions in single capillaries and among multiple capillaries under the conditions of rest and forepaw stimulation-induced activation in the somatosensory cortex.

2 Materials and Methods

2.1 Animal Preparation

All experimental protocols were approved by the Institutional Animal Care and Use Committee. A total of eight male transgenic Wistar rats (250–480 g) were used for the experiments. The animals were anesthetized with 2% isoflurane for surgery and 1.4% for experiments. Intubation was performed for mechanical ventilation. The femoral artery was catheterized with a PE50 tube for monitoring the arterial blood pressure and blood gas sampling, and the femoral vein was catheterized with a PE10 tube for drug administration. The rectal temperature was maintained at 37°C. The animal was fixed on a stereotactic frame, and the left somatosensory cortex was exposed by removing the skull with a dental drill, leaving a layer of thinned dura. The exposed area was covered with warm saline (37°C), and the experiments were initiated approximately 1 h after surgery. Heart rate and arterial blood pressure measurements were recorded with the aid of data acquisition software (AcqKnowledge, Biopac Systems, Inc., Goleta, CA), and end-tidal CO_2 and blood gas levels were maintained within physiologic limits.

2.2 *Image Acquisition and Analysis*

Qdot 605 amino (PEG) quantum dots (1 μM in saline, Invitrogen) were intravenously injected into the animals (1 mL/kg). To obtain a 3D structure of the cortical microvasculature, GFP-expressing RBCs and Qdot plasma were simultaneously imaged using a two-photon microscope (TCS SP5MP, Leica Microsystems, Germany) with a Ti:Sapphire laser (MaiTai HP, Spectra-Physics, CA). The excitation wavelength was 900 nm, and the emission signal was separated by a beam splitter at 560/10 nm and detected through band-pass filters of 525/50 nm and 610/75 nm for GFP and Qdot, respectively. A single image consisted of 1,024 × 1,024 or 512 × 512 pixels, and the volume image was acquired with a step size of 2.5–5 μm. For activation experiments, the right forepaw was electrically stimulated (1.5–1.7 mA, 6 Hz) [16, 17], and volume imaging was conducted before stimulation (pre-stimulus resting state) and during stimulation (activation state). The vessel diameter was compared for both RBC and plasma images.

3 Results and Discussions

The images of the microvasculature that were captured based on Qdot plasma signals showed detectable signals up to a depth of 0.6 mm from the cortical surface, which was in agreement with our previous report [3, 15]. In contrast, the images obtained with the GFP-RBC signals showed a detectable signal up to only 0.4 mm from the cortical surface. The difference in the detectable depth could be due to a differing optical property of the brain tissue for green and red wavelength ranges and/or different characteristics of the fluorescent substances themselves (GFP vs. Qdot). The quantum efficiency of Qdot is higher relative to organic dye [18], which may enhance the depth detection for Qdot plasma imaging. Based on this result, the following analysis was performed for volume images obtained up to a depth of 0.4 mm from the surface where the image contained both RBC and plasma signals.

Comparison of the vessel width showed a thicker width of Qdot plasma than that of the GFP-RBCs (Fig. 23.1). Among the vessels investigated, the plasma layer was observed to be 2–5 μm thick for venous vessels and approximately 1 μm for arterial vessels. In capillaries (<7 μm), no detectable difference in the RBC and plasma width was observed. Additionally, a variable ratio of RBC and plasma distribution was found among multiple capillaries, which indicates dynamic changes of hematocrit in single capillaries as well as among multiple capillaries. Following neural activation, the distribution of RBCs varied among the adjacent capillary networks (Fig. 23.2), which may represent the redistribution of RBCs in the activated region.

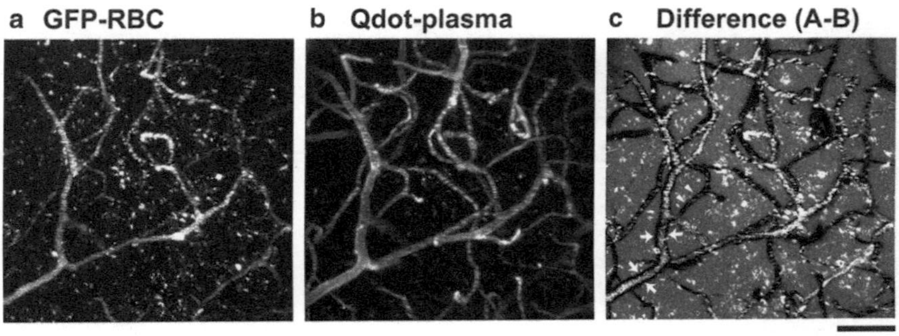

Fig. 23.1 A representative volume image of cortical microvasculature obtained with two-photon microscopy in the cerebral cortex of transgenic rats based on the fluorescent signals of GFP-RBC (**a**) and Qdot plasma (**b**). For visualization purposes, a maximum intensity projection is displayed for the images obtained over a depth of 60–200 μm from the cortical surface. (**c**) Differential image. This image was created by subtracting the plasma image (**b**) from the RBC image (**a**). The *dark area* (*arrows*) apparent in the vessel edge indicates the thicker width of the plasma (i.e., plasma layer). In contrast, no clear plasma layer was found in the capillaries (*arrow head*). The *white spots* that are seen in the extravascular area in the RBC image were due to non-GFP cells' auto-fluorescence, which has a similar emission spectrum to GFP. Scale bar: 50 μm

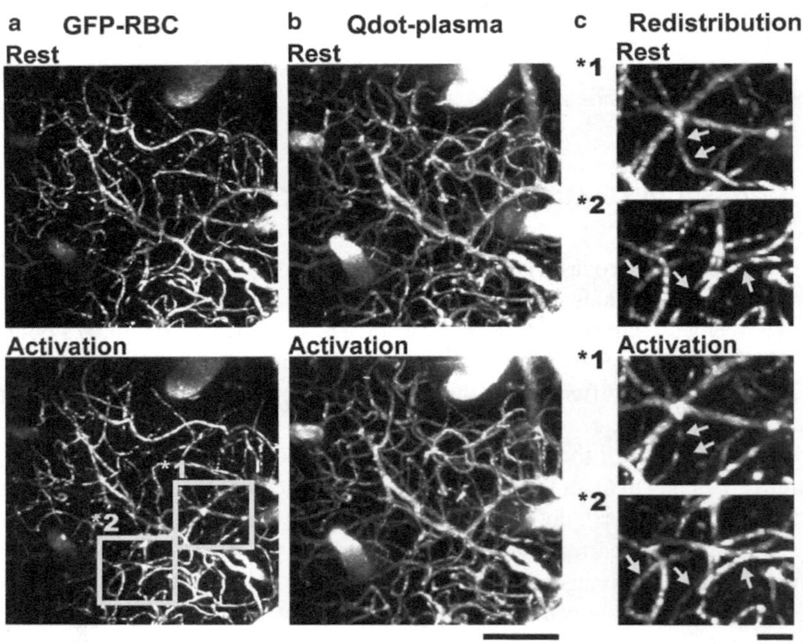

Fig. 23.2 Comparison of the cortical microvasculature during rest (*upper*) and activation (*bottom*) states. The forepaw stimulation-induced activation provoked a change of RBC (**a**) and plasma (**b**) distribution in the vessels. Scale bar: 100 μm. (**c**) Enlarged RBC image. RBC redistribution due to activation was seen in two adjacent capillary networks (regions **1* and **2*). In region **1*, RBC occupation in this capillary (*arrows*) was dense at rest but became sparse during activation. In contrast, the capillaries (*arrows*) in region *2 had increased RBC flux due to activation relative to the resting state. These observations indicate the RBC redistribution among the inter-capillaries in the adjacent capillary networks. Scale bar: 20 μm

These observations strongly indicate that there is local regulation for RBC flow at the single capillary level and/or among multiple-capillary units, which requires further investigation in relation to the activity of surrounding cells, such as neurons, astroglias, and vascular cells.

4 Summary

The present study described 3D imaging of cortical microvasculature based on GFP-expressing RBCs and Qdot-stained plasma fluorescence with two-photon microscopy. The GFP and Qdot images showed an imaging depth up to 0.4 and 0.6 mm, respectively. The plasma layer was thicker in venous vessels relative to arterial vessels of the same size. In response to neural activation, redistribution of RBC and plasma occupancy was detected in individual single capillaries and among multiple capillaries. The findings showed that the animal model presented may be useful in further investigations of the mechanism that controls RBC flow in single capillaries and multiple-capillary units in healthy and diseased brains.

Acknowledgments The authors thank Dr. Junko Taniguchi for help with the experiments. This work was partly supported by Special Coordination Funds for Promoting Science and Technology (K.M.).

References

1. Denk W, Strickler JH, Webb WW (1990) Two-photon laser scanning fluorescence microscopy. Science 248:73–76
2. Helmchen F, Denk W (2005) Deep tissue two-photon microscopy. Nat Methods 2:932–940
3. Park SH, Masamoto K, Hendrich K et al (2008) Imaging brain vasculature with BOLD microscopy: MR detection limits determined by in vivo two-photon microscopy. Magn Reson Med 59:855–865
4. Yoshihara K, Takuwa H, Kanno I et al (2012) 3D analysis of intracortical microvasculature during chronic hypoxia in mouse brain. Adv Exp Med Biol 765: 357–363
5. Kleinfeld D, Mitra PP, Helmchen F et al (1998) Fluctuations and stimulus-induced changes in blood flow observed in individual capillaries in layers 2 through 4 of rat neocortex. Proc Natl Acad Sci USA 95:15741–15746
6. Zipfel WR, Williams RM, Webb WW (2003) Nonlinear magic: multiphoton microscopy in the biosciences. Nat Biotechnol 21:1369–1377
7. Schaffer CB, Friedman B, Nishimura N et al (2006) Two-photon imaging of cortical surface microvessels reveals a robust redistribution in blood flow after vascular occlusion. PLoS Biol 4:e22
8. Dobbe JG, Streekstra GJ, Atasever B et al (2008) Measurement of functional microcirculatory geometry and velocity distributions using automated image analysis. Med Biol Eng Comput 46:659–670
9. Drew PJ, Blinder P, Cauwenberghs G et al (2010) Rapid determination of particle velocity from space-time images using the Radon transform. J Comput Neurosci 29:5–11

10. Kamoun WS, Chae SS, Lacorre DA et al (2010) Simultaneous measurement of RBC velocity, flux, hematocrit and shear rate in vascular networks. Nat Methods 7:655–660
11. Autio J, Kawaguchi H, Saito S, Aoki I, Obata T, Masamoto K, Kanno I (2011) Spatial frequency-based analysis of mean red blood cell speed in single microvessels: investigation of microvascular perfusion in rat cerebral cortex. PLoS One 6:e24056
12. Tomita M, Osada T, Schiszler I et al (2008) Automated method for tracking vast numbers of FITC-labeled RBCs in microvessels of rat brain in vivo using a high-speed confocal microscope system. Microcirculation 15:163–174
13. Unekawa M, Tomita M, Tomita Y et al (2010) RBC velocities in single capillaries of mouse and rat brains are the same, despite 10-fold difference in body size. Brain Res 1320:69–73
14. Tomita M, Tomita Y, Unekawa M et al (2011) Oscillating neuro-capillary coupling during cortical spreading depression as observed by tracking of FITC-labeled RBCs in single capillaries. Neuroimage 56:1001–1010
15. Masamoto K, Obata T, Kanno I (2010) Intracortical microcirculatory change induced by anesthesia in rat somatosensory cortex. Adv Exp Med Biol 662:57–61
16. Masamoto K, Kim T, Fukuda M et al (2007) Relationship between neural, vascular, and BOLD signals in isoflurane-anesthetized rat somatosensory cortex. Cereb Cortex 17:942–950
17. Kim T, Masamoto K, Fukuda M et al (2010) Frequency-dependent neural activity, CBF, and BOLD fMRI to somatosensory stimuli in isoflurane-anesthetized rats. Neuroimage 52: 224–233
18. Resch-Genger U, Grabolle M, Cavaliere-Jaricot S et al (2008) Quantum dots versus organic dyes as fluorescent labels. Nat Methods 5:763–775

Chapter 24
The Effect of Basic Assumptions on the Tissue Oxygen Saturation Value of Near Infrared Spectroscopy

Andreas Jaakko Metz*, Martin Biallas, Carmen Jenny, Thomas Muehlemann, and Martin Wolf

Abstract Tissue oxygen saturation (StO_2), a potentially important parameter in clinical practice, can be measured by near infrared spectroscopy (NIRS). Various devices use the multi-distance approach based on the diffusion approximation of the radiative transport equation [1, 2]. When determining the absorption coefficient (μ_a) by the slope over multiple distances a common assumption is to neglect μ_a in the diffusion constant, or to assume the scattering coefficient (μ_s') to be constant over the wavelength. Also the water influence can be modeled by simply subtracting a water term from the absorption. This gives five approaches A1–A5. The aim was to test how these different methods influence the StO_2 values. One data set of 30 newborn infants measured on the head and another of eight adults measured on the nondominant forearm were analyzed. The calculated average StO_2 values measured on the head were (mean ± SD): A1: 79.99 ± 4.47%, A2: 81.44 ± 4.08%, A3: 84.77 ± 4.87%, A4: 85.69 ± 4.38%, and A5: 72.85 ± 4.81%. The StO_2 values for the adult forearms are: A1: 58.14 ± 5.69%, A2: 73.85 ± 4.77%, A3: 58.99 ± 5.67%, A4: 74.21 ± 4.76%, and A5: 63.49 ± 5.11%. Our results indicate that StO_2 depends strongly on the assumptions. Since StO_2 is an absolute value, comparability between different studies is reduced if the assumptions of the algorithms are not published.

Keywords Absorption coefficient • Methodology • Near-infrared spectroscopy • Tissue oxygen saturation

---

*Andreas Jaakko Metz is the member of the Ph.D. Program imMed

A.J. Metz (✉) • M. Biallas • C. Jenny • T. Muehlemann • M. Wolf
Biomedical Optics Research Laboratory, Division of Neonatology, University Hospital Zurich, Zurich, Switzerland

Zurich Center for Integrative Physiology, University Zurich, Frauenklinikstrasse 10, 8091 Zürich, Switzerland
e-mail: andreas.metz@usz.ch

W.J. Welch et al. (eds.), *Oxygen Transport to Tissue XXXIV*, Advances in Experimental Medicine and Biology 765, DOI 10.1007/978-1-4614-4989-8_24,

1 Introduction

Tissue oxygen saturation (StO_2) has a great potential to become an important clinical parameter, especially in neonatology [3, 4]. It is related to the oxygen metabolism in the tissue on an absolute scale. Slightly different approaches are used to calculate StO_2, depending on the manufacturer. This is reflected in different naming, e.g., tissue oxygenation index for the NIRO (Hamamatsu Photonics, Japan) [5] or regional oxygen saturation for the INVOS (Somanetics Corp., USA) or Critikon (Johnson & Johnson, UK). Studies have been published, which compared the values obtained from the three different devices and found differences between INVOS and Critikon [6] and agreement between the NIRO and INVOS [7, 8]. However, both found *unacceptable* baseline differences. Several reasons were given as explanation: Differences in the technical setup, the effect of extracranial blood flow and differences in the algorithm.

However, the influence of the algorithm itself has to our knowledge not been evaluated. Our aim was to test the influence of basic assumptions of the multi-distance approach [1], which is similar to spatially resolved spectroscopy [2]. Using the different approaches on the same data sets excludes the instrumentation or extracranial blood flow as a source of differences.

2 Methods

2.1 Subjects

Data sets from two different studies have been evaluated. First, 30 newborn infants have been studied previously in our group with the aim to identify precision of NIRS [9]. Second, eight adult subjects (all male, age range 26–45, median 29.5) were investigated within a still ongoing study. Both studies were approved by the ethical committee of the Kanton of Zurich and informed consent was obtained prior to the study.

2.2 Protocol

Neonatal group. The frontal and temporal cerebral region was measured four times for approximately 1 min. The sensor was repositioned between the measurements [9].

Adult group. Five repeated measurements per subject were taken from the non-dominant forearm, near to musculus brachioradialis. The sensor was fixated with an elastic bandage around the forearm. Each measurement took 1 min, in between measurements the bandage was completely removed and the sensor was repositioned to approximately the same place as before.

2.3 NIRS Measurement

The neonatal group was measured with the MCPII, which is described in detail elsewhere [10]. It uses three wavelengths (750, 800, and 875 nm) at distances of 1.25 and 2.5 cm.

The adult group was assessed by a novel continuous wave NIRS device, the *OxyPrem*, which is similar to previous wireless sensors [11]. It measures light attenuation at 760 and 870 nm, at distances of 1.5 and 2.5 cm.

2.4 Theory

Tissue oxygen saturation was calculated by a self-calibrating multi-distance approach [1] based on the diffusion approximation of the radiative transport equation and using two sources and two detectors. The light intensity decreases with the distance. This relation is linear (semi-infinite boundary condition).

$$\ln(dc(r)r^2) = r\mathrm{Sl}_{dc}(\mu_a, \mu_s') + \mathrm{In}_{dc}'(D, K_{dc}), \tag{24.1}$$

$dc(r)$ is the average light intensity as a function of distance r, Sl_{dc} the slope of the intensity loss and In_{dc}' the intercept. μ_a and μ_s' are the absorption and the reduced scattering coefficient, respectively. K_{dc} is a constant. The diffusion constant D equals

$$D = \frac{1}{3\mu_a + 3\mu_s'} \cong \frac{1}{3\mu_s'}. \tag{24.2}$$

μ_a is often neglected because tissue scattering is much larger than absorption (μ_s'). However, here we distinguish between simplified and exact diffusion constant (as seen below).

When evaluating (24.1) at two distances r_L and r_S and subtracting them, the slope can be calculated from the ratio of the measured intensities.

$$\mathrm{Sl}_{dc} = \frac{\frac{1}{2}\ln\left(\frac{dc_1(r_L)dc_2(r_L)}{dc_1(r_S)dc_2(r_S)}\right) + 2\ln\left(\frac{r_L}{r_S}\right)}{r_L - r_S}, \tag{24.3}$$

where r_L is the longer source–detector distance and r_S the shorter one, respectively. Equation (24.3) is a special self-calibrating form, whereby the use of two source–detector pairs [giving the four intensity values $dc_{1,2}(r_L, r_S)$] the coupling factors between the tissue and source/detector cancel out [1]. Then μ_a can be calculated as

$$\mu_a = \mathrm{Sl}_{dc}^2 D. \tag{24.4}$$

When the absorption is determined at least at two wavelengths, concentrations of oxygenated ($[O_2Hb]$) and deoxygenated hemoglobin ($[HHb]$) and the tissue oxygen saturation can be calculated. We used the absorption coefficients from Matcher et al. [12], averaged over the measured intensity spectrum of each light source. Coefficients for scattering were taken from Matcher et al. [13] for the adult arm and from ISS OxyPlex measurements on 36 term infants [14] for the neonates, extrapolated to 750, 800, and 875 nm (3.81, 3.49, and 3.01[cm^{-1}]).

$$\begin{bmatrix}[\mathrm{HHb}]\\ [\mathrm{O_2Hb}]\end{bmatrix} = \mathbf{A}^{-1}\begin{bmatrix}\mu_a(\lambda_1)\\ \mu_a(\lambda_2)\end{bmatrix},\quad \mathbf{A} = \begin{bmatrix} a_{\mathrm{HHb},\lambda_1} & a_{\mathrm{O_2Hb},\lambda_1}\\ a_{\mathrm{HHb},\lambda_2} & a_{\mathrm{O_2Hb},\lambda_2}\end{bmatrix} \tag{24.5}$$

where $a_{i,j}$ is the absorption coefficient for $i=([\mathrm{HHb}], [\mathrm{O_2Hb}])$ at the wavelength j. StO_2 is calculated as $[\mathrm{O_2Hb}]/([\mathrm{O_2Hb}]+[\mathrm{HHb}])$. We examine five different assumptions A1–A5 for the determination of the absorption:

A1.
$$\mu_a = -\frac{\mu_s'}{2} + \sqrt{\frac{1}{4}\mu_s'^2 + \frac{1}{3}\mathrm{Sl}_{dc}^2} - a_{\mathrm{H_2O},\lambda}\, 55.5\,\mathrm{M}\frac{p_{\mathrm{H_2O}}}{100\%}, \tag{24.6}$$

A2.
$$\mu_a = -\frac{\mu_s'}{2} + \sqrt{\frac{1}{4}\mu_s''2 + \frac{1}{3}\mathrm{Sl}_{dc}^2}, \tag{24.7}$$

A3.
$$\mu_a = \frac{\mathrm{Sl}_{dc}^2}{3\mu_s'} - a_{\mathrm{H_2O},\lambda}\, 55.5\,\mathrm{M}\frac{p_{\mathrm{H_2O}}}{100\%}, \tag{24.8}$$

A4.
$$\mu_a = \frac{\mathrm{Sl}_{dc}{}^2}{3\mu_s'}, \tag{24.9}$$

A5.
$$StO_2 = \frac{a_{\mathrm{HHb},\lambda_1} - a_{\mathrm{HHb},\lambda 2}\left(\frac{\mathrm{Sl}_{dc}(\lambda_1)}{\mathrm{Sl}_{dc}(\lambda_2)}\right)^2}{(a_{\mathrm{HHb},\lambda_1} - a_{\mathrm{O_2Hb},\lambda_1}) - (a_{\mathrm{HHb},\lambda_2} - a_{\mathrm{O_2Hb},\lambda_2})\left(\frac{\mathrm{Sl}_{dc}(\lambda_1)}{\mathrm{Sl}_{dc}(\lambda_2)}\right)^2}, \tag{24.10}$$

Equations (24.6) and (24.7) use the exact diffusion constant, while (24.8)–(24.10) use the simplified one. In (24.10), μ_s' is assumed to be constant over the wavelength. Hence, it cancels out in StO_2 calculation as shown. Equations (24.6) and (24.8) are accounting for water in tissue. Here $a_{H2O,\lambda}$ is the absorption of water at the wavelength λ in 1/(M*cm) and p_{H2O} is the amount of water in the tissue. We used 70% for the adult forearm and 90% for the neonatal head. Water contains approximately 55.5 mol atoms/l.

2.5 Statistics

Between-subject variability and within-subject variability were determined using *R* (version 2.6.1, *R* Development Core Team, Austria) with its linear mixed effects function LME. StO_2 was the random variable and subject the factor.

Table 24.1 StO_2, within-subject variability and between-subject variability for 30 newborn infants measured on the head for the five different assumptions A1–A5

	A1	A2	A3	A4	A5
$StO_2 \pm SD$ [%]	79.99±4.47	81.44±4.08	84.77±4.87	85.69±4.38	72.85±4.81
Var_{bet} [%]	4.2	3.84	4.64	4.16	4.56
Var_{within} [%]	2.76	2.55	2.73	2.51	2.83

Table 24.2 StO_2, within-subject variability and between-subject variability for eight adults measured on the forearm for the five different assumptions A1–A5

	A1	A2	A3	A4	A5
$StO_2 \pm SD$ [%]	58.14±5.69	73.85±4.77	58.99±5.67	74.21±4.76	63.49±5.11
Var_{bet} [%]	5.54	4.65	5.52	4.64	4.98
Var_{within} [%]	2.96	2.43	2.95	2.42	2.60

3 Results

For the neonatal head measurements the mean $StO_2 \pm$ standard deviation (SD), the within-subject variability (Var_{within}) and the between-subject variability (Var_{bet}) are given in Table 24.1. In Table 24.2 the values for the adult group are shown.

In both adults and the neonates assumption A5 deviates in value ~10%. In neonates including a water term (A1 vs. A2, A3 vs. A4) has a minor effect on StO_2, but the use of the exact or simplified diffusion constant (A1 vs. A3, A2 vs. A4) induces a change in StO_2 by 5%. In contrast, on the adult arm, the water term makes a large difference of ~15%, while the diffusion constant assumption does induce smaller changes. For both groups, between-subject variability and within-subject variability are smaller when not including the water term (A2 and A4). Both variables are ~0.3% larger when additionally assuming μ_s' to be constant (A5 against A2, A4).

4 Discussion and Conclusion

Our results show, that slight differences in the assumptions have a relevant influence on the final StO_2 value. This difference is also dependent on the measured tissue. The water term seems to have a smaller influence in neonates than the tissue homogeneity ($\mu_a \ll \mu_s'$). This may reflect the influence of the cerebral spinal fluid in the brain [15]. Since the water term only induces a small correction of StO_2 we believe that the water correction is more or less correct. However, the variability within and in between subjects is smaller when not including the water term, although only by ~0.2%.

Regarding the arm tissue of the adults, the concentration of lipid is higher and the water concentration is lower than for the neonatal head. While the diffusion constant assumption does not affect the StO_2 value, the water term makes a difference of ~15%. Since no real reference value exists for StO_2, it is not possible to state if one assumption is more valid than another. From a mathematical point of view, the water term has no relevant influence if the slope (24.3) is much larger than the water term. Hence, the ratio between the long and short distances is much smaller than 1. If the ratio is close to 1, the slope will be small and the water term (usually in the order of 10^{-2}) dominates. This means the ratio is closer to 1 when measuring the adult arm. This may be due to the lipid concentration in the arm, which has not been taken into account, or due to the 70% water assumption, which may be too high, or both. We calculated the body mass index (BMI) for the subjects, which correlated with the change in StO_2 (data not shown), i.e., the higher the BMI and hence the lipid concentration, the higher the change of StO_2 when taking water into account. The latter is supported by the fact that not subtracting the water lowers the variability. The additional assumption A5 lowers the StO_2 values, compared to A2 and A4. This suggests that this assumption is not valid, neither in the neonatal head nor in the adult arm.

In conclusion, we investigated the effect of the assumptions μ_s', $\mu_a \ll \mu_s'$ = constant over the wavelengths and the water contribution and their combinations when using the multi-distance approach of StO_2 calculation. We found significant differences in StO_2 and its variability, depending on the assumptions made and the tissue investigated.

Acknowledgments This work was financially supported by the Zurich Center of Integrative Human Physiology (ZIHP), University of Zurich, Switzerland. The authors would like to thank Raphael Zimmermann for very helpful discussions.

References

1. Hueber DM, Fantini S, Cerussi AE et al (1999) New optical probe designs for absolute (self-calibrating) NIR tissue hemoglobin measurements. Proc SPIE 3597:618–631
2. Matcher SJ, Kirkpatrick P, Nahid K et al (1995) Absolute quantification methods in tissue near infrared spectroscopy. Proc SPIE 2359:486–495
3. van Bel F, Lemmers P, Naulaers G (2008) Monitoring neonatal regional cerebral oxygen saturation in clinical practice: value and pitfalls. Neonatology 94(4):237–244
4. Wolf M, Greisen G (2009) Advances in near-infrared spectroscopy to study the brain of the preterm and term neonate. Clin Perinatol 36(4):807–834
5. Suzuki S, Takasaki S, Ozaki T et al (1999) A tissue oxygenation monitor using NIR spatially resolved spectroscopy. Proc SPIE 3597:582–592
6. McKeating EG, Monjardino JR, Signorini DF et al (1997) A comparison of the Invos 3100 and the Critikon 2020 near-infrared spectrophotometers as monitors of cerebral oxygenation. Anaesthesia 52(2):136–140
7. Thavasothy M, Broadhead M, Elwell C et al (2002) A comparison of cerebral oxygenation as measured by the NIRO 300 and the INVOS 5100 Near-Infrared Spectrophotometers. Anaesthesia 57(10):999–1006
8. Yoshitani K, Kawaguchi M, Tatsumi K et al (2002) A comparison of the INVOS 4100 and the NIRO 300 near-infrared spectrophotometers. Anesth Analg 94(3):586–590
9. Jenny C, Biallas M, Trajkovic I et al (2011) Reproducibility of cerebral tissue oxygenation saturation measurements by near infrared spectroscopy in newborn infants. J Biomed Opt 16(9): 097004

10. Haensse D, Szabo P, Brown D et al (2005) New multichannel near infrared spectrophotometry system for functional studies of the brain in adults and neonates. Opt Express 13(12): 4525–4538
11. Muehlemann T, Haensse D, Wolf M (2008) Wireless miniaturized in-vivo near infrared imaging. Opt Express 16(14):10323–10330
12. Matcher SJ, Elwell CE, Cooper CE et al (1995) Performance comparison of several published tissue near-infrared spectroscopy algorithms. Anal Biochem 227(1):54–68
13. Matcher SJ, Cope M, Delpy DT (1997) In vivo measurements of the wavelength dependence of tissue-scattering coefficients between 760 and 900 nm measured with time-resolved spectroscopy. Appl Opt 36(1):386–396
14. Arri SJ, Muehlemann T, Biallas M et al (2011) Precision of cerebral oxygenation and hemoglobin concentration measurements in neonates measured by near-infrared spectroscopy. J Biomed Opt 16(4):047005
15. Wolf M, Keel M, Dietz V et al (1999) The influence of a clear layer on near-infrared spectrophotometry measurements using a liquid neonatal head phantom. Phys Med Biol 44(7): 1743–1753

Chapter 25
The Effect of Sudden Depressurization on Pilots at Cruising Altitude

Thomas Muehlemann, Lisa Holper, Juergen Wenzel, Martin Wittkowski, and Martin Wolf

Abstract The standard flight level for commercial airliners is ~12 km (40 kft; air pressure: ~ 200 hPa), the maximum certification altitude of modern airliners may be as high as 43–45 kft. Loss of structural integrity of an airplane may result in sudden depressurization of the cabin potentially leading to hypoxia with loss of consciousness of the pilots. Specialized breathing masks supply the pilots with oxygen. The aim of this study was to experimentally simulate such sudden depressurization to maximum design altitude in a pressure chamber while measuring the arterial and brain oxygenation saturation (SaO_2 and StO_2) of the pilots. Ten healthy subjects with a median age of 50 (range 29–70) years were placed in a pressure chamber, breathing air from a cockpit mask. Pressure was reduced from 753 to 148 hPa within 20 s, and the test mask was switched to pure O_2 within 2 s after initiation of depressurization. During the whole procedure SaO_2 and StO_2 were measured by pulse oximetry, respectively near-infrared spectroscopy (NIRS; in-house built prototype) of the left frontal cortex. During the depressurization the SaO_2 dropped from median 93% (range 91–98%) to 78% (62–92%) by 16% (6–30%), while StO_2 decreased from 62% (47–67%) to 57% (43–62%) by 5% (3–14%). Considerable drops in oxygenation were observed during sudden depressurization. The inter-subject variability was high, for SaO_2 depending on the subjects' ability to preoxygenate before the depressurization. The drop in StO_2 was lower than the one in SaO_2 maybe due to compensation in blood flow.

T. Muehlemann • L. Holper • M. Wolf, Ph.D. (✉)
Biomedical Optics Research Laboratory, Division of Neonatology, Department of Obstetrics and Gynecology, University Hospital Zurich, 8091 Zurich, Switzerland
e-mail: martin.wolf@usz.ch

J. Wenzel • M. Wittkowski
Department of Flight Physiology, DLR-Institute of Aerospace Medicine, Cologne, Germany

W.J. Welch et al. (eds.), *Oxygen Transport to Tissue XXXIV*, Advances in Experimental Medicine and Biology 765, DOI 10.1007/978-1-4614-4989-8_25,

Keywords Depressurization • High altitude • Near-infrared spectroscopy • Pilot • Tissue oxygen saturation

1 Introduction

The standard flight level for commercial airliners is 10–12 km (= 42 kft). The maximum certification altitude of modern airliners may be as high as 43–45 kft. The outside air pressure at the 42 kft altitude corresponds to ~170 hPa. This pressure is too low for human subjects to maintain consciousness and consequently the air pressure inside the cabin is kept at a higher level of at least 753 hPa, which corresponds to an altitude of 2.5 km or 8 kft. Loss of structural integrity of an airplane, i.e., a hole, may result in sudden depressurization of the cabin potentially leading to hypoxia, e.g., during Qantas flight QF30 from Hong-Kong to Melbourne in July 2008 an oxygen tank exploded at a cruising altitude of 29 kft and tore a large hole into the airliner's pressure hull, which led to a sudden depressurization. The passengers and crew used oxygen masks, and the pilots immediately started a descent to safe levels of air pressure and nobody was harmed.

The main danger during depressurization is loss of consciousness of the pilots. In cases of sudden depressurization, to protect pilots against hypoxia, specialized breathing masks supply the pilots with oxygen. In some regulations the Pilot Flying has to wear a breathing mask when the plane is at very high altitude.

The first aim of this study was to experimentally simulate such a sudden depressurization to maximum design altitude in a pressure chamber and to determine, whether a paradigm of switching the gas inside the mask to 100% oxygen within 2 s of the depressurization would enable the pilots to maintain consciousness. The second aim was to measure the arterial oxygen saturation (SaO_2) and tissue oxygen saturation (StO_2) of the brain during the depressurization.

2 Subjects and Methods

Ten healthy subjects with a median age of 50 (range 29–70) years, weight of 82 (70–110) kg, height of 181 (170–190) cm and body mass index of 25.0 (22.1–31.8) were included in the study, after giving written informed consent. The study was approved by the local ethical committee.

The subjects were placed in a pressure chamber breathing air from a modified cockpit mask. Sensors of different NONIN based pulse oximeters were attached to the fingers of the subject. The pulse oximeters measured SaO_2. A wireless near-infrared spectroscopy (NIRS) instrument was attached to the left forehead of the subject using an elastic bandage. This instrument was in-house built at the University Hospital Zurich and described in detail previously [1]. It measured StO_2.

Inside the chamber the pressure was set to 753 hPa with the subjects breathing pure oxygen for 30 min as protection against Decompression Sickness (DCS); subsequently they were switched to air at this level for at least 3 min to allow for pulmonary equilibration. The pressure inside the chamber was then suddenly reduced to 148 hPa within 20 s. The test mask was switched to pure oxygen within 2 s after initiation of depressurization. During the whole process SaO_2 and cerebral StO_2 were measured continuously and recorded.

Significances were tested using paired Wilcoxon signed rank test and linear and nonlinear correlations were tested using Pearson's correlation coefficient and Spearman's rho, respectively. Statistics were calculated using SPSS version 19.

3 Results

All subjects remained conscious throughout the experiment.

The results are displayed in Fig. 25.1, which shows the maximal decrease in SaO_2 and cerebral StO_2 during depressurization and compares it to the initial level. Figure 25.2 compares the change in SaO_2 to the change in cerebral StO_2.

During the depressurization the SaO_2 dropped significantly ($p=0.005$) from median 93% (range 91–98%) to 78% (62–92%) by 16% (6–30%), while StO_2 decreased significantly ($p=0.008$) from 62% (47–67%) to 57% (43–62%) by 5% (3–14%) (Fig. 25.1). The decrease in SaO_2 was significantly correlated with the initial value of the SaO_2 at 753 hPa ($p=0.025$ for both Pearson and Spearman). The change in StO_2 was not significantly related to the initial StO_2 value.

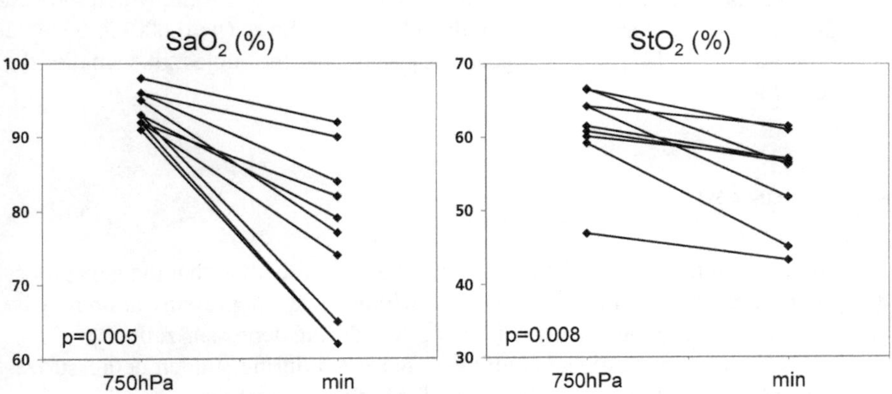

Fig. 25.1 The figure on the *left* displays the change in arterial oxygen saturation (SaO_2) from the initial level at 753 hPa and the minimal value obtained for each subject during the depressurization. Please note the tendency that subjects with an initially lower level of SaO_2 tend to desaturate more during depressurization. The *right panel* displays the change in cerebral tissue oxygen saturation (StO_2) measured by NIRS. In one subject, the data of the NIRS measurement was discarded due to a technical problem

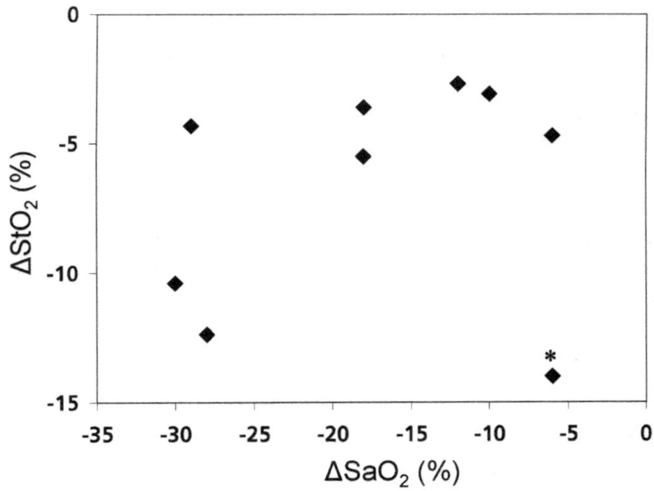

Fig. 25.2 Change in the arterial oxygen saturation (SaO_2) compared to the change in cerebral tissue oxygen saturation (StO_2) during the depressurization. In principle it would be expected that the size of the two changes is related, i.e., the points should be clustered near the diagonal from *left bottom* to *right top*. However, there were outliers and there was no significant relation between the two. One outlier marked with *asterisk* can be explained: the subjects hyperventilated strongly, which leads to cerebral vasoconstriction and lower oxygen supply

The correlation between the changes in SaO_2 and cerebral StO_2 was not significant, neither linearly nor nonlinearly (Fig. 25.2). There was also no correlation between age, height, weight, and body mass index and the initial and minimal SaO_2 and cerebral StO_2 or the change in these parameters, except for the height, which showed a highly significant correlation with the change in StO_2 ($p=0.00043$, $r=0.92$ (Pearson) and $p=0.013$, $\rho=0.78$ (Spearman). The taller the subject, the smaller the change in StO_2.

4 Discussion

All subjects maintained consciousness. This is a first indication that the new procedure to provide 100% O_2 in the pilot's mask within 2 s of a depressurization may be effective in preventing unconsciousness of pilots during depressurization.

Considerable drops in oxygenation were observed during sudden depressurization. The inter-subject variability was high for both SaO_2 and StO_2.

For SaO_2 the size of the drop depended significantly on the initial SaO_2 level. Thus, subjects who had a lower initial SaO_2 also had a larger decrease in SaO_2. This means that the level of oxygenation that a subject is able to maintain at 753 hPa varies and is an important determinant, which affects the desaturation during depressurization. A higher initial preoxygenation is an advantage during depressurization, when the pulmonary air vigorously expands and impedes breathing. In addition, the subjects maintaining a higher initial level of SaO_2 may also show a higher respiratory drive enabling them to more efficient gas exchange in the early phase of depressurization.

There is a strong variability in the initial values of StO_2. In other publications also relatively wide ranges of StO_2 values were observed in healthy subjects, e.g., between 63.3% and 79.1% using a NIRO 300 (Hamamatsu Photonics, Hamamatsu, Japan) [2]. Our initial values were lower, which may be explained by the higher altitude of 8 kft, i.e., 753 hPa, at which our subjects were measured, while the subjects in the mentioned publication [2] were presumably measured at a significantly lower altitude. There is one outlier with a low StO_2 of 46.9%. It is difficult to explain this low value. There are no visible technical or demographic differences between this measurement and others and consequently there is no obvious reason for classifying this measurement as an artifact. Possible explanations are also discussed below.

There was also a high variability in the decrease in StO_2 during the depressurization. In contrast to SaO_2, the size of the decrease was not related to the initial level of StO_2. However, there was a highly significant correlation of the change in StO_2 and the height of the subject, i.e., in larger subjects the brain desaturated much less than in smaller subjects. This correlation is surprising, especially since the change in SaO_2 is unrelated to the height. It may indicate that anatomical features play a role, affecting either the oxygenation of the brain or the principle of the measurement (discussed also below).

The StO_2 decreased considerably less than the SaO_2 during depressurization. In principle, it would also be expected that the size of the change in the cerebral StO_2 is proportional to the change in SaO_2. Although to some degree this seems visible in Fig. 25.2, this correlation is by far not significant and there are two important outliers. One outlier (Fig. 25.2) can be explained: the subject hyperventilated strongly, which leads to cerebral vasoconstriction and lower oxygen supply [3]. In previous studies it was found that the size of the changes in StO_2 depends on the algorithm used to calculate StO_2 and, depending on this change, was slightly to considerably smaller than changes in SaO_2 [4]. In principle, there are several possible explanations, why the change in StO_2 was smaller than the change in SaO_2:

- The drop in SaO_2 was short and occurred only for a few seconds (full width at half minimum <10 s). Since the brain contains a large volume of blood, short changes in SaO_2 are diluted in a pool of blood, which corresponds to an integration over a certain period of time and thus quite naturally to a reduction in the change of StO_2. Analogously, one of the pulse oximeters, which averaged the SaO_2 values, showed smaller changes in SaO_2.
- The brain may compensate for the lack of oxygenation by increasing the cerebral blood flow. This would lead to a smaller size of the change in StO_2. In principle, we measured also the total hemoglobin correlation, which should correlate with the blood flow. However, the total hemoglobin measurement was not reliable in several subjects and thus, there is no evidence that a change in blood flow occurred, although this seems plausible.

- Due to a reduced diffusion pressure for O_2 during the period with low oxygenation, the cerebral metabolic rate of O_2 may have been reduced. This would again lead to a smaller change in StO_2.
- The NIRS instrument was originally designed for neonates and sheep, whose extracerebral tissue layers are thinner compared to adult human subjects. The NIRS instrument consequently employed relatively short source–detector distances of 1.5 and 2.5 cm, which do not penetrate tissue as deeply as longer distances. Thus, there may be a contribution of extracerebral tissue to our StO_2 values. Own unpublished Monte Carlo simulations showed that this contribution is probably small. In addition, we employed a multidistance approach which removes the influence of superficial tissue [5]. Still it cannot be excluded that superficial tissue may have influenced our StO_2 values and in particular the correlation with the height of the subjects may indicate that there is an influence of anatomical features, e.g., height affects head circumference, i.e., head curvature [6], but not skull thickness [7]. Head curvature has no influence on StO_2 [8]. It is difficult to estimate how such an influence of superficial tissue affects our values. In our opinion, it seems plausible that a contribution of superficial tissue would lead to a higher variability between subjects both concerning the size of the change and the initial values. But it seems implausible to us that this would explain the smaller size of the change in StO_2 compared to SaO_2 during depressurization or why taller subjects would show smaller changes in StO_2.
- There may be shifts in the relation of the arterial and venous blood volume that affect the StO_2 significantly [9].

It is difficult to estimate which of these reasons is the most relevant for explaining our data and it is likely that several of them contributed to the results.

5 Conclusions

All subjects remained conscious during the depressurization. Considerable drops in SaO_2 and cerebral StO_2 were observed during sudden depressurization. The intersubject variability was high and for SaO_2 it depended on the oxygenation level before depressurization, which indicates that the preoxygenation is an important factor. The decrease in StO_2 was smaller than the one in SaO_2, maybe due to compensation in blood flow.

Acknowledgments The investigations described here were performed in conjunction with a study supported by Airbus Germany and Zodiac Oxygen Systems, France.

References

1. Muehlemann T, Haensse D, Wolf M (2008) Wireless miniaturized in-vivo near infrared imaging. Opt Express 16(14):10323–10330

2. Quaresima V, Ferrari M, Torricelli A et al (2005) Bilateral prefrontal cortex oxygenation responses to a verbal fluency task: a multichannel time-resolved near-infrared topography study. J Biomed Opt 10(1):11012
3. Tisdall MM, Taylor C, Tachtsidis I et al (2009) The effect on cerebral tissue oxygenation index of changes in the concentrations of inspired oxygen and end-tidal carbon dioxide in healthy adult volunteers. Anesth Analg 109(3):906–913
4. Wolf M, von Siebenthal K, Keel M et al (2000) Tissue oxygen saturation measured by near infrared spectrophotometry correlates with arterial oxygen saturation during induced oxygenation changes in neonates. Physiol Meas 21(4):481–491
5. Choi J, Wolf M, Toronov V et al (2004) Noninvasive determination of the optical properties of adult brain: near-infrared spectroscopy approach. J Biomed Opt 9:221–229
6. Bale SJ, Amos CI, Parry DM et al (1991) Relationship between head circumference and height in normal adults and in the nevoid basal cell carcinoma syndrome and neurofibromatosis type I. Am J Med Genet 40(2):206–210
7. Lynnerup N (2001) Cranial thickness in relation to age, sex and general body build in a Danish forensic sample. Forensic Sci Int 117(1–2):45–51
8. Wallace D, Barbieri B, Hintz SR (2000) Neonatal cerebral oxygenation measurements and the effects of curvature on frequency domain multiple distance near infrared spectroscopy. In Biomedical Optical Spectroscopy and Diagnostics. Lit (ed) Vol 38 of OSA Trends in optics and photonics, paper SuF1. http://www.opticsinfobase.org/abstract.cfm?URL: BOSD-2000-SuF1
9. Tachtsidis I, Tisdall M, Delpy DT et al (2008) Measurement of cerebral tissue oxygenation in young healthy volunteers during acetazolamide provocation: a transcranial Doppler and near-infrared spectroscopy investigation. Adv Exp Med Biol 614:389–396

Chapter 26
Hypoxia in the Diabetic Kidney Is Independent of Advanced Glycation End-Products

Lina Nordquist, Per Liss, Angelica Fasching, Peter Hansell, and Fredrik Palm

Abstract Sustained hyperglycemia is closely associated with increased risk to develop nephropathy. We have previously reported alterations in the intrarenal oxygen metabolism already after the early onset of diabetes. Furthermore, formation of advanced glycation end-products (AGE) is postulated as a major contributor to diabetic nephropathy. We therefore investigated the possible relationship between altered oxygen metabolism and AGE in diabetic kidneys.

Normoglycemic and streptozotocin-diabetic rats with and without chronic treatment with aminoguanidine (AGE inhibitor; 600 mg/kg bw/24 h in drinking water) or L-N^6-(1-Iminoethyl)lysine (L-NIL, iNOS inhibitor, 1 mg/kg bw/24 h in drinking water) were studied 2 weeks after induction of diabetes. Glomerular filtration rate (GFR) was estimated by inulin clearance, oxygen tension (pO_2) and interstitial pH by microelectrodes and regional renal blood flow (RBF) by laser-Doppler. Histological changes were evaluated on fixed tissue.

Glomerular hyperfiltration was unaffected by aminoguanidine, whereas L-NIL normalized GFR in diabetic rats. pO_2 and interstitial pH, but not RBF, were lower in both kidney cortex and medulla compared to control rats, but was unaffected by both chronic treatments. Urinary protein excretion was higher in diabetic rats and unaffected by L-NIL, whereas aminoguanidine paradoxically increased this parameter. Damage scores were similar in all groups.

L. Nordquist, Ph.D. (✉) • A. Fasching • P. Hansell
Division of Integrative Physiology, Departments of Medical Cell Biology, Uppsala University, Biomedical Center, Husargatan3, Box 571, 75123 Uppsala, Sweden
e-mail: Lina.Nordquist@mcb.uu.se

P. Liss
Radiology, Oncology and Radiation Science, Uppsala University, Uppsala, Sweden

F. Palm
Division of Integrative Physiology, Departments of Medical Cell Biology, Uppsala University, Biomedical Center, Husargatan3, Box 571, 75123 Uppsala, Sweden

Department of Medical and Health Sciences, Linköping University, Linköping, Sweden

W.J. Welch et al. (eds.), *Oxygen Transport to Tissue XXXIV*, Advances in Experimental Medicine and Biology 765, DOI 10.1007/978-1-4614-4989-8_26,

In conclusion, diabetes-induced alterations in intrarenal oxygen metabolism are independent of the AGE pathway, and precede any morphological changes. These findings highlight the early stage of diabetes as being a metabolic disorder also in the kidney.

Keywords AGE • Diabetes • Kidney • Oxygen tension

1 Introduction

Diabetic nephropathy is a major cause of morbidity and mortality. Approximately 30% of all type 1 diabetic patients will eventually develop diabetic nephropathy [1], and 20% diagnosed with end-stage renal failure [2]. Diabetes is the most ubiquitous reason patients require dialysis and/or kidney transplant [3]. There are numerous hypotheses for the mechanisms involved in onset and progression of diabetes-induced renal disturbances. Early after onset of diabetes, the kidney displays functional defects such as increased glomerular filtration rate (GFR) and oxygen consumption (QO_2), resulting in decreased oxygen tension (pO_2) in the diabetic kidney [4, 5]. The diabetes-induced renal hypoxia has been postulated as a major contributor to the development of diabetic nephropathy [6].

A unifying mechanism was recently suggested by Nishikawa and coworkers [7]. According to this hypothesis, increased reactive oxygen species (ROS) at the level of the mitochondria are linked to at least three major pathways to diabetes-induced damage, namely activation of the polyol pathway, increased levels of advanced glycation end products (AGE) and activation of protein kinase C (PKC). Of these pathological pathways, however, it is unclear which pathways induce what alteration. We have previously investigated ROS, as well as polyol pathway, and found that inhibition of either pathway prevents the diabetes-induced intrarenal hypoxia [4, 5].

As of today, there is no specific inhibitor of AGE accumulation commercially available. However, aminoguanidine inhibits both AGE formation and inducible nitric oxide synthase (iNOS) [8]. Both AGE and iNOS are thought to participate in the development of diabetic nephropathy [9]. Aminoguanidine has been shown to ameliorate several of the complications commonly associated with prolonged diabetes, including glomerulosclerosis and medullary pathology [10, 11]. However, it remains to be determined if these beneficial effects originate from prevention of the intrarenal hypoxia. We therefore investigated the effect of chronic treatment with aminoguanidine to prevent AGE formation, or L-N^6-(1-Iminoethyl)lysine (L-NIL) to inhibit iNOS, on kidney hypoxia in the diabetic rat kidney.

2 Methods

All chemicals were from Sigma-Aldrich (St. Louis, MO, USA) and of highest grade available if not otherwise stated. Male Wistar–Furth rats, weighing 280–300 g, were purchased from B&K Universal (Sollentuna, Sweden) and had free access to water and rat chow (R3, Ewos, Södertälje, Sweden) throughout the study. All experiments were performed in accordance with the National Institutes of Health guidelines for use and care of laboratory animals and approved by the local Animal Care and Use Committee.

Diabetes mellitus was induced by an intravenous injection of streptozotocin (STZ, 45 mg/kg). Blood glucose concentrations were determined from blood samples obtained from the cut tip of the tail in all animals (MediSense, Bedford, MA, USA). Animals were considered diabetic if blood glucose concentrations increased to ≥15 mmol/l within 48 h after STZ-injection. Animals were divided into five experimental groups: Untreated, normoglycemic controls, normoglycemic controls receiving aminoguanidine in the drinking water (600 mg/kg bw/24 h), untreated diabetics, diabetics receiving aminoguanidine in drinking water (600 mg/kg bw/24 h), and diabetics receiving L-NIL (1 mg/kg bw/24 h) throughout the course of diabetes. All animals were subjected to measurements of renal blood flow, pO_2, interstitial pH, and urinary excretion of electrolytes and proteins 14 days after induction of diabetes as previously described [5]. Picrosirius red staining and microscopic evaluation was performed as previously described [12]. Three slices from each kidney was scored by the same blinded scorer and the average score from each kidney considered as one experiment in the statistical analysis.

Statistical analysis: All values are given as means ± SEM. Multiple comparisons were performed using one way analysis of variance followed by Fisher's protected least significant difference test (Statview, Abacus Concepts, Berkeley, CA). Nonparametric data was analyzed using Kruskal–Wallis' test for multiple comparisons (GraphPad Software Inc.). $P<0.05$ was considered statistically significant.

3 Results

Diabetic animals gained less weight compared to age-matched normoglycemic controls, although both chronic treatments increased body weights of the diabetic animals. All diabetic groups displayed hyperglycemia compared to controls, and both treatments reduced blood glucose levels in diabetes (Table 26.1). MAP and hematocrit (Hct) were similar in all groups.

Untreated and aminoguanidine-treated diabetic animals displayed elevated GFR compared to controls, whereas L-NIL treatment normalized the diabetes-induced glomerular hyperfiltration (Fig. 26.1). Cortical and medullary pO_2 (Fig. 26.2) and interstitial pH (Fig. 26.3) were all lower in diabetic animals and none of the chronic treatments had any effect. Cortical and medullary RBF were similar in all investigated groups (Fig. 26.4).

Urine production was higher in all diabetic groups compared to controls, and none of the treatments affected this parameter (Table 26.2). Urinary Na^+ excretion was higher in untreated diabetic animals, whereas neither of the treated diabetic groups differed from the untreated control group. Urinary K^+ excretion was higher in untreated and aminoguanidine-treated diabetic animals, whereas diabetic animals receiving L-NIL did not differ from controls. Urinary excretion of osmotically

Table 26.1 Body weight (BW), blood glucose, mean arterial blood pressure (MAP) and hematocrit (Hct) in normoglycemic control and diabetic rats with and without chronic treatment with the AGE inhibitor aminoguanidine (AG) or the iNOS inhibitor L-N[6]-(1-Iminoethyl)lysine (L-NIL)

	Treatment	*N*	BW (g)	Blood glucose (mmol/l)	MAP (mmHg)	Hct (%)
Control	–	9	335±2	6.5±0.3	130±3	44±1
	AG	8	324±3	5.5±0.2	119±5	44±0
Diabetes	–	8	271±7*	26.7±0.7*	119±5	44±1
	AG	9	287±4*#	20.5±0.7*#	130±2	43±1
	L-NIL	6	301±3*#	19.6±1.0*#	125±4	44±1

*Denotes $P<0.05$ when compared to untreated normoglycemic control
#Denotes $P<0.05$ when compared to untreated diabetic group

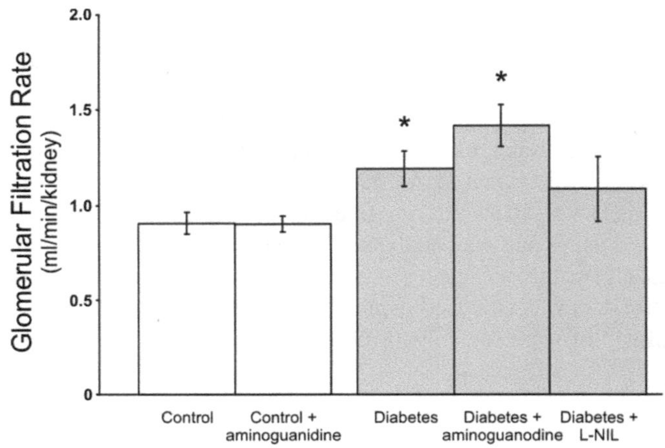

Fig. 26.1 Glomerular filtration rate in normoglycemic controls and diabetic rats with and without the chronic treatment of aminoguanidine or L-N[6]-(1-Iminoethyl)lysine (L-NIL). *Asterisk* denotes $P<0.05$ vs. untreated control group

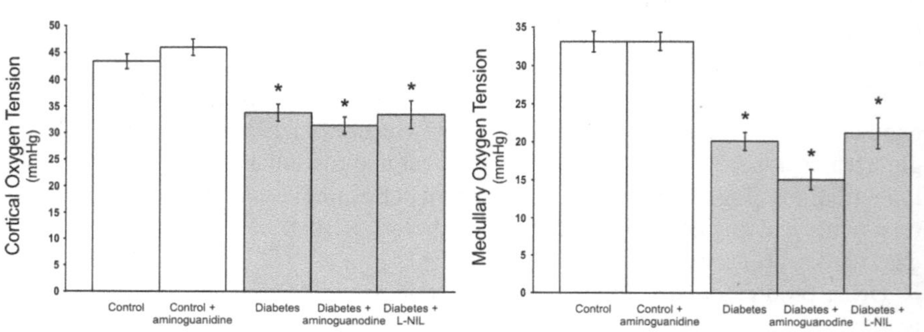

Fig. 26.2 Oxygen tension in kidney cortex (*left*) and medulla (*right*) in normoglycemic controls and diabetic rats with and without the chronic treatment of aminoguanidine or L-N[6]-(1-Iminoethyl) lysine (L-NIL). *Asterisk* denotes $P<0.05$ vs. untreated control group

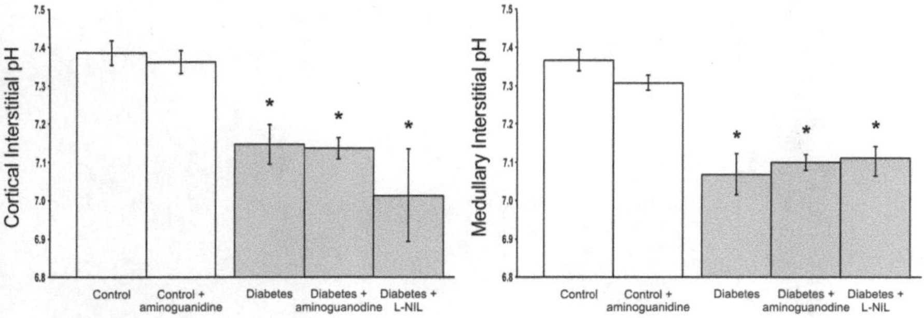

Fig. 26.3 Interstitial pH in kidney cortex (*left*) and medulla (*right*) in normoglycemic controls and diabetic rats with and without the chronic treatment of aminoguanidine or L-N^6-(1-Iminoethyl) lysine (L-NIL). *Asterisk* denotes $P<0.05$ vs. untreated control group

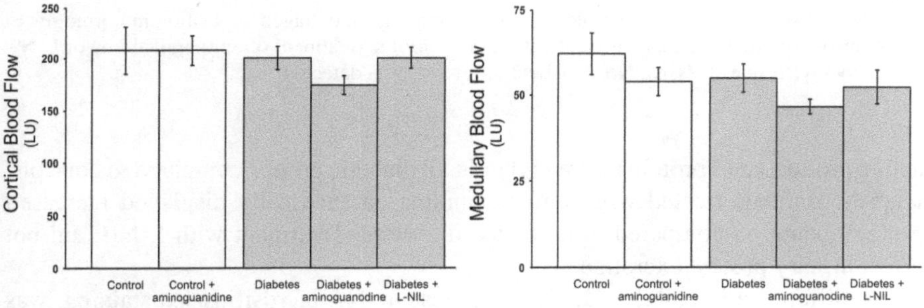

Fig. 26.4 Blood flow in kidney cortex (*left*) and medulla (*right*) in normoglycemic controls and diabetic rats with and without the chronic treatment of aminoguanidine or L-N^6-(1-Iminoethyl) lysine (L-NIL)

Table 26.2 Urine production and urinary excretion of Na^+, K^+, osmotic active products and protein in normoglycemic control and diabetic rats with and without chronic treatment with the AGE inhibitor aminoguanidine (AG) or the iNOS inhibitor L-N^6-(1-Iminoethyl)lysine (L-NIL)

	Treatment	N	Urine flow (ml/min)	Na^+ excretion (pmol/min)	K^+ excretion (pmol/min)	Protein excretion (µg/min)
Control	–	9	0.9±0.2	48±18	84±45	44±4
	AG	9	0.9±0.1	42±7	153±37	45±5
Diabetes	–	8	14±4*	156±42*	344±99*	69±7*
	AG	9	14±2*	82±20	419±78*	113±9*#
	L-NIL	6	12±3*	17±3#	258±127	63±9*

*Denotes $P<0.05$ when compared to untreated normoglycemic control

#Denotes $P<0.05$ when compared to untreated diabetic group

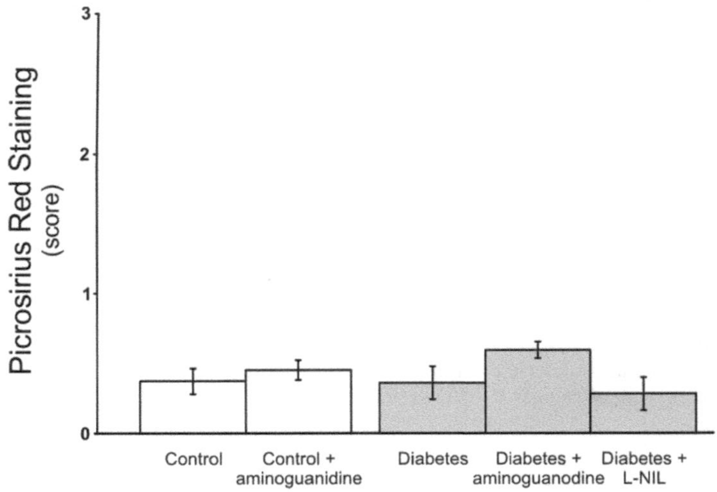

Fig. 26.5 Histological score from picrosirius red staining for collagen deposition in normoglycemic controls and diabetic rats with and without the chronic treatment of aminoguanidine or L-N^6-(1-Iminoethyl)lysine (L-NIL). No statistical difference was detected

active products and proteins were higher in all diabetic groups compared to controls. Diabetic animals treated with aminoguanidine paradoxically displayed increased protein excretion compared to untreated diabetics. Treatment with L-NIL did not affect urinary protein excretion.

No difference in collagen deposition, measured as picrosirius red staining, was detected between any of the investigated groups (Fig. 26.5). Aminoguanidine treatment to normoglycemic controls did not affect any of the investigated parameters.

4 Discussion

The main finding of this study is that inhibition of AGE did not ameliorate diabetes-induced renal tissue hypoxia, hyperfiltration, proteinuria, or renal interstitial acidosis. A 14-day period of sustained hyperglycemia impaired cortical as well as medullary pO_2, without affecting oxygen delivery (i.e., unaltered RBF). In the present study, we showed that these alterations were not prevented by AGE inhibition. Several studies have shown that diabetic kidneys have reduced tissue pO_2, which is linked to increased QO_2 [4]. We have previously shown that the diabetic kidney has increased lactate levels, which is reflected in lower intrarenal interstitial pH [5]. Neither inhibition of AGE formation nor iNOS activity had any effect on the interstitial pH in the diabetic kidney. Furthermore, diabetes-induced urinary protein excretion was not prevented by inhibition of either AGE or iNOS. Actually, AGE inhibition increased the proteinuria about twofold compared to the untreated diabetic rats, but the mechanism for this detrimental effect on kidney function is currently unknown. However, we can conclude that the AGE pathway is not involved in the diabetes-induced increase in kidney QO_2. This is in accordance with a previous study on diabetic retinopathy [13]. Taken together, this implies that the AGE pathway is not responsible for the alterations in intrarenal metabolism observed in rats with early diabetes mellitus. Alterations in the intrarenal oxygen metabolism and pO_2 therefore appear to either precede AGE formation or occur in parallel with activation of the AGE pathway.

There was no difference in collagen deposition between any of the groups in this study, which is not surprising since only very minor fibrotic changes will occur after 2 weeks of diabetes in rats. Inhibition of AGE in diabetic rats has been reported to prevent structural as well as functional manifestations of diabetic nephropathy [14], and especially collagen deposition in long-term diabetes [15]. Thus, the lack of histological alterations in the present study is likely due to the short time of diabetes.

AGE are the result of nonenzymatic glycation, where a glucose carbonyl group reacts with a free amino group. AGE will modify proteins, alter extracellular matrix function, and stimulate production of cytokines and ROS-specific receptors. In 1993, Mitsuhashi et al. showed AGE accumulation in kidney cortex from STZ diabetic rats [16]. Formation of AGE induces oxidative stress from the mitochondria in the kidney [17], and AGE-modified plasma proteins cause vascular pathology [14]. ROS per se can induce AGE formation, making AGE a possible cause as well as a consequence of ROS formation [18].

In a type 2 diabetes model, the renoprotective effects of angiotensin receptor and calcium channel blockers, lipid- (bezafibrate) or glucose-lowering (pioglitazone) agents as well as the HIF activator cobalt chloride were all shown to reduce intrarenal AGE accumulation [19]. In 2004, a placebo-controlled, double-blind multicenter trial with 690 type-1 diabetic patients led to the first clinical report showing attenuation of renal diabetes complications by AGE inhibition [20]. These results implied an important role for AGE in progression of diabetic nephropathy, but there is currently still no drug available for these patients.

In the present study, chronic iNOS inhibition reduced the diabetes-induced glomerular hyperfiltration. It should be noted that site-specific, excessive nitric oxide (NO) production has been shown to contribute to diabetic glomerular hyperfiltration [21]. Renal iNOS is activated in diabetic rats, predominantly by inflammatory cells. Of the four pathways in the unifying hypothesis, two (PKC inhibition and inhibition of the polyol pathway) may prevent glucose-induced iNOS enhancement [22]. Acute iNOS inhibition did not reduce diabetic hyperfiltration [23], and it should be noted that our finding is in disagreement with the previous study by Schwartz et al. [24]. However, approximately 40 times higher doses of aminoguanidine were used in that study, which might provide an explanation for the discrepancy.

Blood glucose and body weight increased in both treatment groups. This may be explained by beneficial effects on residual insulin secretion [25]. iNOS-derived NO may induce beta cell dysfunction and silencing of iNOS has been reported to counteract islet apoptosis [26].

In conclusion, early alterations in intrarenal oxygen metabolism commonly observed in diabetic rats are independent of the AGE pathway, and precede histological changes. These findings highlight the early stages of diabetes as a metabolic disorder also in the kidney, and imply that such alterations may provide an early indication for the development of chronic kidney disease.

Acknowledgments This study was supported by the Swedish Medical Research Council (10840), The Swedish Society for Medical Research, The Magnus Bergvall Foundation and NIH K99/R00 grant (DK077858).

References

1. Schjoedt KJ, Hansen HP, Tarnow L et al (2008) Long-term prevention of diabetic nephropathy: an audit. Diabetologia 51:956–961
2. The Diabetes Control and Complications Trial Research Group (1993) The effect of intensive treatment of diabetes on the development and progression of long-term complications in insulin-dependent diabetes mellitus. N Engl J Med 329:977–986
3. Yip JW, Jones SL, Wiseman MJ et al (1996) Glomerular hyperfiltration in the prediction of nephropathy in IDDM: a 10-year follow-up study. Diabetes 45:1729–1733
4. Palm F, Cederberg J, Hansell P et al (2003) Reactive oxygen species cause diabetes-induced decrease in renal oxygen tension. Diabetologia 46:1153–1160
5. Palm F, Hansell P, Ronquist G et al (2004) Polyol-pathway-dependent disturbances in renal medullary metabolism in experimental insulin-deficient diabetes mellitus in rats. Diabetologia 47:1223–1231
6. Miyata T, de Strihou CY (2010) Diabetic nephropathy: a disorder of oxygen metabolism? Nat Rev Nephrol 6:83–95
7. Nishikawa T, Edelstein D, Du XL et al (2000) Normalizing mitochondrial superoxide production blocks three pathways of hyperglycaemic damage. Nature 404:787–790
8. He C, Sabol J, Mitsuhashi T et al (1999) Dietary glycotoxins: inhibition of reactive products by aminoguanidine facilitates renal clearance and reduces tissue sequestration. Diabetes 48: 1308–1315
9. Inagi R, Yamamoto Y, Nangaku M et al (2006) A severe diabetic nephropathy model with early development of nodule-like lesions induced by megsin overexpression in RAGE/iNOS transgenic mice. Diabetes 55:356–366
10. Kern TS, Engerman RL (2001) Pharmacological inhibition of diabetic retinopathy: aminoguanidine and aspirin. Diabetes 50:1636–1642
11. Wilkinson-Berka JL, Kelly DJ, Koerner SM et al (2002) ALT-946 and aminoguanidine, inhibitors of advanced glycation, improve severe nephropathy in the diabetic transgenic (mREN-2)27 rat. Diabetes 51:3283–3289
12. Junqueira LC, Bignolas G, Brentani RR (1979) Picrosirius staining plus polarization microscopy, a specific method for collagen detection in tissue sections. Histochem J 11:447–455
13. Pulido JS (1996) Experimental nonenzymatic glycosylation of vitreous collagens occurs by two pathways. Trans Am Ophthalmol Soc 94:1029–1072
14. Brownlee M (1992) Glycation products and the pathogenesis of diabetic complications. Diabetes Care 15:1835–1843
15. Kelly DJ, Gilbert RE, Cox AJ et al (2001) Aminoguanidine ameliorates overexpression of prosclerotic growth factors and collagen deposition in experimental diabetic nephropathy. J Am Soc Nephrol 12:2098–2107
16. Mitsuhashi T, Nakayama H, Itoh T et al (1993) Immunochemical detection of advanced glycation end products in renal cortex from STZ-induced diabetic rat. Diabetes 42:826–832

17. Rosca MG, Mustata TG, Kinter MT et al (2005) Glycation of mitochondrial proteins from diabetic rat kidney is associated with excess superoxide formation. Am J Physiol Renal Physiol 289:F420–430
18. Baynes JW (1991) Role of oxidative stress in development of complications in diabetes. Diabetes 40:405–412
19. Miyata T, Izuhara Y (2008) Inhibition of advanced glycation end products: an implicit goal in clinical medicine for the treatment of diabetic nephropathy? Ann New York Acad Sci 1126: 141–146
20. Bolton WK, Cattran DC, Williams ME et al (2004) Randomized trial of an inhibitor of formation of advanced glycation end products in diabetic nephropathy. Am J Nephrol 24:32–40
21. Bank N, Aynedjian HS (1993) Role of EDRF (nitric oxide) in diabetic renal hyperfiltration. Kidney Int 43:1306–1312
22. Noh H, Ha H, Yu MR et al (2002) High glucose increases inducible NO production in cultured rat mesangial cells. Possible role in fibronectin production. Nephron 90:78–85
23. Veelken R, Hilgers KF, Hartner A et al (2000) Nitric oxide synthase isoforms and glomerular hyperfiltration in early diabetic nephropathy. J Am Soc Nephrol 11:71–79
24. Schwartz D, Schwartz IF, Blantz RC (2001) An analysis of renal nitric oxide contribution to hyperfiltration in diabetic rats. J Lab Clin Med 137:107–114
25. Zhao Z, Zhao C, Zhang XH et al (2009) Advanced glycation end products inhibit glucose-stimulated insulin secretion through nitric oxide-dependent inhibition of cytochrome c oxidase and adenosine triphosphate synthesis. Endocrinology 150:2569–2576
26. Bai-Feng L, Yong-Feng L, Ying C (2010) Silencing inducible nitric oxide synthase protects rat pancreatic islet. Diabetes Res Clin Pract 89:268–275

Chapter 27
Tumor Oxygen Measurements and Personalized Medicine

Paul Okunieff, Walter O'Dell, Mei Zhang, Lurong Zhang, and David Maguire

Abstract Tumor hypoxia is probably the most important not yet measurable factor that predicts the outcome of cancer therapy. Hypoxic tumors are resistant to radiation, chemotherapy, and surgery. They signal tumor cells to grow, invade, survive cytotoxic-factor assault, and increase metastatic activity. Therapies aimed at reversing hypoxia-related treatment resistance or normalizing hypoxia are proven effective with level 1 evidence. The weak link remains the lack of satisfactory methods of measurement of tumor oxygenation.

Keywords Personalized medicine • Tumor hypoxia • Tumor resistance

1 Introduction

Personalized medicine involves tailoring treatment to the specific sensitivities of the tumor. To this end, molecular genetics are now routinely available on tumors as are circulating or histochemical molecular diagnostics. Availability of gene markers of tumor resistance and prognosis inform the design of therapeutics. For example, HER2/neu receptor status in breast cancers identifies patients with a previously poor prognosis who now have a "good" prognosis due to the excellent response to herceptin therapy. In the absence of this agent, HER2/neu promotes tumor growth and survival signaling pathways. Because many tumors do not express the receptor, without that molecular marker it would have been difficult or impossible to perform a successful randomized clinical adjuvant trial with herceptin [1]. Similarly, with few exceptions, the increase in survivals with even the best molecular targeted

P. Okunieff (✉) • W. O'Dell • M. Zhang • L. Zhang • D. Maguire
Department of Radiation Oncology, University of Florida,
P.O. Box 103633, Gainesville, FL 32610, USA
e-mail: pokunieff@ufl.edu

W.J. Welch et al. (eds.), *Oxygen Transport to Tissue XXXIV*, Advances in Experimental Medicine and Biology 765, DOI 10.1007/978-1-4614-4989-8_27,

agents has been small and usually has required meta-analyses to confirm benefit. The key to the success of personalized medicine, therefore, is the ability to detect prognostic and diagnostic markers.

Hypoxia is among the first and arguably still the most powerful modifiers of treatment outcome. Yet, our ability to effectively measure hypoxia in vivo, and particularly during therapy, remains elusive. This is in part because hypoxia changes rapidly with time and often changes during a course of therapy. Nevertheless, there are a large number of randomized studies showing clinical benefits even in the absence of the ability to select patients with hypoxic tumors.

We review and support our statement that hypoxia is likely the most important modifier of cancer treatment response. We demonstrate the availability of therapies to overcome hypoxia and their proven effectiveness. Finally, we review methods that might be employed in the clinic for measuring hypoxia and evaluate their potential.

2 Role of Hypoxia in Tumor Resistance to Therapy

The impact of low oxygen on tumor resistance was among the first discoveries in cancer medicine [2, 3]. For example, there is a dose modifying factor of 2.5–3.0 between proliferating cells irradiated at <2.5 mmHg compared to those irradiated at >10 mmHg. Thus, cell killing at 60 Gy, which is usually curative for euoxic tumors, would require >150 Gy for hypoxic cells (60 times 2.5 or 3). Numerically, this is a huge effect, converting a very successful treatment that might have a surviving fraction of 10^{-9} or 10^{-10} into an unsuccessful one with >1% of cells still surviving [4]. The effect on most chemotherapeutic regimens would be similar, since most chemotherapy agents require oxygen to fuel oxidation reactions with DNA (a few chemotherapeutic agents have cytotoxicity only in hypoxia and should improve a chemotherapy regimen if included for hypoxic tumors). Metabolic processes that enable tumor resistance are associated with hypoxia. For example, inflammation associated with hypoxia and reoxygenation injury stimulates the production of growth factors and growth factor receptors. Signaling through HIF-1, considered a therapeutic target by many, would be reduced or eliminated if the tumor was rendered euoxic. Likewise, hypoxia reperfusion injury selects for survival signaling proteins such as p53, which then provide tumors with mechanisms for resistance and malignant progression [5]. Therefore, overcoming hypoxia alone could target many signaling pathways associated with resistance to therapy.

3 Role of Hypoxic Tumors on Local Control and Survival

The role of hypoxic tumors on local control and survival has been reviewed extensively [6]. Briefly, clinical data have universally confirmed the impact of hypoxia on outcomes [7–14]. The “gold standard” method for measurement of hypoxia is the

Eppendorf electrode. Though it has some well-established limitations, it has been used both experimentally and clinically and has provided us with documentation of the role that oxygen plays in patient outcomes. Different metrics have been used in different reports to define a hypoxic tumor. Some have chosen the percent of measurements in the lowest oxygen bin (<2.5 mmHg). Others have used median pO_2 or other thresholds. Independent of metrics, studies have shown extremely large differences in outcomes between hypoxic and non-hypoxic tumors in all types of tumors, including breast cancers [7], head and neck cancers [8, 9], sarcomas [10], uterine cervix [11], and prostate cancers [12, 13]. Importantly, for most cancer types, including lung cancers [15], approximately half the tumors are judged to be hypoxic [7–16]. Thus, not all patients' tumors will be hypoxic on any given treatment day. Some efforts have been made to assess hypoxia during a course of treatment with inconsistent results. For example, some studies predict constant levels of hypoxia until radiation doses reach a high level. Others find that oxygenation improves shortly after a few radiation doses [17]. The best human data on this subject are from ^{18}F-misonidazole (MISO) positron emission tomography (PET) imaging studies [15, 18]. All studies, however, support the notion that the level of hypoxia cannot be well predicted and must be measured.

Simpler approaches to estimating the degree of hypoxia also exist. Prominent among the methods is delivery of ^{18}F-MISO or a similar agent known to bind to DNA in hypoxic cells [19]. Other labeling approaches (e.g., immunohistochemistry or oxidative assays [20, 21]) are sure to be of use, although there is little supporting human data. Other approaches include histochemical measurement of HIF-1 and downstream elements in the HIF-1 pathway. HIF-1's role and its reliable association with hypoxia are controversial. Although HIF-1 staining is not usually microanatomically associated with regions of hypoxia, and the expression of its downstream signaling proteins, such as Glut-1, VEGF, Ca9, and NDRG1, do not consistently overlap in the same regions on tumor immunopathology, the levels of these proteins commonly scale with the level of hypoxia [22–25]. Because hypoxia is very transient and these proteins are very labile, it is important to be conscious of the need to immediately (flash) freeze biopsy specimens.

4 Impact of Therapies Aimed at Tumor Hypoxia

In contrast to the complexity of measuring hypoxia, altering the level of hypoxia can be achieved by a variety of often very inexpensive and nontoxic therapies. Many approaches have been used in clinical studies, including normobaric 100% oxygen or carbogen (95% O_2 5% CO_2) breathing daily with radiation therapy; hyperbaric oxygen breathing; or approaches to reduce metabolic rate, such as hypothermia. To sensitize hypoxic regions, nitroimidazoles have been used as oxygen mimetics, as have agents aimed at both altering metabolism and tumor perfusion, such as nicotinamide. Without dynamic oxygen measurements, it is hard to know how great an impact these have had on hypoxia. Individual randomized studies have not always achieved statistically significant clinical results, but have almost always shown numerical benefit, and meta-analysis is uniformly positive for local disease, distant metastases, and overall survival [26]. Remarkably, given this level of clinical evidence, the measures to overcome hypoxia are rarely employed in the USA, although they are standard in some countries [27]. Given the usual 50% incidence of significant hypoxia, the lost benefit for individual patients who might have been treated personally to achieve euoxic tumor outcomes is indeed a great disadvantage.

5 Limitations of Available Assays for Measuring Hypoxia

Optimally, measurement techniques for tumor oxygen would be logistically simple and easy to interpret and not require expert technical support. The technique should be fast, nontoxic, and noninvasive. It should provide imaging and be performed serially during treatment. It should also document that the intervention actually altered the oxygen state of the tumor. Lastly, it would need to be cheap and satisfactorily calibrated. While Eppendorf electrode technology is the accepted "gold standard," requirement of multiple instrumentations for the tumor and limited range of the needle as well as the required technical skill of the operator make using this technology cumbersome and questionable for routine practice. Even in small studies, multiple measures over a course of treatment have never been performed using the electrode and would be hard to recommend. This method, however, is probably the only one that measures tissue oxygen levels quantitatively. Imaging methods featuring PET have shown some promise. Many agents are blood flow sensitive or have the capacity to detect extremely hypoxic regions. For example ^{18}F-MISO has been used successfully during the course of treatment [15, 16]. Studies using ^{18}F-MISO suggest that a tumor with substantial hypoxia often becomes euoxic during a course of treatment [19]. There are also numerous blood flow-sensitive and oxygen-sensitive ^{1}H magnetic resonance susceptibility imaging [28, 29] and spectroscopy (MRS) techniques [30] that have potential clinical application. ^{19}F MRS operates at frequencies similar to ^{1}H, and can be covalently bound to molecules with high oxygen shifts. ^{19}F-hexafluorobenzene has been used in animal models [31, 32]. Similarly, electron spin resonance (ESR) imaging techniques can measure oxygen concentrations [33]. Although heroic efforts have been made to bring these techniques into the clinic, they are limited to the most superficial tumors and suffer from lack of reproducible data. While this technology has potential, it is not yet ready to be used clinically. Finally, near-infrared optical (NIRO) spectroscopy techniques are under development and meet most requirements we defined for routine tumor oxygen measurement. NIRO exploits the differential absorption of near-infrared photons by deoxy- and oxyhemoglobin [34]. While it does not measure tissue oxygenation directly, NIRO can measure capillary hemoglobin oxygen saturation, which is related to the local tissue oxygen level. The required technologies are already used to measure neonatal brain oxygenation [35] and to minimize brain damage during cardiac bypass surgery [36]. The light can penetrate several centimeters through the skin and can be placed invasively for deeper tumor probes.

Surrogate measurements of oxygen provide the best immediate opportunity to identify patients with hypoxic tumors. Numerous molecular responses to hypoxia are known, and these responses provide powerful histological markers to predict outcome. As discussed earlier, some biomarkers are easily measured in pathological specimens. However, repeated measurement of biopsy specimens is rarely feasible even for superficial tumors. The most reliable surrogate of oxygenation is blood flow. There are many noninvasive methods for measurement of tumor blood flow using CT, PET, and MRI [37]. All have substantial expense but allow for serial imaging. In proton therapy, a convenient treatment by-product is ^{15}O-water. This molecule has been used with PET imaging successfully in the clinic for measurement of tumor blood flow [38]. While the number of proton therapy facilities has steadily increased, and the technology to measure ^{15}O-water blood flow during proton therapy has been long demonstrated [39], its potential for clinical benefit has yet to be tested. Given the level one evidence of benefit for oxygen-targeted therapies in cancer, it is realistic to hope that some of these tests will become reimbursable procedures and clinically feasible.

Tumor metabolism tends to be oxygen limited. Thus, simple reductions or increments of inspired oxygen greatly affect the ratio of high energy to low energy phosphates that can be measured using techniques, such as ^{31}P-MRS [40]. Tests of this sort could be performed during tumor imaging. The ability to alter the breathing mixture for a patient can help assure that the target tumor is oxygen limited and clarify the degree of hypoxia [28]. Noninvasive dynamic studies of tumor metabolism are instructive as they confirm the notion that tumors can survive a seriously hypoxic environment, such as tourniquet ischemia, in the absence of both glucose and oxygen and in a very acidotic environment. The implication of these observations is ominous as we try to better understand the capacity for hypoxia to select an aggressive tumor phenotype that can withstand extremely demanding treatment conditions.

6 Summary

There are few if any modifiers of treatment response and assays predictive of prognosis more important that hypoxia. Hypoxia drives many signaling pathways, which are themselves the target of therapies against cancer. Given that hypoxia is the root of many deleterious aspects of tumor physiology, malignant progression, treatment resistance, and allowing for the plentiful and easily implemented opportunities to overcome its nefarious effects, it is unfortunate that we do not yet have a satisfactory method of measuring hypoxia. Methods are increasingly available to measure oxygen or a surrogate. Developing an optimal method requires simplicity, accuracy, reproducibility, economy, and a capacity to serially image tumors deep seated in the body. The methods currently available include direct measures of oxyhemoglobin saturation and mechanized Clark-type electrodes, neither of which is yet ready for routine use. Future work, including biological markers, either endogenously expressed (i.e., cytokines) or delivered to a patient immediately before biopsy, present an opportunity to directly or indirectly measure tumor oxygenation. Future hopes also lie in molecular imaging techniques, including PET, MRI/MRS, NIRO, and related methods for measuring blood flow or metabolism.

Acknowledgments We thank Kate Casey-Sawicki for editing and preparing this manuscript for publication.

References

1. Perez EA, Reinholz MM, Hillman DW et al (2010) HER2 and chromosome 17 effect on patient outcome in the N9831 adjuvant trastuzumab trial. J Clin Oncol 28:4307–4315
2. Warburg O (1956) On respiratory impairment in cancer cells. Science 124:269–270
3. DeBerardinis RJ, Lum JJ, Hatzivassiliou G et al (2008) The biology of cancer: metabolic reprogramming fuels cell growth and proliferation. Cell Metab 7:11–20
4. Hall EJ, Giaccia AJ (2006) Radiobiology for the radiologist, 6th edn. Lippincott Williams & Wilkins, Philadelphia, pp 85–105
5. Graeber TG, Osmanian C, Jacks T et al (1996) Hypoxia-mediated selection of cells with diminished apoptotic potential in solid tumours. Nature 379:88–91
6. Vaupel P, Mayer A (2007) Hypoxia in cancer: significance and impact on clinical outcome. Cancer Metastasis Rev 26:225–239
7. Vaupel P, Briest S, Höckel M (2002) Hypoxia in breast cancer: pathogenesis, characterization and biological/therapeutic implications. Wien Med Wochenschr 152:334–342
8. Gatenby RA, Kessler HB, Rosenblum JS et al (1988) Oxygen distribution in squamous cell carcinoma metastases and its relationship to outcome of radiation therapy. Int J Radiat Oncol Biol Phys 14:831–838
9. Nordsmark M, Bentzen SM, Rudat V et al (2005) Prognostic value of tumor oxygenation in 397 head and neck tumors after primary radiation therapy. An international multi-center study. Radiother Oncol 77:18–24
10. Nordsmark M, Alsner J, Keller J et al (2001) Hypoxia in human soft tissue sarcomas: adverse impact on survival and no association with p53 mutations. Br J Cancer 84:1070–1075
11. Hockel M, Schlenger K, Aral B et al (1996) Association between tumor hypoxia and malignant progression in advanced cancer of the uterine cervix. Cancer Res 56:4509–4515
12. Movsas B, Chapman JD, Hanlon AL et al (2002) Hypoxic prostate/muscle pO2 ratio predicts for biochemical failure in patients with prostate cancer: preliminary findings. Urology 60:634–639
13. Wang JZ, Li XA, Mayr NA (2006) Dose escalation to combat hypoxia in prostate cancer: a radiobiological study on clinical data. Br J Radiol 79:905–911
14. Rampling R, Cruickshank G, Lewis AD et al (1994) Direct measurement of pO2 distribution and bioreductive enzymes in human malignant brain tumors. Int J Radiat Oncol Biol Phys 29:427–431
15. Eschmann S-M, Paulsen F, Reimold M et al (2005) Prognostic impact of hypoxia imaging with 18F-misonidazole PET in non-small cell lung cancer and head and neck cancer before radiotherapy. J Nucl Med 46:253–260
16. Rischin D, Hicks RJ, Fisher R et al (2006) Prognostic significance of [18F]-misonidazole positron emission tomography-detected tumor hypoxia in patients with advanced head and neck cancer randomly assigned to chemoradiation with or without tirapazamine: a substudy of Trans-Tasman Radiation Oncology Group Study 98.02. J Clin Oncol 24:2098–2104
17. Karar J, Maity A (2009) Modulating the tumor microenvironment to increase radiation responsiveness. Cancer Biol Ther 8:1994–2001

18. Koh WJ, Rasey JS, Evans ML et al (1992) Imaging of hypoxia in human tumors with [F-18] fluoromisonidazole. Int J Radiat Oncol Biol Phys 22:199–212
19. Mortensen LS, Buus S, Nordsmark M et al (2010) Identifying hypoxia in human tumors: a correlation study between ^{18}F-FMISO PET and the Eppendorf oxygen-sensitive electrode. Acta Oncol 49:934–940
20. Varia MA, Calkins-Adams DP, Rinker LH et al (1998) Pimonidazole: a novel hypoxia marker for complementary study of tumor hypoxia and cell proliferation in cervical carcinoma. Gynecol Oncol 71:270–277
21. Hlatky L, Ring CS, Sachs RK (1989) Detection of an intrinsic marker in hypoxic cells. Cancer Res 49:5162–5166
22. Chou L-W, Wang J, Chang P-L et al (2011) Hyaluronan modulates accumulation of hypoxia-inducible factor-1 alpha, inducible nitric oxide synthase, and matrix metalloproteinase-3 in the synovium of rat adjuvant-induced arthritis model. Arthritis Res Ther 13:R90
23. Said HM, Supuran CT, Hageman C et al (2010) Modulation of carbonic anhydrase 9 (CA9) in human brain cancer. Curr Pharm Des 16:3288–3299
24. Cangul H (2004) Hypoxia upregulates the expression of the NDRG1 gene leading to its over-expression in various human cancers. BMC Genet 5:27
25. Ding I, Okunieff P, Salnikow K et al (2003) A new intrinsic hypoxia marker in esophageal cancer. Adv Exp Med Biol 540:227–233
26. Horsman MR (1995) Nicotinamide and other benzamide analogs as agents for overcoming hypoxic cell radiation resistance in tumours. A review. Acta Oncol 34:571–587
27. Overgaard J (2011) Hypoxic modification of radiotherapy in squamous cell carcinoma of the head and neck – a systematic review and meta-analysis. Radiother Oncol 100(1):22–32
28. Kotas M, Schmitt P, Jakob PM et al (2009) Monitoring of tumor oxygenation changes in head-and-neck carcinoma patients breathing a hyperoxic hypercapnic gas mixture with a noninvasive MRI technique. Strahlenther Onkol 185:19–26
29. Fan B, Wang X-Y, Yang X-D et al (2011) Blood oxygen level-dependent MRI for the monitoring of neoadjuvant chemotherapy in breast carcinoma: initial experience. Magn Reson Imaging 29:153–159
30. Pacheco-Torres J, López-Larrubia P, Ballesteros P et al (2011) Imaging tumor hypoxia by magnetic resonance methods. NMR Biomed 24:1–16
31. Diepart C, Magat J, Jordan BF et al (2011) In vivo mapping of tumor oxygen consumption using (19)F MRI relaxometry. NMR Biomed 24:458–463
32. Salmon HW, Siemann DW (2004) Utility of 19F MRS detection of the hypoxic cell marker EF5 to assess cellular hypoxia in solid tumors. Radiother Oncol 73:359–366
33. Epel B, Haney CR, Hleihel D et al (2010) Electron paramagnetic resonance oxygen imaging of a rabbit tumor using localized spin probe delivery. Med Phys 37:2553–2559
34. Finlay JC, Foster TH (2004) Hemoglobin oxygen saturations in phantoms and in vivo from measurements of steady-state diffuse reflectance at a single, short source-detector separation. Med Phys 31:1949–1959
35. Austin T, Gibson AP, Branco G et al (2006) Three dimensional optical imaging of blood volume and oxygenation in the neonatal brain. Neuroimage 31:1426–1433
36. Murkin JM, Adams SJ, Novick RJ et al (2007) Monitoring brain oxygen saturation during coronary bypass surgery: a randomized, prospective study. Anesth Analg 104:51–58
37. Taber KH, Black KJ, Hurley RA (2005) Blood flow imaging of the brain: 50 years experience. J Neuropsychiatry Clin Neurosci 17:441–446
38. Lodge MA, Jacene HA, Pili R et al (2008) Reproducibility of tumor blood flow quantification with 15O-water PET. J Nucl Med 49:1620–1627
39. Okunieff P, Lee J, Vaupel P (1992) Measurement of human tumor blood flow: a positron technique using an artifact of high energy radiation therapy. Adv Exp Med Biol 317: 169–176
40. Dewhirst MW, Poulson JM, Yu D et al (2005) Relation between pO2, 31P magnetic resonance spectroscopy parameters and treatment outcome in patients with high-grade soft tissue sarcomas treated with thermoradiotherapy. Int J Radiat Oncol Biol Phys 61:480–491

Chapter 28
Wavelet Cross-Correlation to Investigate Regional Variations in Cerebral Oxygenation in Infants Supported on Extracorporeal Membrane Oxygenation

Maria Papademetriou, Ilias Tachtsidis, Martin J. Elliott, Aparna Hoskote, and Clare E. Elwell

Abstract Extracorporeal membrane oxygenation can potentially affect cerebral blood flow dynamics and consequently influence cerebral autoregulation. We applied wavelet cross-correlation (WCC) between multichannel cerebral oxyhemoglobin concentration (HbO_2) and mean arterial pressure (MAP), to assess regional variations in cerebral autoregulation. Six infants on veno-arterial (VA) ECMO were studied during sequential changes in the ECMO flows. WCC between MAP and HbO_2 for each flow period and each channel was calculated within three different frequency (wavelet scale) bands centered around 0.1, 0.16, and 0.3 Hz chosen to represent low frequency oscillations, ventilation, and respiration rates, respectively. The group data showed a relationship between maximum WCC and ECMO flow. During changes in ECMO flow, statistically significant differences in maximum WCC were found between right and left hemispheres. WCC between HbO_2 and MAP provides a useful method to investigate the dynamics of cerebral autoregulation during ECMO. Manipulations of ECMO flows are associated with regional changes in cerebral autoregulation which may potentially have an important bearing on clinical outcome.

Keywords Cerebral oxygenation • Infants

The original version of this chapter was revised. An erratum to this chapter can be found at DOI 10.1007/978-1-4614-4989-8_53

M. Papademetriou • I. Tachtsidis • C.E. Elwell (✉)
Medical Physics and Bioengineering Department, University College London, Malet Place Engineering Building, Gower Street, London, WC1E 6BT, UK
e-mail: celwell@medphys.ucl.ac.uk

M.J. Elliott • A. Hoskote
Cardiothoracic Unit, Great Ormond Street Hospital for Children, London, UK

W.J. Welch et al. (eds.), *Oxygen Transport to Tissue XXXIV*, Advances in Experimental Medicine and Biology 765, DOI 10.1007/978-1-4614-4989-8_28,

1 Introduction

Extracorporeal membrane oxygenation (ECMO) is a life support system for infants with cardiorespiratory failure. Neurological complications are the largest cause of morbidity and mortality in these patients, with the reported frequency of abnormal neuroimaging ranging from 28 to 52 % [1]. Initiation of ECMO involves cannulation of the major great vessels—right common carotid artery and internal jugular vein—which may cause lateralizing cerebrovascular injury. ECMO infants suffer from hypoxia, asphyxia, and hypercarbia which can disrupt cerebral autoregulation, leaving the cerebral microcirculation vulnerable to alterations in blood pressure [2].

Methods to assess the status of autoregulation by considering the relationship between spontaneous fluctuations in MAP and cerebral blood flow (CBF) surrogates, such as (HbO_2) measured by NIRS, in either the time or frequency domain using Fourier transforms were reported extensively in the literature [3]. These, conventional methods suffer from the big drawback of averaging out all the potential useful time information, hence treating cerebral autoregulation as a stationary, linear process.

Recent studies have emphasized that cerebral autoregulation is a dynamic process [4]. The continuous wavelet transform (CWT) possesses the ability to construct a time–frequency representation of a signal that offers time and frequency localization. Latka et al. used CWT to compute a synchronization index between CBF and ABP signals [5]. Wavelet cross-correlation (WCC) was introduced by Rowley et al. as the cross-correlation between CWT coefficients of two time series [6]. Spectral analysis using wavelets provides a framework for analysis of nonstationary effects in cerebral hemodynamics, thus overcoming the restrictions intrinsic to earlier methods.

Previously we used a dual-channel NIRS system and showed the presence of oscillations related to vasomotion, respiration, and heart rates [7]. Preliminary results using multichannel NIRS indicate regional variation in cerebral oxygenation [8]. Here, we investigate the use of WCC as a method to study the concordance between multisite cerebral HbO_2 and mean arterial pressure in order to assess regional variations in cerebral oxygenation in neonates supported on ECMO.

2 Methods

2.1 Subjects and Instrumentation

A total of six veno-arterial (VA) ECMO patients, age range 1–16 days, were monitored during sequential changes in the ECMO flows. Alterations in the ECMO flows refer to successive decrease in the ECMO flow by 10 % from the initial flow, approximately every 10 min, down to 70 % of the initial flow followed by successive increase back to baseline (Fig. 28.2b).

A multichannel NIRS system (ETG-100, Hitachi Medical Ltd., Japan) was used to measure changes in oxy-(HbO_2), deoxy-(HHb), and total hemoglobin (HbT) concentrations at 5 Hz. A novel cap was constructed to accommodate the optical sources and detectors (interoptode distance = 3 cm), allowing data to be collected from 12 channels. Multimodal data were collected synchronously including systemic parameters (arterial blood pressure [ABP], heart rate [HR], and arterial oxygen saturation [SpO_2]) and ECMO circuit parameters (venous oxygen saturation [SvO_2], arterial saturation at the cannula [SaO_2]).

2.2 Data Analysis

Mean arterial pressure (MAP) was obtained by trapezoid integration of ABP every 0.2 s, equivalent to sampling frequency of 5 Hz. The time series of MAP and HbO_2 were divided into sections representing each ECMO flow period (Fig. 28.3c). Each section of data was then high and low pass filtered at 0.008 and 1 Hz.

An approximate relationship between the scale α in the wavelet domain and frequency in the Fourier transform, f_α, can be computed as [5]:

$$f_\alpha = \frac{f_c}{\alpha \cdot \delta t}, \tag{28.1}$$

where f_c is the center frequency and δ_τ is the sampling period.

Wavelet analysis was performed on HbO_2 data. WCC and synchronization index, γ, were used as methods to investigate the relation between MAP and HbO_2. The complex Morlet wavelet was used to calculate the CWT coefficients for MAP and HbO_2 using the MatLab wavelet toolbox function *cwt*. The central frequency (f_c) and bandwidth (f_b) of the complex Morlet wavelet were both chosen as 1 in order to be in agreement with previous methods [5, 6]. A scale range with unit spacing from 5 to 100, representing frequencies 0.008–1 Hz was used to obtain two complex time series, $W_{MAP}(a,t)$ and $W_{HbO_2}(a,t)$ for each flow period A–G and across each of the 12 channels.

The WCC between MAP and HbO_2 in each channel and for each flow was obtained using the equation below [6]:

$$\overline{\text{WCC}} = \frac{\left|R_{X,Y}(W_{\text{MAP}}, W_{\text{HbO}_2}\alpha,\tau)\right|}{\sqrt{\left|R_{X,X}(W_{\text{MAP}},\alpha,0)\cdot R_{X,X}(W_{\text{HbO}_2}\alpha,0)\right|}}, \tag{28.2}$$

in which $R_{X,Y}(s1, s2, \alpha, \tau)$ denotes the cross-correlation of the wavelet coefficients of the series $s1$ and $s2$ at a scale α and for a relative time shift τ and $R_{X,X}(s1, \alpha, 0)$ denotes the autocorrelation of the time series $s1$ for zero time shift. WCC(α, τ) represents the cross-spectral power in the two time series (shifted relative to each other by τ) as a fraction of the total power in the two time series. WCC ranges from 0 to 1. At a given wavelet scale, WCC = 1 would indicate that the coefficients of the two wavelet transforms are related to each other by a simple scaling factor, suggesting strong synchronization at this frequency [6].

The phase difference between the two time series, MAP and HbO_2 was also calculated using the circular mean, $\overline{\Delta\Phi(\alpha)}$, of the instantaneous phase difference between the two signals $\Delta\Phi(\alpha, \tau)$ over the duration of a test segment [5]

$$\overline{\Delta\Phi(\alpha)} = \tan^{-1}\left(\frac{\sum_t \sin(\Delta\phi(\alpha,t))}{\sum_t \cos(\Delta\phi(\alpha,t))}\right). \tag{28.3}$$

For each time series pair at each flow period and for each channel, the maximum value of $WCC(\alpha, \tau)$ was found within three scale bands: $\alpha_i = 5 < \alpha < 20$ ($f_{\alpha i} = 0.25$ Hz $< f_a < 1$ Hz), $\alpha_{ii} = 20 < \alpha < 40$ ($f_{\alpha ii} = 0.13$ Hz $< f_\alpha < 0.25$ Hz), $\alpha_{iii} = 40 < \alpha < 80$ ($f_{\alpha iii} = 0.06$ Hz $< f_\alpha < 0.13$ Hz). These bands were chosen to overlap with respiration rate (RR), ventilation rate (VR), and slow M-waves, respectively. The maximum circular mean, $\Delta\Phi_{max}$, were also calculated within each scale band, for each flow period and each channel. Student's *t*-test was then used to analyze the statistical significance of the differences in the group mean of each of these variables between channels.

3 Results

Figure 28.1 shows a set of typical WCC contours obtained from two patients at baseline flow and minimum flow. For patient 1 WCC shows no distinct peaks at baseline flows indicating no correlation between MAP and HbO_2. At minimum flow, peaks in the WCC contours are shown at scales 15 ($f_\alpha = 0.33$ Hz), 29 ($f_\alpha = 0.17$ Hz) and a relatively weaker peak at scale 55 ($f_\alpha = 0.09$ Hz). WCC for patient 2 at baseline flow shows a relatively weak peak at a scale 34 ($f_\alpha = 0.15$ Hz). As with patient 1 correlation between MAP and HbO_2 becomes stronger at minimum flow with the peak at scale 34 spreading to higher Mayer-waves related scales and another peak occurring at scale 10 ($f_\alpha = 0.5$ Hz). These peaks appear shifted from zero time lag in agreement with Rowley et al. [6].

In general, WCC between MAP and HbO_2 revealed three distinct peaks within three scale regions. The first peak typically occurs at a scale of around 14 (0.36 Hz), the second at a scale around 30 (0.16 Hz) and the third at a scale around 50 (0.1 Hz). These peaks could correspond to the RR, VR, and Mayer-waves, respectively.

Figure 28.2 shows the group data for the mean of the maximum WCC, WCC^i_{max}, within scale band $\alpha_i = 5 < \alpha < 20$ ($f^i_a = 0.25$ Hz $< f_\alpha < 1$ Hz) at each flow period and across the 12 channels. By convention a value of WCC below 0.5 indicates no correlation between MAP and HbO_2 [5]. A + sign indicates that HbO_2 lags MAP, i.e. $\Delta\Phi > 0$, only for WCC > 0.5. A – sign is used to indicate that HbO_2 is leading MAP,

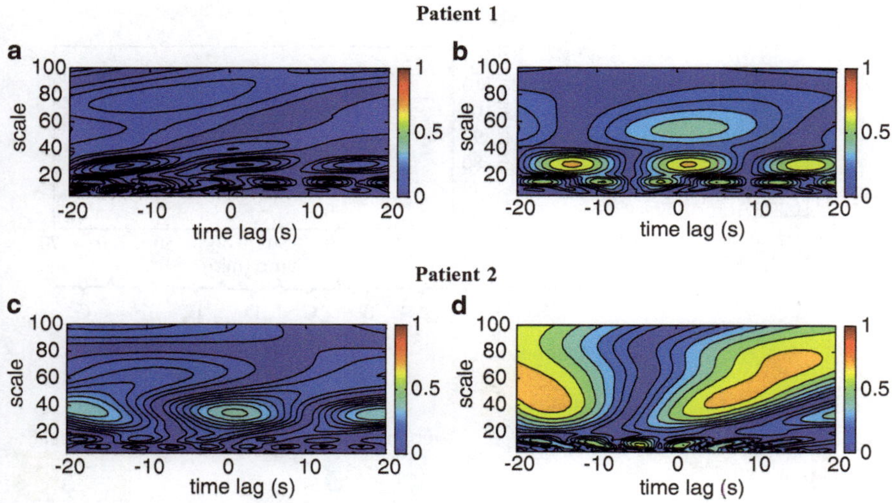

Fig. 28.1 Wavelet cross-correlation (WCC) between MAP and HbO_2 for two ECMO patients. Low correlation is shown at baseline ECMO flows (**a** and **c**) and high correlation around scales 16 and 30 for patient 1 at minimum flow (**b**) and around scales 16 and 40–80 for patient 2 (**d**) at minimum flow

i.e. $\Delta\Phi < 0$, where WCC > 0.5. There are statistically significant differences ($p < 0.05$) in mean WCC^i_{max} across all flows between symmetrical channels most likely positioned on the right and left parietal lobes (Fig. 28.3d). WCC^i_{max} for all flows in the three channels positioned on the left parietal lobe is below 0.5 suggesting no correlation between MAP and HbO_2 in these channels. A general increase in WCC^i_{max} was observed with decrease in flow across all channels. WCC^i_{max} is highest either at flow period E or F. a^i_{max} across flow changes for all channels ranges from 9 to 17 (0.29–0.56 Hz) (Fig. 28.3e). Most of the channels show a shift in a^i_{max} to a lower scale when the highest WCC^i_{max} is reached (flow period E).

4 Discussion and Conclusions

WCC between HbO_2 and MAP provides a useful method to investigate the dynamics of cerebral autoregulation. Cerebral autoregulation on ECMO is poorly studied, since there have been no easy noninvasive methods to study and interpret complex cerebral physiological process. Our results showed a relationship between WCC and ECMO flow in the grouped data of six patients. These differences were statistically significant between right and left hemispheres, especially when the flows were weaned sequentially. Modest manipulations of ECMO flows are associated with regional changes in cerebral autoregulation which may potentially have an important bearing on clinical outcome.

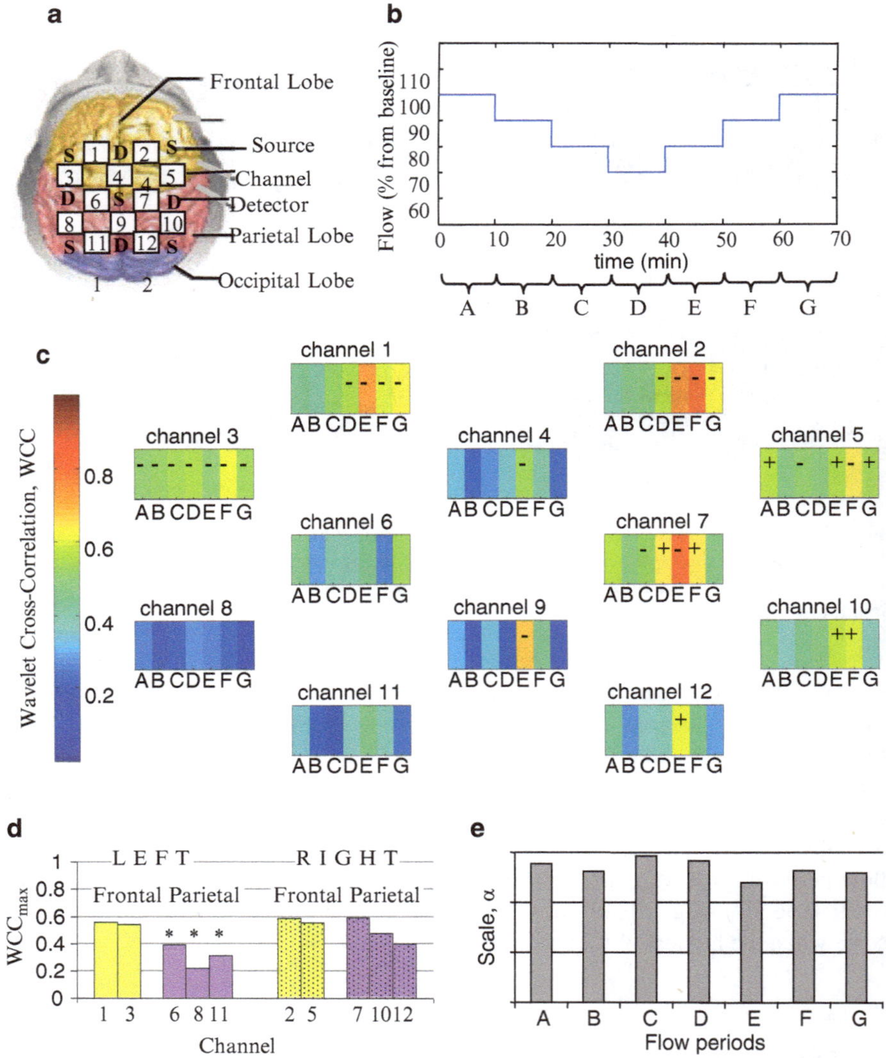

Fig. 28.2 Group WCC on between MAP and HbO_2 within scale band $\alpha_i = 5 < \alpha < 20$ ($f_{a_i} = 0.25$ Hz$< f_\alpha <$1 Hz): (**a**) channel arrangement; (**b**) sequence of flow changes ($A = 100$ %, $B = 90$ %, $C = 80$ %, $D = 70$ %, $E = 80$ %, $F = 90$ %, $G = 100$ %); (**c**) WCC^i_{max} at all flow periods across all channels; (**d**) mean WCC^i_{max} across all flow periods of channels on the right side and symmetrical channels on the left side; (**e**) mean of scale at WCC^i_{max}, a^i_{max}, for each flow period across all channels. +/- denotes HbO_2 lagging/leading MAP for $WCC^i_{max} > 0.5$. (*asterisk*) Statistical significant difference between symmetrical channels on right and left hemispheres ($p < 0.05$)

Acknowledgment This work was supported by Hitachi Medical Ltd., Japan.

References

1. Bulas D, Glass P (2005) Neonatal ECMO: neuroimaging and neurodevelopmental outcome. Semin Perinatol 29:58–65
2. Liem D, Hopman J, Oeseburg B, de Haan A, Festen C, Kollee L (1995) Cerebral oxygenation and hemodynamics during induction of extracorporeal membrane oxygenation as investigated by near infrared spectrophotometry. Pediatrics 95:555–561
3. Czosnyka M, Brady K, Reinhard M, Smielewski P, Steiner L (2009) Monitoring cerebrovascular autoregulation: facts, myths and missing links. Neurocrit Care 10:373–86
4. Panerai RB, Moody M, Eames PJ, Potter JF (2005) Cerebral blood flow velocity during mental activation: interpretation with different models of the passive pressure–velocity relationship. J Appl Physiol 99:2352–2362
5. Latka M, Turalska M, Latka MG, Kolodziej W, Latka D, West BJ (2005) Phase dynamics in cerebral autoregulation. Am J Heart Circ Physiol 289:H2272–H2279
6. Rowley AB, Payne SJ, Tachtsidis I, Ebden MJ, Whiteley JP, Gavaghan DJ, Smith M, Elwell CE, Delpy DT (2007) Synchronisation between arterial blood pressure and cerebral concentration investigated by wavelet cross-correlation. Physiol Meas 28:161–173
7. Papademetriou MD, Tachtsidis I, Leung TS, Elliott MJ, Hoskote A, Elwell CE (2010) Cerebral and peripheral tissue oxygenation in infants and children supported on ECMO for cardio-respiratory failure. Adv Exp Med Biol 662:447–453
8. Papademetriou MD, Tachtsidis I, Leung TS, Elliott MJ, Hoskote A, Elwell CE (2011) Regional cerebral oxygenation measured by multichannel near-infrared spectroscopy (optical topography) in an infant supported on venoarterial extracorporeal membrane oxygenation. J Thorac Cardiovasc Surg 141:e31–e33

Chapter 29
Association of the Red Cell Distribution Width with Red Blood Cell Deformability

Kushang V. Patel, Joy G. Mohanty, Bindu Kanapuru, Charles Hesdorffer, William B. Ershler, and Joseph M. Rifkind

Abstract The red cell distribution width (RDW) is a component of the automated complete blood count (CBC) that quantifies heterogeneity in the size of circulating erythrocytes. Higher RDW values reflect greater variation in red blood cell (RBC) volumes and are associated with increased risk for cardiovascular disease (CVD) events. The mechanisms underlying this association are unclear, but RBC deformability might play a role. CBCs were assessed in 293 adults who were clinically examined. RBC deformability (expressed as the elongation index) was measured using a microfluidic slit-flow ektacytometer. Multivariate regression analysis identified a clear threshold effect whereby RDW values above 14.0% were significantly associated with decreased RBC deformability (β=–0.24; p=0.003). This association was stronger after excluding anemic participants (β=–0.40; p=0.008). Greater variation in RBC volumes (increased RDW) is associated with decreased RBC deformability, which can impair blood flow through the microcirculation. The resultant hypoxia may help to explain the previously reported increased risk for CVD events associated with elevated RDW.

Keywords (MeSH) Ektacytometer • Erythrocyte count • Erythrocyte deformability • Erythrocyte indices

---

K.V. Patel (✉)
Laboratory of Epidemiology, Demography, and Biometry, National Institute on Aging/NIH, 7201 Wisconsin Ave, Suite 3C309, Gateway Building, Bethesda, MD 20814, USA
e-mail: patelku@mail.nih.gov

J.G. Mohanty • J.M. Rifkind
Molecular Dynamics Section, National Institute on Aging/NIH, Bethesda, MD, USA

B. Kanapuru • C. Hesdorffer
Clinical Research Branch, National Institute on Aging/NIH, Bethesda, MD, USA

W.B. Ershler
Institute for Advanced Studies in Aging, Gaithersburg, MD, USA

W.J. Welch et al. (eds.), *Oxygen Transport to Tissue XXXIV*, Advances in Experimental Medicine and Biology 765, DOI 10.1007/978-1-4614-4989-8_29,

1 Introduction

The red cell distribution width (RDW) is a component of the complete blood count (CBC) that quantifies heterogeneity in the size of circulating erythrocytes. Nearly all modern automated blood cell counters report the RDW as the coefficient of variation of red blood cell (RBC) volume, which is computed by dividing the standard deviation of RBC volume by the mean corpuscular volume (MCV) and multiplying this quantity by 100. Therefore, higher values of RDW reflect greater variation in the population distribution of RBC volumes.

Several epidemiologic studies have shown that higher RDW is strongly associated with increased risk for cardiovascular disease (CVD) events (e.g., myocardial infarction, stroke) and mortality in middle-aged and older adults, independent of hemoglobin concentration and nutritional status [1–7]. These associations have been observed in a variety of settings, including in the general community-dwelling population [3, 4], in patients with clinically significant CVD [1, 2, 5, 7], and among hospitalized patients [5–7]. The mechanisms underlying these associations are unclear; however, it is conceivable that RBC deformability might play a role as elevated RDW is associated with increased inflammation and decreased levels of antioxidants [8], which can reduce RBC deformability and survival. Accordingly, we sought to determine whether higher RDW values are associated with decreased RBC deformability.

2 Methods

2.1 *Study Population*

Blood was collected in 293 community-dwelling adults participating in the Baltimore Longitudinal Study of Aging (BLSA), an ongoing prospective cohort study. Participants in the current study were 32–98 years of age, were examined between May 2009 and March 2010, and provided written informed consent. The protocol of the BLSA was reviewed and approved by the Institutional Review Board of the National Institute on Aging, Intramural Research Program.

2.2 *RBC Measures*

Fasting blood draws were completed in the morning. A Sysmex XE-2100 automated blood cell counter was used to obtain the CBC, which included the following Erythrocyte indices: hemoglobin concentration, MCV, mean corpuscular hemoglobin concentration (MCHC), and RDW. In each whole blood sample, RBC deformability was measured using a microfluidic RheoScan-D slit-flow ektacytometer (Rheo Meditech, Seoul, South Korea). The methodological details of this instrument

Table 29.1 Characteristics of the study sample ($N = 293$)

Characteristic	Mean (SD) or *n* (%)
Age in years, mean (SD)	71.1 (13.3)
Women, *n* (%)	136 (46.4)
Hemoglobin in g/dL, mean (SD)	13.7 (1.3)
MCV in fL, mean (SD)	90.3 (4.7)
MCHC in g/dL, mean (SD)	33.5 (1.1)
RDW %, mean (SD)	13.6 (1.0)
RBC deformability (elongation index), mean SD	0.32 (0.02)

have previously been published [9]. Briefly, blood (~6 μl) was suspended (~1/100 dilution) by slowly mixing it in 600 μl of the highly viscous PVP360 solution (viscosity ~30 cP) provided in the test kits from Rheo Meditech, and then 500 μl of this solution were loaded onto the sample reservoir of the microfluidic chip. During operation of the ektacytometer, the vacuum generating mechanism allowed RBCs to flow through the micro-channel at a range of shear stresses, while the elliptical diffraction patterns of the flowing cells were generated by a laser beam (wavelength 635 nm from a 1.5 mW laser diode). The elliptical diffraction patterns of the flowing RBCs, at different shear stress levels, are projected on a screen and captured by a CCD-video camera. The image data are then analyzed by an ellipse-fitting software. Deformability of RBCs was expressed as the elongation index (EI), which is defined as $(L-W)/(L+W)$, where L and W are the major and minor axes of the ellipse, respectively, at various shear stress values (0–20 Pa). However, for comparative analysis of the deformability values of multiple blood samples, EI values (mean and standard deviation of three measurements) at 3 Pa were used, which is approximately at the halfway point on the EI-shear stress curve [10].

2.3 *Statistical Analysis*

Means and standard deviations (SD) were used to describe the distribution of participants' age and RBC measurements (Table 29.1). Multivariable linear regression analysis was used to test the association of RDW with RBC deformability (Table 29.2). The elongation index values (RBC deformability) were first standardized into Z-scores and then were regressed on age, sex, hemoglobin concentration, MCV, MCHC, and RDW. An initial exploratory graphical analysis showed a clear nonlinear association between RBC deformability and RDW with an inflection point near an RDW value of 14.0%; a threshold above which mortality risk is substantially increased [3, 4]. Spline functions are often used in epidemiology to fit a nonlinear dose–response curve. Therefore, linear splines were added to the regression analysis with a knot arbitrarily set at RDW = 14.0% (Fig. 29.1 and Table 29.2). All analyses were completed using Stata/SE statistical software (version 10.1; StataCorp LP, College Station, Texas).

Table 29.2 Association of RDW with RBC deformability (elongation index)[a], adjusting for age, sex, and other Erythrocyte indices ($N=293$)

	Regression coefficient	95% Confidence interval	p-Value
Age in years	0.006	–0.002, 0.013	0.152
Men (vs. women)	–0.327	–0.522, –0.131	0.001
Hemoglobin in g/dL	0.081	–0.003, 0.165	0.060
MCV in fL	0.104	0.083, 0.125	<0.001
MCHC in g/dL	–0.228	–0.329, –0.127	<0.001
RDW %			
Spline below 14.0%	–0.027	–0.224, 0.171	0.791
Spline above 14.0%	–0.242	–0.401, –0.082	0.003

[a]*Note.* RBC deformability is expressed as the elongation index and it was standardized into a Z-score (mean=0, SD=1) for this regression analysis. The model adjusted $R^2=0.37$

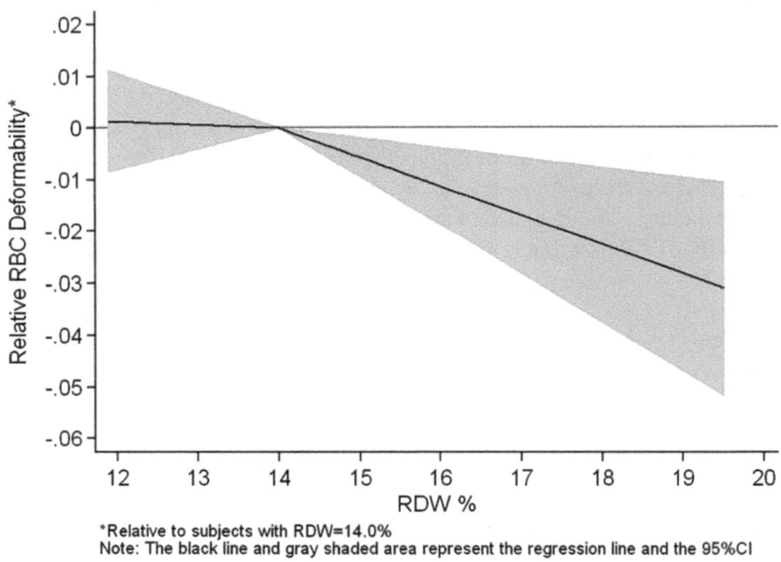

Fig. 29.1 Association of RDW with RBC deformability (elongation index), adjusted for age, sex, hemoglobin concentration, MCV, and MCHC ($N=293$)

3 Results

The characteristics of the study sample are shown in Table 29.1. Participants had a mean age of 71.1 years and 46.4% were women. The distributions of the Erythrocyte indices are consistent with normative data from population studies of adults.

Figure 29.1 illustrates the relationship between RBC deformability and RDW. Compared to participants with an RDW value of 14.0%, there was no difference in

RBC deformability among those with RDW <14.0%; however, participants with RDW above 14.0% had significantly decreased RBC deformability. The detailed results of this regression model are shown in Table 29.2. While there was no association of the participants' age with RBC deformability, men had significantly decreased RBC deformability compared with women. Higher hemoglobin concentration was associated with increased RBC deformability, although the p-value for this association was 0.06. As expected, lower MCV and higher MCHC were associated with decreased RBC deformability, reflecting the effects that reduced cell membrane and increased intracellular hemoglobin concentration can have on deformability. After excluding 51 participants with anemia, the effects of age, sex, and the RBC indices on deformability were essentially the same as those shown in Table 29.2, including for RDW whereby values above 14.0% were significantly associated with decreased RBC deformability (regression coefficient = –0.40; $p = 0.008$).

4 Conclusions

In this study of 293 community-dwelling adults, greater variation in RBC volumes (increased RDW) was associated with decreased RBC deformability, which can impair blood flow through the microcirculation [11]. The resultant hypoxia may help to explain the previously reported increased risk for CVD events associated with elevated RDW, particularly among those with RDW values above 14.0%. To our knowledge, this is first study to examine RDW with erythrocyte deformability. Future studies are needed to replicate the current findings as well as to investigate other aspects of hemorheology with RDW and identify the mechanisms through which RDW is associated with adverse health outcomes.

Acknowledgment This research was supported by the Intramural Research Program of the National Institute on Aging, National Institutes of Health.

References

1. Felker GM, Allen LA, Pocock SJ et al (2007) Red cell distribution width as a novel prognostic marker in heart failure. J Am Coll Cardiol 50:40–47
2. Tonelli M, Sacks F, Arnold M et al (2008) Relation between red blood cell distribution width and cardiovascular event rate in people with coronary disease. Circulation 117:163–168
3. Patel KV, Ferrucci L, Ershler W, Longo DL, Guralnik JM (2009) Red blood cell distribution width and the risk of death in middle-aged and older adults. Arch Intern Med 169:515–523
4. Patel KV, Semba RD, Ferrucci L et al (2010) Red cell distribution width and mortality in older adults: a meta-analysis. J Gerontol A Biol Sci Med Sci 65A:258–265
5. Lippi G, Filippozzi L, Montagnana M et al (2009) Clinical usefulness of measuring red cell distribution width on admission in patients with acute coronary syndromes. Clin Chem Lab Med 47:353–357

6. Wang F, Pan W, Pan S et al (2011) Red cell distribution width as a novel predictor of mortality in ICU patients. Ann Med 43:40–46
7. Zalawadiya SK, Zmily H, Farah J et al (2011) Red cell distribution width and mortality in predominantly African American population with decompensated heart failure. J Card Fail 17:292–298
8. Semba RD, Patel KV, Ferrucci L et al (2010) Serum antioxidants and inflammation predict red cell distribution width in older women. Clin Nutr 29:600–604
9. Shin S, Hou JX, Suh JS, Singh M (2007) Validation and application of a microfluidic ektacytometer (RheoScan-D) in measuring erythrocyte deformability. Clin Hemorheol Microcirc 37:319–328
10. Baskurt OK, Boynard M, Cokelet GC et al (2009) New guidelines for hemorheological laboratory techniques. Clin Hemorheol Microcirc 42:75–97
11. Simchon S, Jan KM, Chien S (1987) Influence of reduced red cell deformability on regional blood flow. Am J Physiol 253:H898–H903

Chapter 30
Kidney Function After In Vivo Gene Silencing of Uncoupling Protein-2 in Streptozotocin-Induced Diabetic Rats

Malou Friederich Persson, William J. Welch, Christopher S. Wilcox, and Fredrik Palm

Abstract Kidney uncoupling protein 2 (UCP-2) increases in streptozotocin-induced diabetes, resulting in mitochondria uncoupling, i.e., increased oxygen consumption unrelated to active transport. The present study aimed to investigate the role of UCP-2 for normal and diabetic kidney function utilizing small interference RNA (siRNA) to reduce protein expression. Diabetic animals had increased glomerular filtration rate and kidney oxygen consumption, resulting in decreased oxygen tension and transported sodium per consumed oxygen. UCP-2 protein levels decreased 2 and 50% after UCP-2 siRNA administration in control and diabetic animals respectively. Kidney function was unaffected by in vivo siRNA-mediated gene silencing of UCP-2. The reason for the lack of effect of reducing UCP-2 is presently unknown but may involve compensatory mitochondrial uncoupling by the adenosine nucleotide transporter.

Keywords Diabetes • Kidney • Mitochondria • siRNA • Uncoupling protein 2

M.F. Persson (✉)
Division of Integrative Physiology, Department of Medical Cell Biology, Biomedical Center, Uppsala University, Box-571 75123 Uppsala, Sweden
e-mail: Malou.Friederich@mcb.uu.se

W.J. Welch • C.S. Wilcox
Division of Nephrology and Hypertension, Department of Medicine, Medical Center, Georgetown University, Washington, DC, USA

F. Palm
Division of Integrative Physiology, Department of Medical Cell Biology, Biomedical Center, Uppsala University, Box-571 75123 Uppsala, Sweden

Division of Nephrology and Hypertension, Department of Medicine, Medical Center, Georgetown University, Washington, DC, USA

Department of Medicine and Health Sciences, Linköping University, Linköping, Sweden

W.J. Welch et al. (eds.), *Oxygen Transport to Tissue XXXIV*, Advances in Experimental Medicine and Biology 765, DOI 10.1007/978-1-4614-4989-8_30,

1 Introduction

Diabetic nephropathy is a common complication in diabetes mellitus and often results in the need for dialysis and/or kidney transplantation. In the year of 2030 the projected prevalence of diabetes mellitus worldwide is 366 million people [1] with 37% expected to develop kidney damage [2].

The mechanisms underlying diabetic nephropathy are presently unclear, but there is a connection between kidney dysfunction and alterations in the energy metabolism, increased oxidative stress, mitochondria dysfunction and increased oxygen consumption (Q_{O2}) [3–5]. The increased oxidative stress directly contributes to increased Q_{O2} resulting in decreased oxygen tension (pO_2) [4]. A major source of reactive oxygen species (ROS) is the mitochondria electron transport chain and normally 0.2% of total Q_{O2} is due to production of superoxide radicals ($O_2^{\cdot-}$) [6] but this increases in diabetes [7]. Cultured renal proximal tubular cells exposed to hyperglycemia display increased membrane potential together with increased production of $O_2^{\cdot-}$. After the initial response, a decrease in the mitochondrial membrane potential, increased Q_{O2} and decreased $O_2^{\cdot-}$production has been reported [8], indicating mitochondria uncoupling. Uncoupling proteins (UCP-2) releases protons independently of ATP-production and therefore decreases membrane potential and $O_2^{\cdot-}$production [9, 10]. In diabetes, up-regulation of kidney UCP-2 results in mitochondria uncoupling which may contribute to increased Q_{O2} [5, 11]. Therefore, we investigated the role of kidney UCP-2 in diabetes by utilizing in vivo gene silencing of UCP-2.

2 Materials and Methods

All animal procedures were approved by the Animal Care and Use Committee. Diabetes was induced in male Sprague Dawley rats (250–300 g, streptozotocin 65 mg/kg bw, tail vein). Blood glucose was measured using a reagent test strip (MediSense, Bedford, MA, USA) in a blood sample obtained from the cut tip of the tail and considered diabetic if it increased ≥15 mM within 24 h of injection and remained elevated. For administration of siRNA, a polyethylene catheter was placed in the carotid artery under isoflurane anesthesia (2% in 40% O_2). A nonfunctional scrambled (scrl siRNA) or a siRNA directed against UCP-2 (id no. 50931, Ambion, Austin, TX, USA) was administered in a volume of 6 ml of 37 °C sterile saline during 6 s via a catheter in the carotid artery which was subsequently ligated. siRNA was administered at day 5 of diabetes and all experiments carried out 2 days thereafter.

Kidney cortexes were homogenized in RIPA buffer (1% tergitol type NP40, 0.5% sodium deoxycholate, 0.1% SDS, 10 mM NaF, 80 mM Tris, pH 7.5). Enzyme inhibitors (phosphatase inhibitor cocktail-2; 10 μl/ml, Sigma–Aldrich, St. Louis, MO, USA and Complete Mini; one tablet/1.5 ml; Roche Diagnostics, Mannheim, Germany) were added before each experiment. Molecular weight separation was performed on 12.5% Tris–HCl gels with Tris/glycine/SDS buffer, the proteins transferred to nitrocellulose membranes and detected with goat anti-rat UCP-2 (1:1,000; Santa Cruz Biotechnology, Santa Cruz, CA, USA) and HRP-conjugated rabbit anti-goat, (1:10,000; Kirkegaard and Perry Laboratories, Gaithersburg, MD, USA). β-actin was detected with mouse anti-rat β-actin antibody (1:10,000, Sigma–Aldrich, St Louis, MO, USA) and secondary HRP-conjugated goat-anti mouse antibody (1:60,000; Kirkegaard and Perry Laboratories, Gaithersburg, MD, USA).

Animals were sedated with sodium thiobutabarbital (Inactin, 120 mg/kg bw nondiabetic animals, 80 mg/kg bw diabetic animals, i.p) and were placed on a heating pad that was servo-controlled at 37 °C. Tracheotomy was performed and polyethylene catheters were placed in the femoral artery and vein to allow monitoring of blood pressure (Statham P23dB, Statham Laboratories, Los Angeles, CA, USA), blood sampling and infusion of saline (5 ml/kg bw/h nondiabetic animals, 10 ml/kg bw/h diabetic animals). The left kidney was exposed by a subcostal flank incision and immobilized. The left ureter and bladder were catheterized to allow for timed urine sampling and urinary drainage, respectively. After surgery, the animal was allowed to recover for 40 min. Kidney pO_2 was measured using Clark-type oxygen electrodes (Unisense, Aarhus, Denmark). Glomerular filtration rate (GFR) and renal blood flow (RBF) were measured by clearance of ^{14}C-inulin and ^{3}H-para-aminohippuric acid (185 kBq bolus followed by 185 kBq/kg bw/h, American Radiolabelled Chemicals, St. Louis, MO, USA). RBF was calculated assuming an extraction of 70%. GFR was calculated as inulin clearance=($[inulin]_{urine}$*urine flow)/$[inulin]_{plasma}$ and RBF with PAH-clearance adjusted for the hematocrit. Total kidney Q_{O2} (μmol/min) was estimated from the arteriovenous difference in O_2 content ($O_2ct = ([Hb]*oxygen\ saturation*1.34 + pO_2*0.003)$). *total RBF. Tubular Na^+ transport (T_{Na+}, μmol/min) was calculated as follows: $T_{Na+} = [P_{Na+}]*GFR-U_{Na}V$, where $[P_{Na+}]$ is plasma Na^+ concentration and $U_{Na}V$ is the urinary Na^+ excretion. T_{Na+} per Q_{O2} was calculated as T_{Na+}/Q_{O2}. Statistical comparisons were performed using one-way analysis of variance with Bonferroni post hoc test were appropriate. $p<0.05$ was considered significant and all values are mean ± SEM.

3 Results

Increased blood glucose and decreased weight were observed with diabetes. Weight loss was also observed after administration of siRNA (Table 30.1). UCP-2 was increased in diabetic rats compared to controls but decreased in both groups by UCP-2 siRNA (Fig. 30.1). Diabetics had increased GFR which was maintained after UCP-2 siRNA (Fig. 30.2). No differences in RBF were detected between any of the groups (data not shown). Q_{O2} increased and T_{Na+}/Q_{O2} decreased in diabetic animals compared to controls (Figs. 30.3 and 30.4). Diabetic animals had decreased kidney pO_2 compared to control animals which was unaffected by UCP-2 siRNA (Fig. 30.5). All parameters investigated were unaffected by scrambled siRNA.

Table 30.1 Body weight and blood glucose

	Bw (g)	BG (mM)
Control	349±7	4.8±0.15
Control + scrambled siRNA	272±7*	4.9±0.35
Control + UCP-2 siRNA	303±6*	5.65±0.47
Diabetes	278±8*	19.3±0.52*
Diabetes + scrambled siRNA	296±9*	18.2±1.1*
Diabetes + UCP-2 siRNA	308±13*	18.2±0.7*

*Denotes $p < 0.05$ vs. untreated control animals

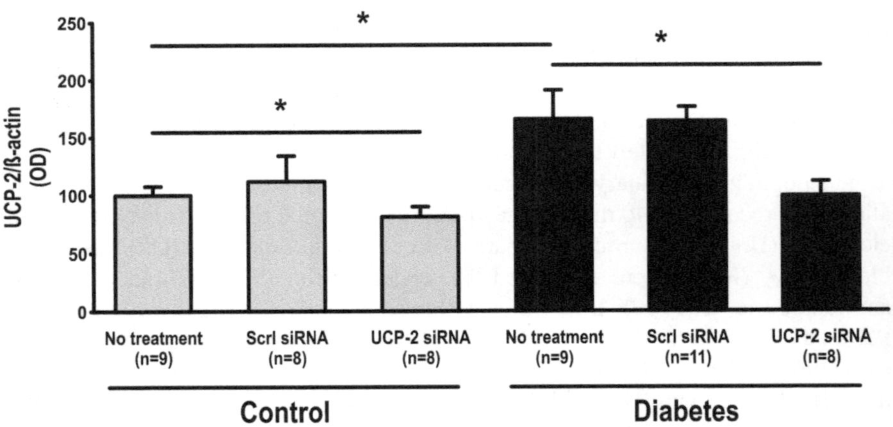

Fig. 30.1 UCP-2 protein expression in kidney cortex. *Scrl* scrambled, *asterisk* denotes $p < 0.05$

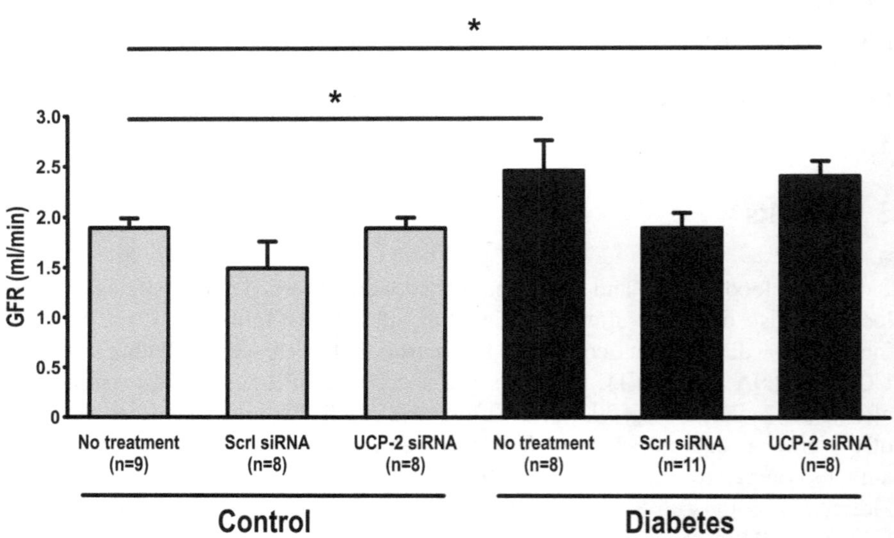

Fig. 30.2 Glomerular filtration. *Scrl* scrambled, *asterisk* denotes $p < 0.05$

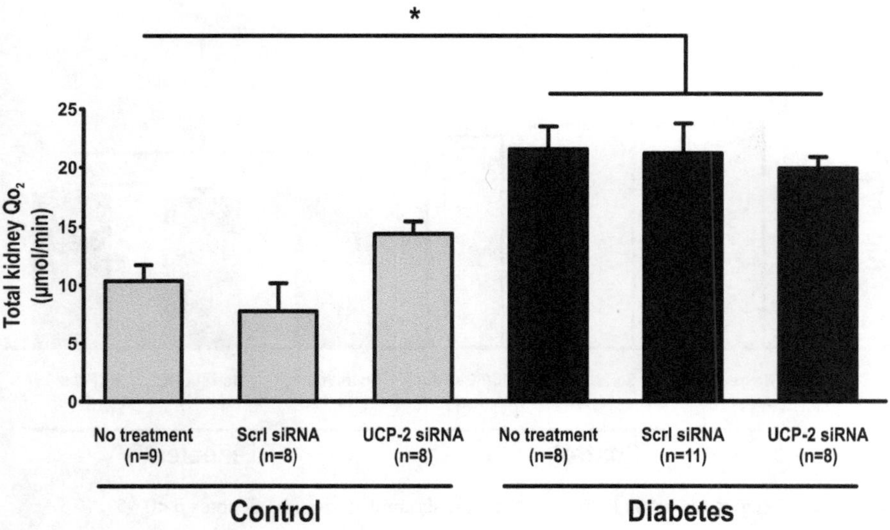

Fig. 30.3 Total kidney oxygen consumption. *Scrl* scrambled, *asterisk* denotes $p<0.05$

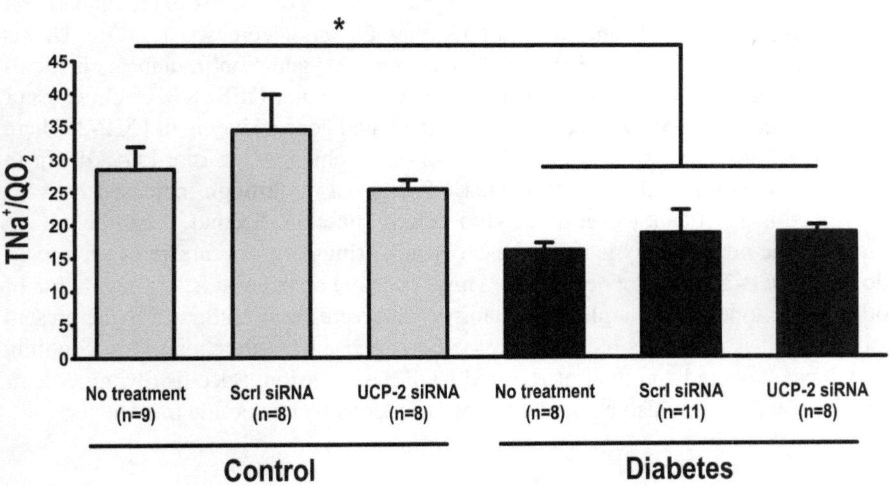

Fig. 30.4 Transported sodium per consumed oxygen. *Scrl* scrambled, *asterisk* denotes $p<0.05$

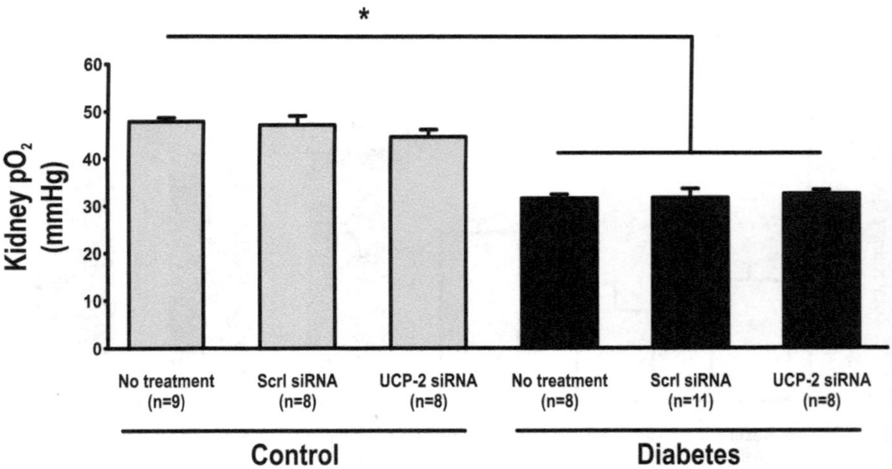

Fig. 30.5 Oxygen tension in kidney cortex. *Scrl* scrambled, *asterisk* denotes $p<0.05$

4 Discussion

The present study demonstrates increased protein levels of UCP-2 in the diabetic rat kidney concomitant with increased total kidney Q_{O2} and decreased $T_{Na+/}Q_{O2}$. This is consistent with the hypothesis that defective renal oxygenation in diabetes is mediated by enhanced expression of UCP-2. However, neither GFR, RBF, pO_2, Q_{O2} nor T_{Na+}/Q_{O2} were affected by selective siRNA-mediated gene silencing of UCP-2. There are several possible explanations. First, we only achieved a partial knockdown in UCP-2. It is possible that the residual UCP-2, possibly through increased activity, prevented a significant effect on in vivo kidney function. Second, it can be argued that diabetic nephropathy is a disease evolving during several years and 48 h knockdown of UCP-2 might not be enough. Third, possible compensatory up-regulation of other mitochondrial uncoupling mechanisms that were not investigated in the present study, but it is also possible the increased compensatory mitochondria uncoupling may have obscured an effect of reduced UCP-2 expression. Specifically, mitochondrial uncoupling can also be mediated by the adenosine nucleotide transporter.

5 Conclusion

The diabetic kidney had defective oxygenation and enhanced expression of UCP-2. siRNA-mediated knockdown of UCP-2 for 48 h in the diabetic kidney did not reverse that defect on kidney function in vivo. The role of UCP-2 and mitochondria uncoupling in diabetes needs to be further evaluated, preferentially using genetically modified mice.

Acknowledgments MPF and FP were supported by a grant from the Swedish Research Council and the Swedish Diabetes Foundation whereas WJW and CSW were supported by a grant from the NIH (DK 49870; DK 36079; HL 68686).

References

1. Wild S, Roglic G, Green A, Sicree R, King H (2004) Global prevalence of diabetes: estimates for the year 2000 and projections for 2030. Diabetes Care 27:1047–1053
2. Ostchega Y, Yoon S, Hughes J, Louis T (2008) Hypertension awareness, treatment and control – continued disparities in adults: United States, 2005–2006, NCHS DATA Brief. http://www.cdc.gov/nchs/data/databriefs/db03
3. Palm F (2006) Intrarenal oxygen in diabetes and a possible link to diabetic nephropathy. Clin Exp Pharmacol Physiol 33:997–1001
4. Palm F, Cederberg J, Hansell P, Liss P, Carlsson PO (2003) Reactive oxygen species cause diabetes-induced decrease in renal oxygen tension. Diabetologia 46:1153–1160
5. Friederich M, Fasching A, Hansell P, Nordquist L, Palm F (2008) Diabetes-induced up-regulation of uncoupling protein-2 results in increased mitochondrial uncoupling in kidney proximal tubular cells. Biochim Biophys Acta 1777:935–940
6. St-Pierre J, Buckingham JA, Roebuck SJ, Brand MD (2002) Topology of superoxide production from different sites in the mitochondrial electron transport chain. J Biol Chem 277:44784–44790
7. Nishikawa T, Edelstein D, Du XL, Yamagishi S, Matsumura T, Kaneda Y, Yorek MA, Beebe D, Oates PJ, Hammes HP, Giardino I, Brownlee M (2000) Normalizing mitochondrial superoxide production blocks three pathways of hyperglycaemic damage. Nature 404:787–790
8. Munusamy S, MacMillan-Crow LA (2009) Mitochondrial superoxide plays a crucial role in the development of mitochondrial dysfunction during high glucose exposure in rat renal proximal tubular cells. Free Radic Biol Med 46:1149–1157
9. Arsenijevic D, Onuma H, Pecqueur C, Raimbault S, Manning BS, Miroux B, Couplan E, Alves-Guerra MC, Goubern M, Surwit R, Bouillaud F, Richard D, Collins S, Ricquier D (2000) Disruption of the uncoupling protein-2 gene in mice reveals a role in immunity and reactive oxygen species production. Nat Genet 26:435–439
10. Negre-Salvayre A, Hirtz C, Carrera G, Cazenave R, Troly M, Salvayre R, Penicaud L, Casteilla L (1997) A role for uncoupling protein-2 as a regulator of mitochondrial hydrogen peroxide generation. Faseb J 11:809–815
11. Friederich M, Nordquist L, Olerud J, Johansson M, Hansell P, Palm F (2009) Identification and distribution of uncoupling protein isoforms in the normal and diabetic kidney. Adv Exp Med Biol 645: 205–212

Chapter 31
Adenosine A2 Receptor-Mediated Regulation of Renal Hemodynamics and Glomerular Filtration Rate Is Abolished in Diabetes

Patrik Persson, Peter Hansell, and Fredrik Palm

Abstract Alterations in glomerular filtration rate (GFR) are one of the earliest indications of altered kidney function in diabetes. Adenosine regulates GFR through tubuloglomerular feedback mechanism acting on adenosine A1 receptor. In addition, adenosine can directly regulate vascular tone by acting on A1 and A2 receptors expressed in afferent and efferent arterioles. Opposite to A1 receptors, A2 receptors mediate vasorelaxation. This study investigates the involvement of adenosine A2 receptors in regulation of renal blood flow (RBF) and GFR in control and diabetic kidneys. GFR was measured by inulin clearance and RBF by a transonic flow probe placed around the renal artery. Measurements were performed in isoflurane-anesthetized normoglycemic and alloxan-diabetic C57BL/6 mice during baseline and after acute administration of 3,7-dimethyl-1-propargylxanthine (DMPX), a selective A2 receptor antagonist. GFR and RBF were lower in diabetic mice compared to control (258 ± 61 vs. 443 ± 33 μl min^{-1} and $1{,}083 \pm 51$ vs. $1{,}405 \pm 78$ μl min^{-1}). In control animals, DMPX decreased RBF by −6%, whereas GFR increased +44%. DMPX had no effects on GFR and RBF in diabetic mice. Sodium excretion increased in diabetic mice after A2 receptor blockade (+78%). In conclusion, adenosine acting on A2 receptors mediates an efferent arteriolar dilatation which reduces filtration fraction (FF) and maintains GFR within normal range in normoglycemic mice. However, this regulation is absent in diabetic mice, which may contribute to reduced oxygen availability in the diabetic kidney.

Keywords Alloxan • C57BL/6 • DMPX • Renal blood flow

P. Persson • P. Hansell • F. Palm, Ph.D. (✉)
Department of Medical Cell Biology, Uppsala University, BMC Box 571, 75123, Uppsala, Sweden
e-mail: fredrik.palm@mcb.uu.se

F. Palm, Ph.D. (✉)
Department of Medical Cell Biology, Uppsala University, BMC Box 571, 75123, Uppsala, Sweden

Department of Medical and Health Sciences, Linköping University, Linköping, Sweden
e-mail: fredrik.palm@mcb.uu.se

W.J. Welch et al. (eds.), *Oxygen Transport to Tissue XXXIV*, Advances in Experimental Medicine and Biology 765, DOI 10.1007/978-1-4614-4989-8_31,

1 Introduction

Sustained hyperglycemia is associated with several complications, including nephropathy, and is the leading cause for end-stage renal disease. The severity of the hyperglycemia is a major predictor for development of diabetes-induced complications according to the Diabetes Control and Complication Trial Research Group [1]. Increased glomerular filtration rate (GFR) is one of the earliest indications of altered kidney function in diabetic patients and is commonly used as a predictor for later development of progressive renal dysfunction [2]. The mechanism mediating diabetes-induced glomerular hyperfiltration has been the subject of extensive research, and potential mechanisms have been proposed. So far, the most accepted hypothesis for the diabetes-induced glomerular hyperfiltration involves an inactivated tubuloglomerular feedback (TGF) mechanism. This inactivation is a result of reduced tubular sodium load to the macula densa due to increased tubular reabsorption secondary to increased tubular glucose load [3]. However, this hypothesis has recently been questioned by studies using adenosine A1 receptor knockout mice. These mice lack the TGF mechanism, but still display diabetes-induced glomerular hyperfiltration [4, 5]. However, they still express the two subtypes of the adenosine A2 receptor, $A2_a$ and $A2_b$. Adenosine is traditionally believed to regulate local organ blood flow, increasing during exercise and therefore increase oxygen supply in relation to metabolic demand. Blood flow regulation in the kidney is more complex since increased blood flow is coupled to increased GFR and subsequently more tubular electrolyte transport. Renal blood flow (RBF) is regulated to a great extent by adenosine. Adenosine A1 and A2 receptors are expressed on both tubules and vessels and therefore involved in both regulation of RBF and tubular electrolyte transport so that the supply of oxygen is fine tuned to match demand. Opposite to the effect of A1, activation of A2 leads to arteriolar vasodilatation. The $A2_b$ is mainly distributed on the afferent arteriole and has low affinity to adenosine while the $A2_a$ is mainly localized to the efferent arteriole and has a higher affinity to adenosine [6–8]. If adenosine concentration in isolated perfused afferent arterioles is gradually increased the initial vasoconstriction is reversed to a vasodilatation. In vivo studies in rat using a nucleoside transport inhibitor, increasing tissue adenosine concentration results in dilatation of both the afferent and efferent arteriole. The vasodilatation was abolished when the selective A2 antagonist 3,7-dimethyl-1-propargylxanthine (DMPX) was given. This dose has no blood pressure effects and adding an A1 antagonist further increased the vasodilator response [9]. These results show that high tissue adenosine concentration will increase RBF opposite to a lower concentration that will constrict the afferent arteriole thereby reducing RBF. Furthermore, the adenosine A1 and A2 receptor expressions are upregulated at both mRNA and protein levels in streptozotocin-diabetic rat with a greater increase in A2 distribution, especially in kidney cortex [10]. Receptor level seems to be the rate limiting step for adenosine function [11]. The aim of the present study is to investigate the involvement of the adenosine A2 receptors in the regulation of GFR and RBF in control and diabetic mice.

2 Methods

Age-matched male C57BL/6 mice were purchased from Charles River (Sulzfeld, Germany). Animals had free access to tap water and pelleted standard mouse chow. All experiments were conducted in accordance with national guidelines and approved by the Animal Care and Use Committee at Uppsala University. Diabetes was induced by a single injection of alloxan (75 mg kg^{-1} body wt) in the tail vein. Animals were considered diabetic if blood glucose concentration increased to >18 mM within 24 h after injection. At 21–28 days after induction of diabetes animals were anesthetized with isoflurane 1.5–2% in 100% oxygen, placed at an operating table with a servo-controlled heating pad maintaining body temperature at 37.5 °C. Catheters were placed in carotid artery and jugular vein for withdrawal of blood samples, blood pressure monitoring and infusion of Ringer solution (0.35 ml h^{-1} for controls and 0.7 ml h^{-1} for diabetics). Bladder was catheterized for urine collection for subsequent analysis. GFR was determined by measurement of [^{3}H]-inulin clearance. ^{3}H-inulin (American Radiolabeled Chemicals, St Louis, MO, USA) was given as a continuous infusion at a constant rate of 2 μCi h^{-1}. Inulin concentration in plasma and urine was determined by standard liquid scintillation technique and GFR was calculated with standard formula. After a 40-min recovery period, baseline data were obtained for a 40-min period. DMPX (Sigma-Aldrich, St Louis, MO, USA) was administered as a single bolus dose 1 mg kg^{-1} and all parameters were followed for another 40-min period. RBF was measured using an ultrasound probe (Transonic Systems, Ithaca, NY, USA) placed around the left renal artery. Kidneys were weighed at the end of each experiment. Urine flows were measured gravimetrically and urinary sodium concentration was determined by flame spectrophotometry (model IL543, Instrumentation Lab, Milan, Italy). Statistical analyses were performed using two-way ANOVA followed by Bonferroni's post hoc test.

3 Results

Diabetic animals displayed sustained hyperglycemia compared to normoglycemic controls (25.6±1.0 vs. 8.3±0.2 mM). Left kidney weight was increased in diabetic animals compared to normoglycemic controls (0.23±0.012 vs. 0.17±0.005 g). Diabetic animals weighed less compared to age-matched normoglycemic controls (26±0.7 vs. 29±0.8 g). MAP was similar at baseline and did not change by A2 blockade (Table 31.1). In diabetic mice, GFR and RBF were lower compared to controls. DMPX increased GFR and reduced RBF in controls, whereas it had no effects in diabetic mice (Figs. 31.1 and 31.2). Filtration fraction (FF) was similar during baseline and DMPX increased FF only in the controls (Table 31.1). Renal vascular resistance (RVR) was higher in diabetic mice compared to control but was unaffected by DMPX. Sodium excretion was increased in diabetic mice by DMPX.

Table 31.1 Mean arterial pressure (MAP), filtration fraction (FF), renal vascular resistance (RVR), sodium excretion (Na^+ E), and hematocrit (Hct) in control and diabetic mice during baseline and after selective A2-receptor inhibition (DMPX)

		MAP	FF	RVR	Na^+ E	Hct
		(mmHg)	–	(mmHg ml^{-1} min^{-1})	(μmol min^{-1})	(%)
Control	Baseline	73±1	0.27±0.01	54±3	45±27	41±0.6
(*n*=11)	DMPX	72±1	0.39±0.03*	56±3	40±23	40±1.0
Diabetes	Baseline	74±2	0.22±0.02	70±9#	41±7	44±0.9
(*n*=10)	DMPX	73±2	0.22±0.04	66±11	73±13*	42±0.8

All values are mean±SEM

#Denotes $P<0.05$ compared to corresponding normoglycemic control

*Denotes $P<0.05$ compared to baseline

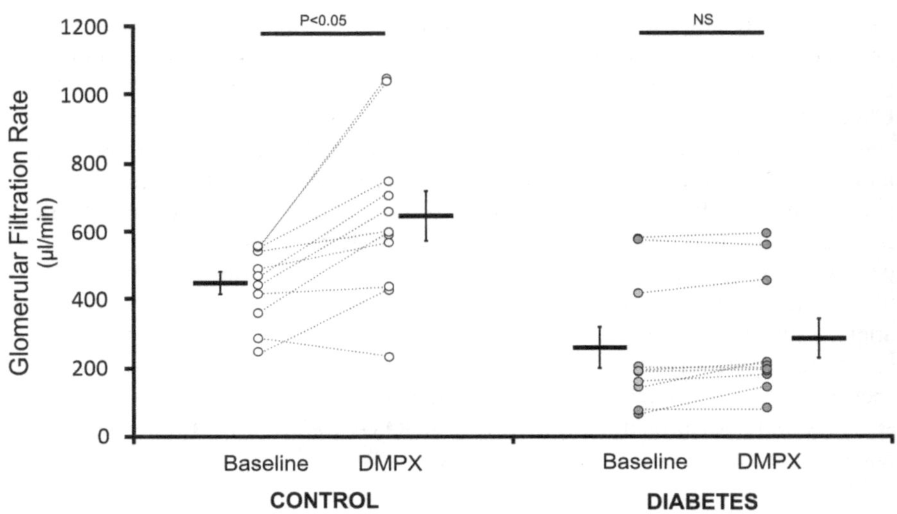

Fig. 31.1 Glomerular filtration rate in control and diabetic mice during baseline and after selective adenosine A2 receptor inhibition (DMPX). All values are mean±SEM

4 Discussion

The main finding from the present study is that A2 is an important regulator of GFR and RBF in control mice but this regulation is abolished in diabetic mice. Adenosine regulates RBF through its effect on vascular tone, mainly constriction of the afferent arteriole by A1 activation and dilatation of efferent arteriole by A2 activation. A major problem in the kidney is to match oxygen supply with demand since increasing RBF also increases GFR with a subsequent increase in tubular transport work, leaving oxygen tension unaltered. Therefore, adenosine also affects tubular transport, by stimulating proximal tubular reabsorption in the well oxygenated cortex via A1 activation and increasing medullary blood flow via A2 activation.

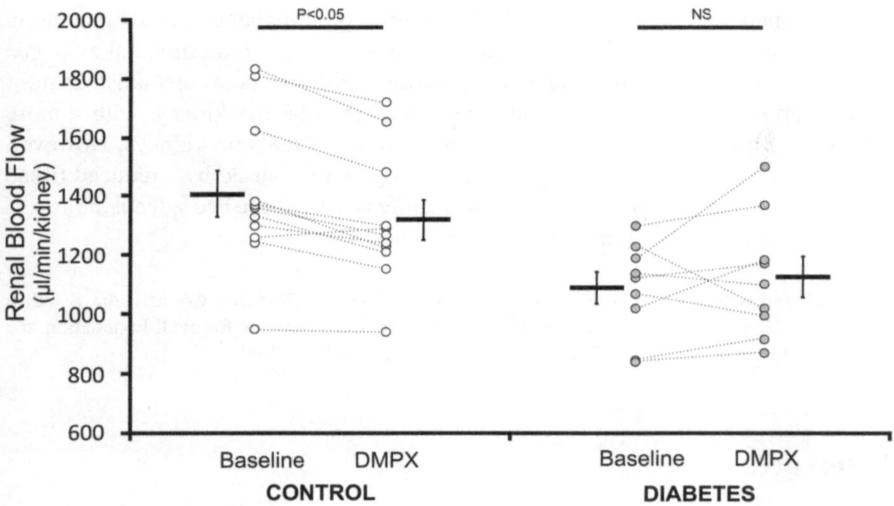

Fig. 31.2 Renal blood flow in control and diabetic mice during baseline and after selective adenosine A2 receptor inhibition (DMPX). All values are mean ± SEM

However, this is a fairly complicated system and a targeted effect of adenosine on the efferent arteriole will bring the same result. It will reduce net filtration pressure and GFR while RBF is maintained. GFR is determined by the glomerular capillary hydrostatic pressure (P_{gc}), pressure in Bowman's space and oncotic pressure in glomerular capillaries. P_{gc} is favoring GFR and determined by the vascular tone and interplay between the afferent and efferent arteriole. Increasing efferent arteriolar diameter will reduce net filtration pressure and GFR, while RBF is remarkably unchanged, which makes it an interesting therapeutic target to improve renal oxygen tension in diabetes. In this study we showed that basal A2 receptor activation in the normal kidney reduces FF and GFR, implicating that the main effect is on the efferent arteriole. This mechanism is probably important in reducing GFR but maintaining RBF, and therefore oxygenation, and is in line with a suggested efferent arteriolar TGF mediated by the A2 receptor [12]. The predominantly efferent arteriolar effect in control mice is most likely explained by the A2 receptor distribution. $A2_b$ receptors have lower affinity for adenosine and are expressed in the afferent arteriole [6], whereas $A2_a$ receptors have higher affinity and are expressed in the efferent arteriole [7]. Adenosine increases medullary blood flow and decreases cortical blood flow and, A2 receptor inhibition prevents the increase in medullary blood flow [13, 14], indicating that A2 receptors are important in shifting blood supply to the medullary region during increased adenosine formation. The observation that A2 receptor blockade has no effect on GFR in diabetic animals implicates that diabetes is inducing changes in the adenosine mediated regulation of GFR that will place extra stress on the diabetic kidney, and potentially be one factor contributing to the intra-renal hypoxia in diabetes. The mechanism for the lack of effect on GFR may include reduced adenosine formation in the diabetic kidney. However, A2 blockade increased excretion of sodium exclusively in diabetic animals indicating that adenosine influences tubular function in these animals. Experimental evidence for alterations of adenosine-mediated regulation of GFR consists of studies demonstrating up regulation of adenosine receptors in the diabetic kidney, with a more pronounced up regulation of A2 receptors in cortex of diabetic kidneys. However, this might be the result of a compensatory upregulation caused by a reduced tissue adenosine concentration. This would be manifested as abolished adenosine-mediated regulation of GFR and RBF during diabetes.

Acknowledgments The work was supported by the Swedish Research Council, the Swedish Society for Medical Research, the Lars Hierta Foundation, the Magnus Bergvall Foundation, the Åke Wiberg Foundation, and NIH/NIDDK K99/R00 grant (DK077858).

References

1. The effect of intensive treatment of diabetes on the development and progression of long-term complications in insulin-dependent diabetes mellitus. The Diabetes Control and Complications Trial Research Group (1993). N Engl J Med 329 (14):977–986
2. Magee GM, Bilous RW, Cardwell CR, Hunter SJ, Kee F, Fogarty DG (2009) Is hyperfiltration associated with the future risk of developing diabetic nephropathy? A meta-analysis. Diabetologia 52(4):691–697
3. Thomson SC, Vallon V, Blantz RC (2004) Kidney function in early diabetes: the tubular hypothesis of glomerular filtration. Am J Physiol Renal Physiol 286(1):F8–F15
4. Sallstrom J, Carlsson PO, Fredholm BB, Larsson E, Persson AE, Palm F (2007) Diabetes-induced hyperfiltration in adenosine A(1)-receptor deficient mice lacking the tubuloglomerular feedback mechanism. Acta Physiol (Oxf) 190(3):253–259
5. Faulhaber-Walter R, Chen L, Oppermann M, Kim SM, Huang Y, Hiramatsu N, Mizel D, Kajiyama H, Zerfas P, Briggs JP, Kopp JB, Schnermann J (2008) Lack of A1 adenosine receptors augments diabetic hyperfiltration and glomerular injury. J Am Soc Nephrol 19(4): 722–730
6. Feng MG, Navar LG (2010) Afferent arteriolar vasodilator effect of adenosine predominantly involves adenosine A2B receptor activation. Am J Physiol Renal Physiol 299(2):310–315
7. Al-Mashhadi RH, Skott O, Vanhoutte PM, Hansen PB (2009) Activation of A(2) adenosine receptors dilates cortical efferent arterioles in mouse. Kidney Int 75(8):793–799
8. Jackson EK, Zhu C, Tofovic SP (2002) Expression of adenosine receptors in the preglomerular microcirculation. Am J Physiol Renal Physiol 283(1):F41–F51
9. Nakamoto H, Ogasawara Y, Kajiya F (2008) Visualisation of the effects of dilazep on rat afferent and efferent arterioles in vivo. Hypertens Res 31(2):315–324
10. Pawelczyk T, Grden M, Rzepko R, Sakowicz M, Szutowicz A (2005) Region-specific alterations of adenosine receptors expression level in kidney of diabetic rat. Am J Pathol 167(2):315–325
11. Yaar R, Jones MR, Chen JF, Ravid K (2005) Animal models for the study of adenosine receptor function. J Cell Physiol 202(1):9–20. doi:10.1002/jcp. 20138
12. Ren Y, Garvin JL, Carretero OA (2001) Efferent arteriole tubuloglomerular feedback in the renal nephron. Kidney Int 59(1):222–229
13. Zou AP, Nithipatikom K, Li PL, Cowley AW Jr (1999) Role of renal medullary adenosine in the control of blood flow and sodium excretion. Am J Physiol 276(3 Pt 2):R790–R798
14. Agmon Y, Dinour D, Brezis M (1993) Disparate effects of adenosine A1- and A2-receptor agonists on intrarenal blood flow. Am J Physiol 265(6 Pt 2):F802–F806

Chapter 32
Can Mitochondrial Cytochrome Oxidase Mediate Hypoxic Vasodilation Via Nitric Oxide Metabolism?

Zimei Rong, Murad Banaji, Tracy Moroz, and Chris E. Cooper

Abstract The brain responds to hypoxia with an increase in cerebral blood flow (CBF). Many mechanisms have been proposed for this hypoxic vasodilation, but none has gained universal acceptance. Although there is some disagreement about the shape of the relationship between arterial oxygen partial pressure (PaO_2) and CBF, it is generally agreed that CBF does not increase until the PaO_2 reaches a threshold value. We used a previously published computational model of brain oxygen transport and metabolism (BRAINSIGNALS) to test possible molecular mechanisms for such a threshold phenomenon. One suggestion has been that a decrease in the metabolism of nitric oxide by mitochondrial cytochrome *c* oxidase (CCO) at low PaO_2 could be responsible for raising NO levels and the consequent triggering of the hypoxic blood flow increase. We tested the plausibility of this mechanism using the known rate constants for NO interactions with CCO. We showed that the shape of the CBF–PaO_2 curve could indeed by reproduced, but only if NO production by the enzyme nitric oxide synthase had a very low Michaelis constant K_m for oxygen. Even then, in the current version of BRAINSIGNALS the NO-induced CBF rise occurs at much lower PaO_2 than is consistent with the in vivo data.

Keywords Cytochrome oxidase • Hypoxic vasodilation • Nitric oxide

Z. Rong (✉)
Department of Biological Sciences, University of Essex, Colchester CO4 3SQ, UK

Department of Medical Physics and Bioengineering, University College London, London WC1E 6BT, UK

University of Essex, 23 Sydney Road, Ilford, Essex, IG6 2ED, UK
e-mail: zrong@essex.ac.uk

M. Banaji • T. Moroz • C.E. Cooper
Department of Biological Sciences, University of Essex, Colchester CO4 3SQ, UK

Department of Medical Physics and Bioengineering, University College London, London WC1E 6BT, UK

W.J. Welch et al. (eds.), *Oxygen Transport to Tissue XXXIV*, Advances in Experimental Medicine and Biology 765, DOI 10.1007/978-1-4614-4989-8_32,

1 Introduction

Hypoxic vasodilation refers to the process whereby an oxygen partial pressure (content) decrease causes blood vessel expansion and consequent increase in blood flow. A key question is to understand the mechanisms bridging the oxygen tension decrease and the blood flow increase. More specifically, what causes blood vessel dilation leading to flow resistance decrease? Although hypoxic vasodilation is a general phenomenon, in this paper we focus on a key oxygen-dependent organ—the brain.

Shimojyo et al. [1] examined the details of the CBF–PaO_2 relationship in the human and concluded that there was no significant change in CBF until PaO_2 reached a 30 mmHg threshold. Gjedde [2] analyzed the experimental data reported by Shimojyo et al. and proposed a formula for CBF as a function of PaO_2, that enabled cerebral metabolic rate of oxygen consumption $CMRO_2$ to remain constant through the hypoxic period. However, human data are necessarily limited in extent and quality. Grubb et al. [3, 4] published high quality CBF–PaO_2 curves in ducks and proposed a functional formula as $CBF = CBF_0 + CBF_1 \times \exp(-k \times PaO_2)$, where CBF_0, CBF_1, and k are constants.

We then wish to use these models to test a specific mechanism whereby a decrease in PaO_2 can trigger a CBF rise. The mechanism we test here is whether mitochondrial cytochrome *c* oxidase can trigger hypoxic vasodilation by controlling the levels of the vasodilator, nitric oxide [5].

2 Mathematical Model

The PaO_2 vasodilation threshold is the PaO_2 value below which CBF starts to increase. For the purposes of this paper, we determined the threshold as the CBF increases more than the noise by simple visual inspection. We digitized the reported experimental results and found that the following threshold values for PaO_2: 40 mmHg in humans by Shimojyo et al. [1], 45 mmHg in humans by Brown et al. [6], 60 mmHg in dogs by Kogure et al. [7] and McDowall [8], 75 mmHg in ducks in normocapnia by Grubb et al. [3], and 55 mmHg in hypocapnia [4]. This led to the generally held view that there is a clear threshold in the CBF–PaO_2 curve [9]; although Ellingsen et al. [10] contested this idea they did not provide detailed experimental evidence to support their claims. It is important to note that the threshold behavior may be partially explained by the fact that, as a consequence of the sigmoidal shape of the hemoglobin–oxygen dissociation curve, when oxygen partial pressure decreases from 100 to 50 mmHg, the oxygen content in blood shows a much smaller decrease.

Our BRAINSIGNALS model [11] which is available online [12] simulates phenomenologically a direct effect of oxygen concentration on blood vessel expansion with no implied mechanism. BRAINSIGNALS is a two component model with submodels of cerebral blood circulation and mitochondrial metabolism [11]. The key variable controlling blood flow is an averaged inner radius of the resistance vessels which is determined from the equilibrium of forces acting on the vessel walls. One of the forces, the muscular force is expressed as a function of stimuli η, which in turn is a function of biophysical and metabolic variables. In the original model, a stimulus η is a function of four quantities: arterial blood pressure, capillary oxygen level, arterial carbon dioxide partial pressure and a quantity representing metabolic demand. In this paper we only consider the effects ascribed to capillary oxygen, maintaining the other components as constant (whilst this is a valid assumption for studies on anesthetized animals with controlled ventilation [8], we are aware that in human volunteer studies hypoxia-induced secondary changes in other inputs [especially $PaCO_2$] need to be taken into account [1]). The relevant equation is:

$$\eta = R_X\left(\frac{v_X}{v_{Xn}} - 1\right) + \cdots X = O_2. \tag{32.1}$$

We also consider a revised version of this equation, where NO is the variable X in this equation (with O_2 indirectly controlling NO levels via its effect on CCO). v_{O_2} and v_{NO} are concentrations of O_2 and NO passed through a first order filter [2] in order to take account of possible time delays in the processes leading to vasodilation. v_{O_2n} and v_{NOn} are the normal values. R_{O_2} and R_{NO} are sensitivity parameters corresponding to the stimuli. As calculated from the balance of the forces on the vessel wall, the radius is a function of the stimulus. This provides one link between the circulatory and metabolic components. The other link between these two components is CBF which is also a function of radius r: $CBF=(P_a-P_v)K_G r^4$. Oxygen delivery from blood to tissue (mitochondria) is proportional to CBF.

Metabolism of O_2 occurs in mitochondria (Fig. 32.1a). The same notation is used as in [11]. We focus on two redox centers in CCO, namely, Cu_A and cytochrome a_3. First, NADH transfers electrons to Cu_{Ao} with a reaction rate f_1, forming Cu_{Ar} and NAD^+. Cu_{Ar} then transfers electrons to cytochrome a_3 with a rate of f_2, forming a_{3r} and Cu_{Ao}. In the final step a_{3r} transfers electrons to oxygen with a rate of f_3. During the reaction, a_{3r} is oxidized to a_{3o} and oxygen is reduced to water.

To model NO mediation of hypoxic vasodilation, we first introduce an NO synthesis pathway via a simple Michaelis–Menten equation representing the activity of the enzyme nitric oxide synthase (NOS).

$$\frac{dNO}{dt} = \frac{V_{mNO} \times O_2}{K_{mNO} + O_2}. \tag{32.2}$$

We modelled the NO interactions with CCO incorporating both the inhibition via bonding to the reduced enzyme and the metabolism via binding to the oxidized enzyme [13]. The mechanism and rates used were from the simplified model of these

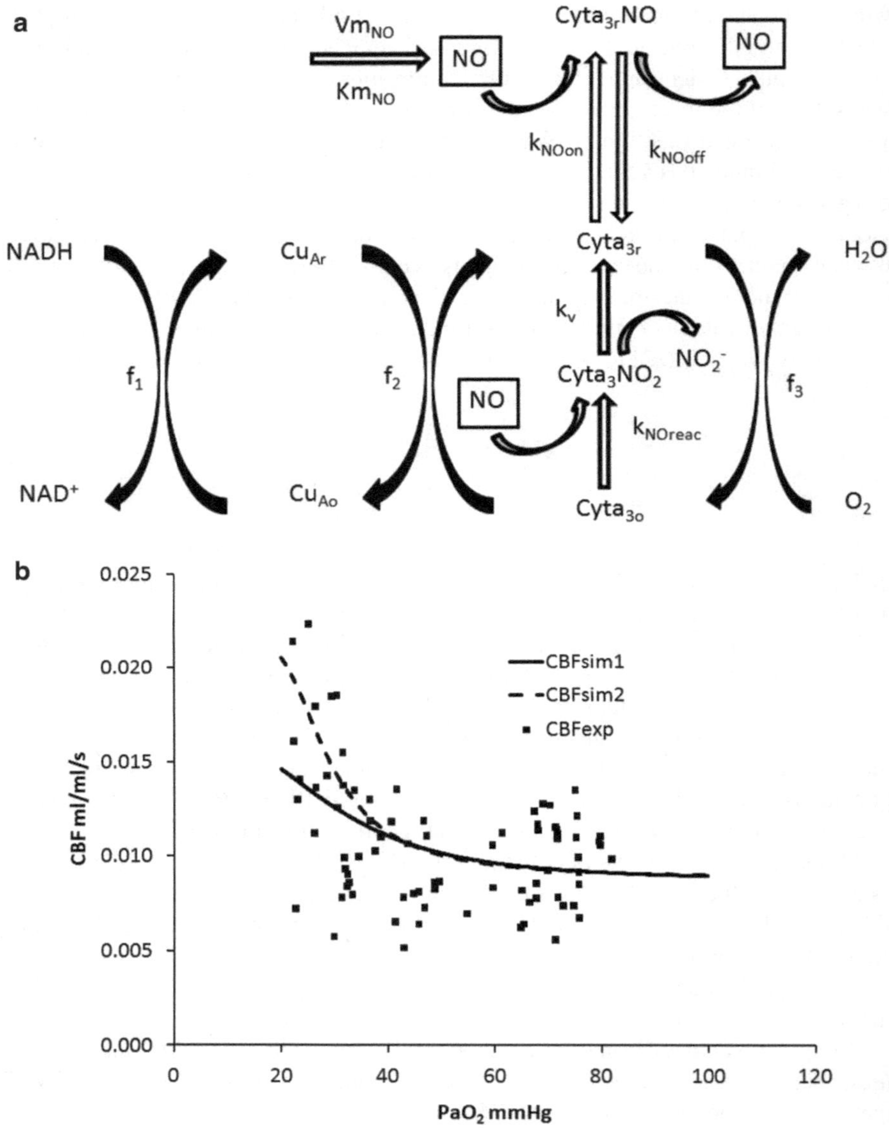

Fig. 32.1 (**a**) Schematic diagram of the mitochondrial submodel. The chemical reaction rates for competitive inhibition are $k_{NOon}=0.04\ nM^{-1}\ s^{-1}$ and $k_{NOoff}=0.16\ s^{-1}$. The chemical reaction rates for uncompetitive inhibition are $k_{NOreac}=2.0\times10^{-4}\ nM^{-1}\ s^{-1}$ and $k_v=0.75\ s^{-1}$. (**b**) Experimental CBF ~ PaO_2 in humans was reported by Shimojyo et al. [1]. The *solid line* is the simulated CBF ~ PaO_2 using the BRAINSIGNALS model with the stimuli as (32.1) and the *dashed line* with the stimuli as (32.5) [11]

interactions by Antunes et al. [14]. First, NO reversibly reacts with reduced CCO in competition with O_2 with a binding rate k_{NOon} and dissociation rate of k_{NOoff}.

$$\frac{\mathrm{dNO}}{\mathrm{d}t} = -k_{\mathrm{NOon}} \times a_{3\mathrm{r}} \times \mathrm{NO} + k_{\mathrm{NO_{NOoff}}} \times a_{3\mathrm{r}}\mathrm{NO}. \tag{32.3}$$

Second, NO reacts with oxidized CCO in two steps. NO reacts with $a_{3\mathrm{o}}$ to form an a_3NO_2 complex with a reaction rate of k_{NOreac}. Then a_3NO_2 dissociates to become $a_{3\mathrm{r}}$ and NO_2^-.

$$\frac{\mathrm{dNO}}{\mathrm{d}t} = -k_{\mathrm{NOreac}} \times a_{3\mathrm{o}} \times \mathrm{NO}. \tag{32.4}$$

As the enzyme converts from the NO-metabolizing oxidized state to the NO inhibited reduced state following a drop in PaO_2, the combination of these processes has been proposed by Palacios-Callender et al. [5] to result in NO-induced hypoxic vasodilation.

3 Results and Discussions

In a first set of simulations we examined whether the BRAINSIGNALS model, without the inclusion of NO, could reproduce the CBF–PaO_2 curve, which we digitized from Shimojyo et al. [1]. It can be seen that the simulated CBF (solid line of Fig. 32.1b) gradually increases when PaO_2 decreases. Although, because of the noise in the experimental data, it is difficult to evaluate agreement between experiment and simulation for PaO_2 values ranging from 50 to 100 mmHg PaO_2 the simulated CBF did not appear to be an optimal fit to the data. Consequently we propose replacing (32.1) with a new equation modelling the PaO_2 component of cerebral blood flow:

$$\eta = R_{\mathrm{O}_2}\left(1 - \frac{\left[1 + \left(\frac{P_{50}}{v_{\mathrm{O}_2}}\right)^{n_h}\right]}{\left[1 + \left(\frac{P_{50}}{v_{\mathrm{O}_2 n}}\right)^{n_h}\right]}\right). \tag{32.5}$$

Equation (32.5) the simulated CBF (dashed line in Fig. 32.1b) appears to match the experimental values better.

Due to the noise in the human data we then turned to the higher quality animal CBF ~ PaO_2 data sets and compared the simulation results with the CBF ~ PaO_2 relationship in dogs using experimental data digitized from McDowall [8]. Again using

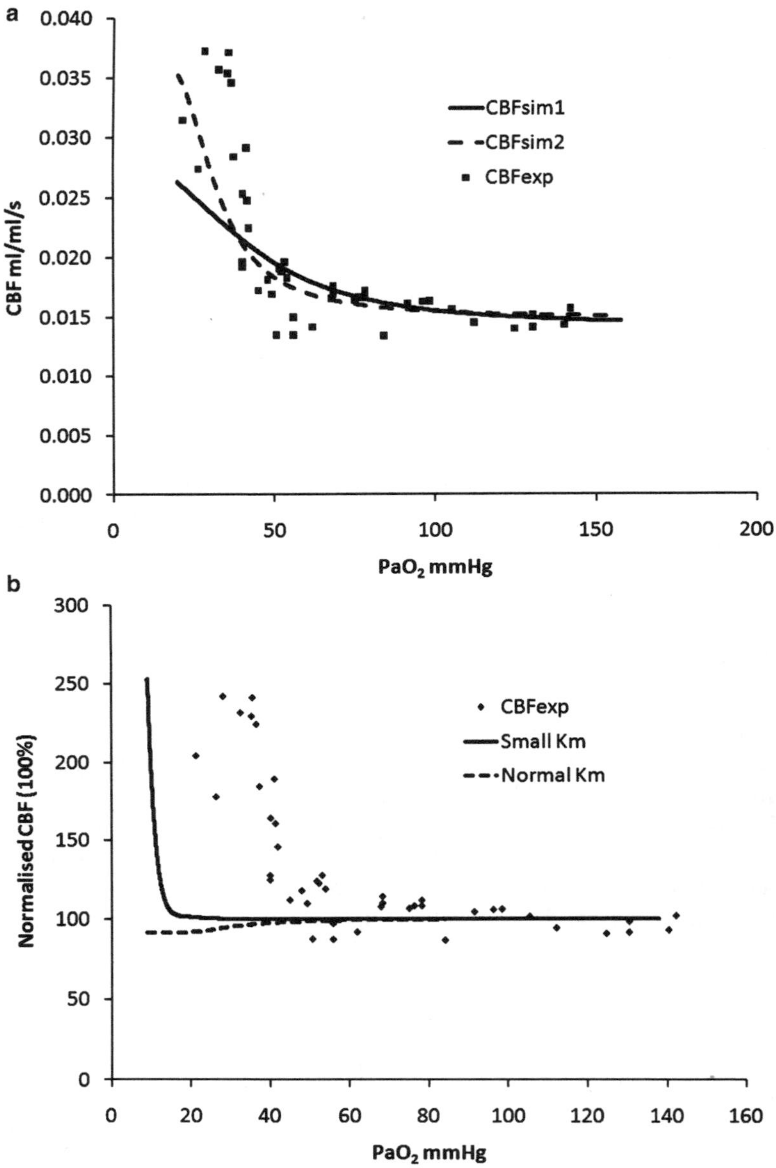

Fig. 32.2 (**a**) Experimental CBF ~ PaO_2 in dogs was reported by McDowall [8]. The *solid line* is the simulated CBF ~ PaO_2 using the BRAINSIGNALS model with the stimuli as (32.1) and the *dashed line* with the stimuli as (32.5) [11]. (**b**) Experimental CBF ~ PaO_2 in humans was reported by Shimojyo et al. [1]. The NOS oxygen K_m was set to 0.012 mM (*dashed line*) and 0.00003 mM (*solid line*)

(32.5) appears on visual inspection to offer a better fit to the experimental data (Fig. 32.2a), especially when PaO_2 is <60 mmHg.

Equation (32.5) still lacks mechanistic detail. However, it is possible to modify our BRAINSIGNALS model to test the specific molecular mechanism of hypoxic vasodilation discussed above, namely, that changes in NO metabolism by CCO trigger the blood flow increase [5]. Figure 32.2b shows the results of such a simulation. Nitric oxide synthase (NOS) is an oxygen-sensitive enzyme. If NOS activity was set to its published oxygen K_m (12 μM), then simulations actually predict an oxygen-dependent *drop* in NO levels. NO concentration decreases from 1 to 0.005 nM when PaO_2 decreases from 140 to 9 mmHg, resulting in hypoxic vasoconstriction, not vasodilation. However, if the NOS K_m is set very low (30 nM) then the drop in NO metabolism by CCO dominates the change in NO levels during hypoxia. NO concentration now rises from 1 to 10 nM when PaO_2 decreases to 10 mmHg. This results in a CBF increase with a sharp threshold. However, the threshold is too low at 15 mmHg.

4 Conclusions and Future Work

NO is one of several possible effectors that can mediate hypoxic vasodilation. This work has shown that it is theoretically possible for mitochondrial cytochrome *c* oxidase to control NO levels in the brain. Decreased NO metabolism at low PaO_2 can indeed trigger a CBF increase. However, this requires an unusually low K_m for oxygen for NOS. Even then the fit to the data is poor. The computational model used—BRAINSIGNALS—has minimal compartmentation, yet O_2 and NO sensing, signalling and metabolism can occur in spatially distinct compartments. Indeed a significant fraction of the sensing mechanism could be localized to a region of the brain [15]; this might explain some of the anomalies. Alternatively, rather than NO metabolism changes being the hypoxic trigger, NO synthesis could increase in hypoxia as suggested by the hemoglobin nitrite reductase theory of hypoxic vasodilation [16]. We are actively exploring these possibilities.

Acknowledgment This work is financially supported by the Leverhulme Trust.

References

1. Shimojyo S, Scheinberg P, Kogure K et al (1968) The effect of graded hypoxia upon transient cerebral blood flow and oxygen consumption. Neurology 18:127–133
2. Gjedde A (2002) Cerebral blood flow change in arterial hypoxemia is consistent with negligible oxygen tension in brain mitochondria. NeuroImage 17:1876–1881
3. Grubb B, Colacino JM, Schmidt-Nielsen K (1978) Cerebral blood flow in birds: effect of hypoxia. Am J Physiol 234 (Heart Circ Physiol 3(3)):H230–H234

4. Grubb B, Jones JH, Schmidt-Nielsen K (1979) Avian cerebral blood flow: influence of the Bohr effect on oxygen supply. Am J Physiol 234 (Heart Circ Physiol 5(5)):H744–H749
5. Palacios-Callender M, Hollis V, Mitchison M et al (2007) Cytochrome *c* oxidase regulates endogenous nitric oxide availability in respiring cells: a possible explanation for hypoxic vasodilation. Proc Natl Acad Sci USA 104:18508–18513
6. Brown MM, Wade JPH, Marshall J (1985) Fundamental importance of arterial oxygen content in the regulation of cerebral blood flow in man. Brain 108:81–93
7. Kogure K, Scheinberg P, Reinmuth OM et al (1970) Mechanisms of cerebral vasodilatation in hypoxia. J Appl Physiol 29:223–229
8. McDowall DG (1966) Interrelationship between blood oxygen tensions and cerebral blood flow. In: Payne JP, Hill DW (eds) Oxygen measurements in blood and tissue and their significance. Churchill, London, pp 205–219
9. Gupta AK, Menon DK, Czosnyka M et al (1997) Threshold for hypoxic cerebral vasodilation in volunteers. Anesth Analg 85:817–820
10. Ellingsen I, Hauge A, Nicolaysen G et al (1987) Changes in human cerebral blood flow due to step changes in P_{AO2} and P_{ACO2}. Acta Physiol Scand 129:157–163
11. Banaji M, Mallet A, Elwell CE et al (2008) A model of brain circulation and metabolism: NIRS signal changes during physiological challenges. PLoS Comput Biol 4(11):e1000212. doi:10.1371/journal.pcbi.1000212
12. BRAINCIRC: an open source modelling environment. http://braincirc.sourceforge.net. Brain signal model: http://www.medphys.ucl.ac.uk/braincirc/download/repos/NIRSmodel.html.
13. Cooper CE (2002) Nitric oxide and cytochrome oxidase: substrate, inhibitor or effector? Trend Biochem Sci 27:33–39
14. Antunes F, Boveris A, Cadenas E (2007) On the biological role of the reaction of NO with oxidized cytochrome *c* oxidase. Antioxid Redox Signal 9:1569–1579
15. Golanov EV, Christenson JRC, Reis DJ (2001) Neurons of a limited subthalamic area mediate elevations in cortical cerebral blood flow evoked by hypoxia and excitation of neurons of the rostral ventrolateral medulla. J Neurosci 21:4032–4041
16. Crawford JH, Isbell TS, Huang Z et al (2006) Hypoxia, red blood cells, and nitrite regulate NO-dependent hypoxic vasodilation. Blood 107:566–574

Chapter 33
Effects of Occlusal Disharmony on Working Memory Performance and Prefrontal Cortex Activity Induced by Working Memory Tasks Measured by NIRS

Kaoru Sakatani, Takeo Tsujii, Teruyasu Hirayama, Youichi Katayama, Tomotaka Takeda, Ai Amemiya, and Keiichi Ishigami

Abstract The effects of artificial occlusal disharmony (AOD) on working memory function and prefrontal cortex (PFC) activity in the elderly were examined. We evaluated working memory function using the modified Sternberg test (ST). We measured activity in the bilateral PFC during ST using near-infrared spectroscopy (NIRS) before and after AOD: the mandibular position was displaced by a splint for 10 min. AOD caused a gradual increase of oxyhemoglobin (oxy-Hb) in the bilateral PFC. The response time of ST (six digits) after AOD was longer than that before AOD. The oxy-Hb increase during ST after AOD was smaller than that before AOD. These results indicate that short-term physical stress caused by AOD decreased working memory function in elderly subjects, associated with a decrease of the evoked PFC activity during working memory function.

Keywords Dementia • NIRS • Occlusal disharmony • Prefrontal cortex • Working memory

---

K. Sakatani (✉) • T. Tsujii
Division of Optical Brain Engineering, Nihon University School of Medicine, 30-1 Oyaguchi-uemachi, Itabasi-ku, Tokyo 173-0032, Japan
e-mail: sakatani@med.nihon-u.ac.jp

T. Hirayama • Y. Katayama
Division of Neurosurgery, Department of Neurological Surgery, Nihon University School of Medicine, Tokyo, Japan

T. Takeda • A. Amemiya • K. Ishigami
Department of Sports Dentistry, Tokyo Dental College, Chiba, Japan

W.J. Welch et al. (eds.), *Oxygen Transport to Tissue XXXIV*, Advances in Experimental Medicine and Biology 765, DOI 10.1007/978-1-4614-4989-8_33,

1 Introduction

Recent studies have shown that chewing can enhance cognitive performance. For example, gum chewing appeared to be of benefit to verbal working memory, immediate episodic long-term memory, language-based attention, and processing speed [1]. These findings suggested that acute occlusal disharmony might attenuate cognitive function, including working memory function. Indeed, reduced mastication caused by occlusal disharmony could be a risk factor for development of dementia in humans [2].

In the present study, we examined the effects of artificial occlusal disharmony (AOD) on working memory, which is a system for actively maintaining and manipulating information, and forms an integral part of the human memory system. We used NIRS to evaluate the effects of AOD on working memory performance and prefrontal cortex (PFC) activity during a working memory task in the elderly.

2 Methods

2.1 *Procedures*

We studied 18 normal elderly subjects (7 males, 11 females; mean age of 66.1 ± 4.8 years). All subjects were healthy, with no psychiatric or neurological disorders. Written informed consent was obtained from each subject on forms approved by the ethical committee of the Nihon University School of Medicine.

We employed the modified Sternberg test as a working memory task. In the Sternberg test, subjects were asked to remember one digit and six digits by turns. There were eight 1-digit trials and eight 6-digit trials. Each trial began with the presentation of one digit or a set of six digits to be encoded for 1 s on a CRT. Then a blank display was inserted for 2 s, followed by the test digit until a response was obtained within 2 s. Subjects held a small box with two buttons side by side. They were required to press the right button if they thought the test digit was contained within the encoded stimulus and to press the left one if not, as quickly and accurately as possible. Similar tasks have been used previously in NIRS experiments and have been demonstrated to activate the LPFC. In order to assess psychological stress levels, subjects were asked to fill in the State-Trait Anxiety Inventory (STAI) before and after AOD.

2.2 *NIRS Measurements and Data Analysis*

We measured cerebral blood oxygenation (CBO) in the bilateral PFC using a two-channel NIRS system (PNIRS-10, Hamamatsu Photonics K.K., Japan), which sends data wirelessly to a PC. The NIRS system uses LEDs of three different wavelengths (735, 810, and 850 nm) as light sources and one photo-diode as a detector; it has two channels. Two AAA batteries allow up to 8 h of continuous measurement for the two-probe operation. The sampling rate was 61.3 Hz (i.e., the sampling time was about 16.3 ms). The NIRS probes were set symmetrically on the forehead; the positioning is similar to the midpoint between electrode positions Fp1/Fp3 (left) and Fp2/Fp4 (right) of the international electroencephalographic 10–20 system.

Three experimental conditions were established: a resting condition, AOD (a position in which experimental horizontal mandibular deviation was maintained by a splint), and control condition (a mandibular rest position maintained by a splint). We monitored the CBO changes continuously by NIRS during (1) resting conditions for 5 min, (2) the working memory task for 5 min, (3) the recovery phase for 1 min, (4) AOD or control condition for 10 min, (5) resting conditions for 5 min, (6) the working memory task for 5 min, and (7) the recovery phase for 5 min.

To analyze PFC activity, we calculated changes in oxy-Hb concentration. The mean control values (measured during the first 10 s) were subtracted from the mean activation values (measured throughout task performance). We compared oxy-Hb changes and STAI score in control condition (splint without displacement of the lower jaw) and AOD (splint with displacement of the lower jaw).

3 Results

STAI in the AOD condition (38.6 ± 1.9) tended to show a higher score than that in the control condition (37.9 ± 1.9, $p = 0.07$). NIRS showed a gradual increase of oxy-Hb in the bilateral PFC during AOD. Figure 33.1a shows a typical example of oxy-Hb changes in the bilateral PFC during AOD. AOD increased oxy-Hb in both the right and left PFC. However, the increase of oxy-Hb in the right PFC was statistically significant ($p < 0.05$), while the increase in the left PFC was not significant ($p = 0.18$) (Fig. 33.2b).

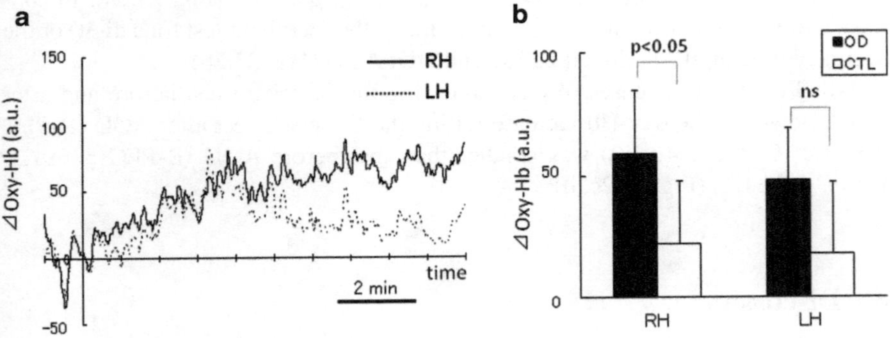

Fig. 33.1 Changes of oxy-Hb during AOD. (**a**) A typical example of oxy-Hb changes during AOD. (**b**) Differences in AOD-induced oxy-Hb changes between the right and left PFC. AOD increased oxy-Hb in both the right and left PFC. However, the increase of oxy-Hb in the right PFC was statistically significant ($p < 0.05$), while the increase in the left PFC was not significant ($p = 0.18$)

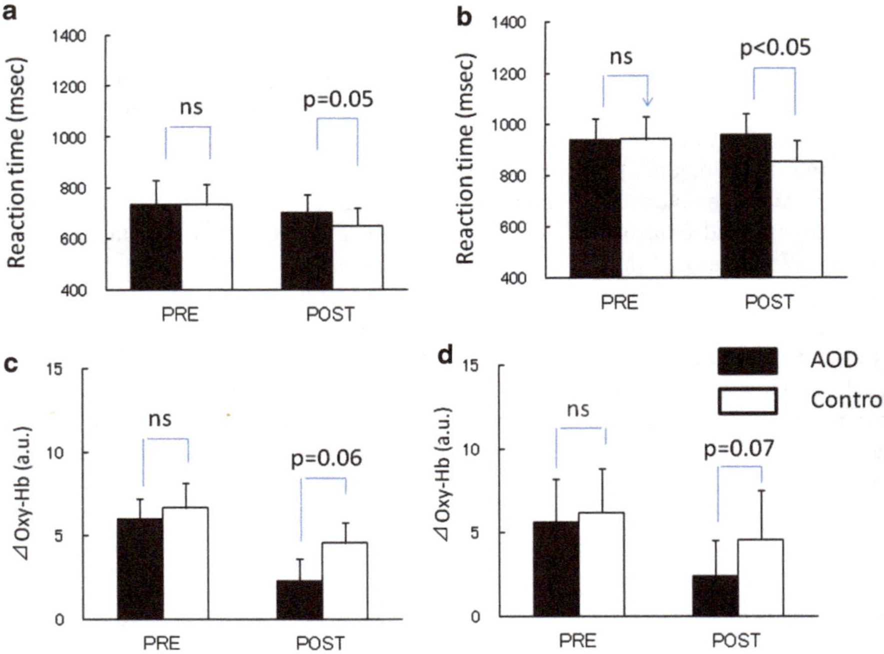

Fig. 33.2 Effects of AOD on working memory. (**a**) Effect of AOD on reaction time during the Sternberg test (one-digit). The reaction time after AOD tended to be slower than that before AOD ($p=0.07$). (**b**) Effect of AOD on reaction time during the Sternberg test (six-digit). The reaction time after AOD was significantly slower than that before AOD ($p<0.05$). (**c** and **d**) Effect of AOD on PFC activity during working memory performance in the left (**c**) and right (**d**) PFC. AOD tended to suppress the oxy-Hb increase during the Sternberg test in the left ($p=0.06$) and right ($p=0.07$) PFC

The response time of the Sternberg test (six digits) after AOD (960.9 ± 83.7 ms) was longer than that before AOD (941.0 ± 84.2 ms, $p<0.05$) (Fig. 33.2a). In contrast, there was no difference in response time of the Sternberg test (one digit) or the accuracy (one digit, six digits) before and after AOD (Fig. 33.2b).

NIRS revealed increases of oxy-Hb during the Sternberg test before and after AOD; however, the oxy-Hb increase during the Sternberg test after AOD (R-PFC 2.3 ± 1.4; L-PFC 2.4 ± 1.2) was smaller than that before AOD (R-PFC 6.0 ± 1.3; L-PFC 5.7 ± 1.1) (Fig. 33.2c, d).

4 Discussion

The subjects felt discomfort or pain during AOD, which could cause stress responses in the brain. Indeed, AOD increased oxy-Hb in the bilateral PFC, indicating that AOD induced neuronal activation of the PFC. The AOD-induced PFC activation could cause activation of the hypothalamic–pituitary–adrenal (HPA) axis, since neuronal networks exist between the PFC and the neuroendocrine centers in the medial hypothalamus [3]. Interestingly, recent studies have demonstrated that the HPA axis is regulated by the PFC, particularly the right PFC. Electroencephalographic studies have shown that a greater right frontal activation is associated with increased heart rate during unpleasant emotional stimuli [4]. Animal experiments demonstrated that lesions to the right or bilateral PFC decrease prestress glucocorticoids (GC) levels and the stress-induced GC response in rats [5]. A recent fMRI study revealed that right dominance of PFC activity during mental stress tasks correlated with changes in salivary-GC levels and heart rate [6]. In addition, employing NIRS, we have demonstrated that right dominant PFC activity during mental tasks caused larger heart rate changes and activation of the HPA axis [7–10].

The present study revealed that AOD decreased working memory performance, associated with a decrease of the evoked PFC activity during working memory performance in elderly subjects. The decreases of working memory performance and evoked PFC activity might be caused by activation of the HPA axis, which increases levels of GC secretion. It has been reported that stress exposure or GC administration impairs working memory, which relies on the integrity of the PFC. Systemic injections of GC impair working memory performance in rats [11] and human subjects [12].

In summary, the present results indicate that short-term physical stress caused by AOD decreased working memory function in elderly subjects, associated with a decrease of the evoked PFC activity during working memory function. We suggest that reduced mastication could be a risk factor for cognitive dysfunction in the elderly, and emphasize the importance of oral care in elderly persons.

Acknowledgments This research was partly supported by Japan Science and Technology Agency, under the Strategic Promotion of Innovative Research and Development Program, and a Grant-in-Aid from the Ministry of Education, Culture, Sports, Sciences and Technology of Japan (B23300247).

References

1. Wilkinson L, Scholey A, Wesnes K (2002) Chewing gum selectively improves aspects of memory in healthy volunteers. Appetite 38:235–236
2. Stein PS, Desrosiers M, Donegan SJ et al (2007) Tooth loss, dementia and neuropathology in the Nun study. J Am Dent Assoc 138:1314–1322
3. Buijs RM, van Eden CG (2000) The integration of stress by the hypothalamus, amygdale and prefrontal cortex: balance between the autonomic nervous system and the neuroendocrine system. Prog Brain Res 126:117–132
4. Waldstein SR, Kop WJ, Schmidt LA et al (2000) Frontal electrocortical and cardiovascular reactivity during happiness and anger. Biol Psychol 55:3–23
5. Sullivan RM, Gratton A (1999) Lateralized effects of medial prefrontal cortex lesions on neuroendcrine and autonomic stress responses in rats. J Neurosci 19:2834–2840
6. Wang J, Rao H, Wetmore GS et al (2005) Perfusion functional MRI reveals cerebral blood flow pattern under psychological stress. Proc Natl Acad Sci U S A 102:17804–17809

7. Tanida M, Sakatani K, Takano R et al (2004) Relation between asymmetry of prefrontal cortex activities and the autonomic nervous system during a mental arithmetic task: Near infrared spectroscopy study. Neurosci Lett 369:69–74
8. Tanida M, Katsuyama M, Sakatani K (2007) Relation between mental stress-induced prefrontal cortex activity and skin conditions: a near infrared spectroscopy study. Brain Res 1184:210–216
9. Tanida M, Katsuyama M, Sakatani K (2008) Effects of fragrance administration on stress-induced prefrontal cortex activity and sebum secretion in the facial skin. Neurosci Lett 432:157–161
10. Sakatani K, Tanida M, Katsuyama M (2010) Effects of aging on activity of the prefrontal cortex and autonomic nervous system during mental stress task. Adv Exp Med Biol 662:473–478
11. Roozendaal B, McReynolds JR, McGaugh JL (2004) The basolateral amygdala interacts with the medial prefrontal cortex in regulating glucocorticoid effects on working memory impairment. J Neurosci 24:1385–1392
12. Wolf OT, Convit A, McHugh PF et al (2001) Cortisol differentially affects memory in young and elderly men. Behav Neurosci 115:1002–1011

Chapter 34
Biological Maintenance of Distal Vein Arterialization

Tadahiro Sasajima and Tomiyasu Koyama

Abstract Eleven weeks after surgery, a fine microvessel network was seen in the feet of patients with *arteriosclerosis obliterans* treated by distal vein arterialization. A possible mechanism for establishment of blood flow to, and biological maintenance of, the graft in the foot, in relation to oxygen consumption rate during walking was investigated, using Krogh's tissue cylinder model. Our calculations showed that the increase in oxygen consumption rate of the muscle when patients walked would reduce the size of the oxygen front in the tissue cylinder, thus producing small, transient hypoxic regions in skeletal foot muscle. Such muscle hypoxia, although localized, could stimulate synthesis of vascular endothelial growth factor and facilitate angiogenesis in the grafted tissue. The architecture of fine microvessel networks observed in the foot by angiography seems consistent with this supposition and, moreover, suggests that the reinstatement of blood vessel networks in the foot tissues after grafting is supported by "normal" biological mechanisms.

Keywords Distal vein arterialization

1 Introduction

Transplanted veins have been used widely as conduits for coronary artery bypass and their adaptation to an arterial environment has long been recognized. Unfortunately, postsurgical complications arise in 20–50 % of vein grafts [1].

T. Sasajima
Department of Vascular Surgery, Asahikawa Medical University, Asahikawa, Japan

T. Koyama (✉)
Hokkaido University, N-1 W-25 2-7-516, Sapporo, Hokkaido, 064-0821, Japan
e-mail: tomkoyamajp@yahoo.co.jp

W.J. Welch et al. (eds.), *Oxygen Transport to Tissue XXXIV*, Advances in Experimental Medicine and Biology 765, DOI 10.1007/978-1-4614-4989-8_34,

In surgery for distal vein arterialization, (DVA), in patients suffering from chronic arterial ischemia caused by *arteriosclerosis obliterans* [2], venous valves are removed and the whole venous system in the sole of the foot is exposed to retrograde perfusion by the arterial pressure. A branch of a healthy artery is grafted to the *vena plantaris* below the ankle and the arterial blood is injected directly into the peripheral vein, without passage through the capillary network. In a previous theoretical study on rat skeletal muscle, the venular network was investigated using Krogh`s tissue cylinder model and a hypothetical venular tube that corresponded to the total length of the venular network in the unit tissue volume being considered. It was concluded that sufficient oxygen would be transported to the peripheral tissue through the venular network at rest [3]. A related question concerning the probable large increase in lymphatic outflow caused by the high venous pressure resulting from retrograde perfusion was explored by consideration of the filterability and total surface area of the hypothetical venular tube. It was suggested that the increase in lymphatic outflow would be tolerable [4]. The present study is an attempt to understand the biological mechanism for maintenance of venular networks under the initially severely unphysiological conditions imposed by DVA. Our interest was focused on the possibility of development of transient, local hypoxic states in the tissue after DVA, since it has been shown that hypoxic tissues increase production of vascular endothelial growth factor (VEGF). Serum VEGF is significantly higher than normal in critically ischemic lower limbs of patients with peripheral arterial disease [5]. A 3-min occlusion of the left anterior descending coronary artery followed by reperfusion caused an expression of VEGF in myocardial microvessels within 2 h in rats and a proliferation of cell nuclear antigen (PCNA) in 24 h [6]. Adenoviral VEGF gene delivery to hindlimb muscles of rabbits led to an enlargement of arterioles and venules, an increase in capillary number, and a decrease in the number of microvessels expressing dipeptidyl peptidase IV [7]. These findings resulted in the conclusion that VEGF induced an arterialization of microvessels.

In the present article we use a patient's foot as an example of the effect of DVA surgery. The possibility of the occurrence of localized tissue hypoxia caused by an increase in oxygen consumption rate is considered in the hypothetical tissue cylinder. It is suggested that the growth and maintenance of peripheral blood vessels may be supported by physiological angiogenic potential of the tissues.

2 Methods

2.1 *Angiogram of Vascular Nets*

An iodinated intravascular X-ray imaging agent was injected into the femoroposterior tibial artery to visualize the patent blood vessels. The X-ray image was recorded with a Siemens Artis FA. Although no exact calibration of the image size could be made it was assumed that the smallest diameter of recognizable microvessels would be <1 pixel when their image could not be made sharp by the edging procedure.

2.2 *Estimation of Oxygen Diffusion Front*

The oxygen diffusion profile in the virtual tissue cylinder was calculated using a radial diffusion equation applied to the terminal portion of the hypothetical venular tube. In the present study a one-dimensional diffusion equation was used for simplicity. The oxygen diffusion front was calculated by the equation, $F=\sqrt{(2\alpha D P_0/760VO_2SM)}$, where α, D, P_0, and VO_2SM represent the solubility coefficient of oxygen in tissue (0.023 ml ml^{-1} × 760 mmHg^{-1}), diffusion coefficient (1.6 × 10^{-5} cm^2 s^{-1}), PO_2 mmHg at the outlet of the venular tube (40 mmHg), and oxygen consumption rate of the skeletal muscle, respectively. The values for VO_2SM at rest, during slow walk (4.0 km h^{-1}), and during fast walk (5.6 km h^{-1}) were estimated to be 0.00089, 0.00307, and 0.00790 ml min^{-1} g^{-1}, respectively (see the Appendix).

3 Results

The foot used as our clinical example was edematous, with the great toe missing and an ulcer present on the back of the foot before the DVA operation. The edematous state and ulcer had almost disappeared 7 days after the surgery. The angiogram of the treated foot of this patient showed many arcuate blood vessels of different diameters 11 weeks after DVA. The diameters of small blood vessels which were recognizable on the angiogram seemed to be <1 pixel which corresponded approximately to 135 μm. The mesh-like vesicular structure resembled the sketch of arcade venules in skeletal muscle made by Engelson et al. [8]. The networks were formed by anastomoses of patent blood vessels and the diameter of the small meshes seemed <1 mm. Thus it appears the angiograms show that the blood circulation of the patient's foot was restored via the retrograde perfusion with the arterial blood.

By inserting the values of oxygen consumption rates by skeletal muscle at rest and during slow and fast walks into the equation, the maximal oxygen diffusion fronts were calculated to be 587, 317, and 197 μm, respectively. These results were assumed to be applicable for radial diffusion of oxygen and are schematically represented in Fig. 34.1 together with the values for hypothetical venular tube, tissue cylinder and oxygen fronts at rest (FI) and for fast walk (FII). It can be seen that the oxygen front is larger than the tissue cylinder at rest but reduces to a level smaller than the tissue cylinder by the increase in oxygen consumption during the fast walk. The reduction in the diffusion front was remarkable and suggests development of a hypoxic area in the tissue cylinder. The oxygen front was small in the *slow* walk but still *larger* than tissue cylinder.

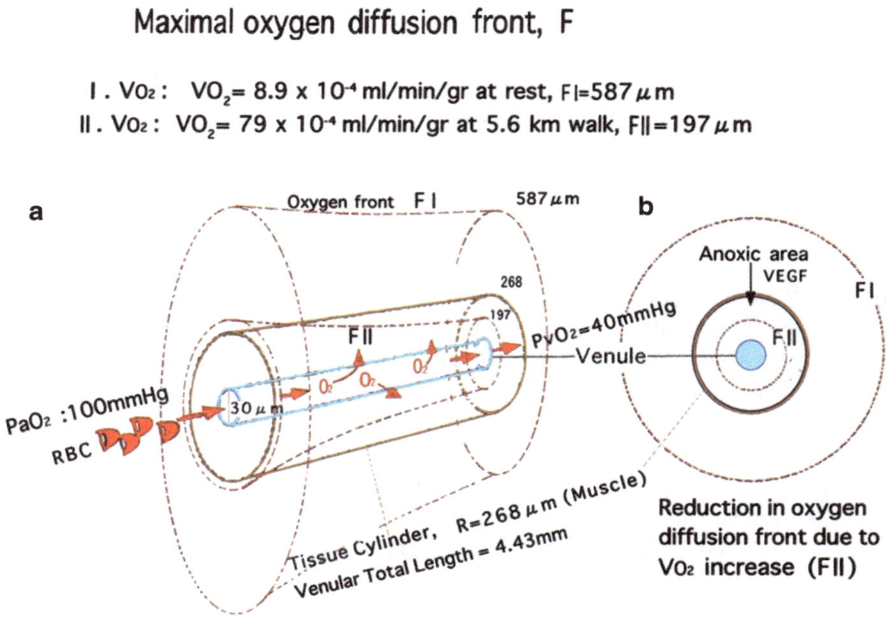

Fig. 34.1 Illustration of the oxygen front. (**a**) The central thin tube and the surrounding thick tube represent the venular tube and the tissue cylinder, respectively. The thick and thin tapered cylinders, shown with *dotted lines*, stand for the oxygen fronts at rest and during the fast walk. (**b**) A cross section of oxygen delivery system at the termination of the hypothetical venular tube. *Circles* from the innermost toward the outermost are venular tube, oxygen front during walk, FII (*dotted line*), tissue cylinder and oxygen front at rest, FI (*dotted line*). The ring area between FII and tissue cylinder rings is hypoxic during fast walking

4 Discussion

The DVA surgery restored the peripheral arcuate microvascular networks in the patient's foot. However, it is not known if the intima of the vessel walls was thickened as it is in arterial microvessels. Nevertheless, the dense networks were patent after surgery and adequate oxygen could be transported to peripheral foot tissues at rest [3]. Patients subjected to DVA surgery start walking after a few days and their foot muscle consumes more oxygen during exercise than at rest. The present calculation suggests that the small increase in oxygen consumption produces a localized anoxic area in the tissue cylinder, which may reduce walking capacity. However, local anoxia may also be beneficial as a physiological stimulus for growth factor expression by endothelial cells, consistent with an earlier suggestion that "local reduction in PO_2 may be important in VEGF induction" [9]. The increased production of VEGF in the distal part of tissue cylinder caused by temporary anoxia when patients walk should stimulate proliferation of endothelial cells and microvessels.

A similar phenomenon may be alternating perfusion of capillary nets in skeletal muscle [10], since capillaries are not constantly perfused at a constant rate but flow fluctuates. The advantage of such an intermittent perfusion system may be the avoidance of hyperoxia and the maintenance of proper tissue condition for angiogenesis [10]. Furthermore, it has been reported that VEGF increases the expression of ephrine B2, a distinct marker of arterial cells. Arterial pressure and shear stress in the venular network will act on the tissues to produce basic fibroblast growth factor, bFGF, and vascular smooth muscle growth factor, VSGF [11]. It seems probable that venular vessels may become gradually arterialized.

5 Conclusion

The venular system in the foot can be perfused with arterial blood introduced retrogradely by arterial graft. The oxygen transport will be adequate at rest but walking will cause localized oxygen deficit in skeletal muscle. The local hypoxia will stimulate VEGF production followed by angiogenesis. Thus, the peripheral microvessels maintain their normal biological function by adaptation to the arterialization of the distal vein.

Acknowledgments Thanks are due to Professor Dr. Ian Silver and Professor Dr Maria Erecinska for their kind reading of the manuscript and valuable suggestions.

Appendix: Estimation of VO_2 SM (Skeletal Muscle) from VO_2WB (Whole Body)

The VO_2 SM has been described only partially, while the oxygen consumption rate for the human whole body and VO_2 WB at rest and during exercise have been well studied. For estimation of VO_2SM during slow and fast walks, numerical estimation was required from available data on VO_2WB during exercise. VO_2WB in a man weighing 70 kg is 250 ml min^{-1} at rest, 700 ml min^{-1} at a mild walk (4.0 km h^{-1}) and 1,400 ml min^{-1} at a fast walk (5.6 km h^{-1}) [12]. The contribution of VO_2SM to the VO_2WB is 25 % at rest [13] and increases in proportion with the increase in work load of the whole body. Therefore, the following relation is given. VO_2SM ratio to VO_2WB (%) = 25 + (VO_2WB − 250) · (80 − 25)/(4,600 − 250). The contribution ratio of VO_2SM to VO_2 WB is 30.7 and 39.5 % for 4 km and 5.6 km h^{-1} walks for 70 kg body weight, respectively. VO_2SM at rest = 0.250 × 250 = 62.5 ml min^{-1} × 70 kg^{-1} = 0 .00089 ml min^{-1} g^{-1}, VO_2SM mild walk = 0.307 × 700 = 215 m min^{-1} × 70 kg^{-1} = 0.00 307 ml min^{-1} g^{-1}, and VO_2SM fast walk = 0.395 × 1,400 = 553 ml min^{-1} × 70 kg^{-1} = 0. 00790 ml min^{-1} g^{-1}. Putting these values into the equation, the maximal oxygen diffusion fronts were roughly estimated.

References

1. Muto A, Model L et al (2010) Mechanisms of vein graft adaptation to the arterial circulation. Circ J 74:1501–1512
2. Azuma N, Inaba M, Akasaka N, Sasajima T et al (2005) Foot salvage achieved by free flap transfer with bypass surgery. Jpn J Vasc Surg 14:151–158
3. Koyama T, Sasajima T (2009) Sufficient oxygen can be transported to resting skeletal muscle via arterialization of the vein: theoretical in a rat model. In: LaManna JC et al (eds) Oxygen transport to tissue, vol XXXII. Springer, NewYork, NY, pp 335–339
4. Koyama T, Sasajima T (2012) Retrograde perfusion of the hind leg in diabetic patients suffering from the arteriosclerosis obliterans: theoretical considerations of Oxygen supply and lymphatic flow based on rat models. Adv Exp Med Biol 737: 259–262
5. Findley CM, Mitchell RG et al (2008) Plasma levels of soluble Tie 2 and vascular endothelial growth factor distinguish critical limb ischemia from intermittent claudication in patients with peripheral arterial disease. J Am Coll Cardiol 52:387–393
6. Xie Z, Gao M, Koyama T (1997) Effects of transient coronary occlusion on the capillary network in the left ventricle of rat. Jpn J Physiol 47:537–543
7. Rissanen TT, Korpisalo P, Markkanen JE et al (2005) Blood flow remodels growing vasculature during vascular endothelial growth factor gene therapy and determines between capillary endothelial growth factor gene therapy and determines between capillary arterization and sprouting angiogenesis. Circulation 112:3937–3946
8. Engelson EL, Schmid-Schonbein GW, Zweifach BM (1985) The microvasculature in skeletal muscle III. Venous network anatomy in normotensive and spontaneously hypertensive rats. Int J Microcirc Clin Exp 4:229–248
9. Breen EC, Johnson EC, Wagner H et al (1996) Local reduction in PO2 may be important in VEGF induction. J Appl Physiol 81(1):355–361
10. Shibata M, Ichioka S, Ando J et al (2005) Nonlinear regulation of capillary perfusion in relation to ambient PO_2 changes in skeletal muscle. Eur J Appl Physiol 94(3):352–355
11. Swift MR, Weinstein BM (2009) Arterial-venous specification during development. Circ Res 104:576–588
12. Lumb AB (2005) Nunn's applied respiratory physiol, 6th edn. Elsevier, Philadelphia, p 240
13. Mottram RF (1955) "Exercise" in human muscle oxygen consumption. J Physiol 128:268–276

Chapter 35
Bayesian STAI Anxiety Index Predictions Based on Prefrontal Cortex NIRS Data for the Resting State

Masakaze Sato, Wakana Ishikawa, Tomohiko Suzuki, Takashi Matsumoto, Takeo Tsujii, and Kaoru Sakatani

Abstract Several distinctive activity patterns have been observed in the brain at rest. The aim of this study was to determine whether the STAI index can be predicted from changes in the oxy- and deoxy-hemoglobin (Hb) concentrations by using two-channel prefrontal cortex (PFC) NIRS data for the ***resting state***. The study population comprised 19 subjects. Each subject performed four trials, each of which consisted of resting with no task for 3 min. Data were acquired using a portable NIRS device equipped with two channels. The prediction algorithm was derived within a Bayesian machine learning framework. The prediction errors for seven subjects were not greater than 5.0. Because the STAI index varied between 20 and 80, these predictions appeared reasonable. The present method allowed prediction of mental status based on the NIRS data at resting condition obtained in the PFC.

Keywords NIRS • Prefrontal cortex • STAI anxiety index

1 Introduction

The brain in the resting state has been an active area of research in brain science [1]. In the resting state, the brain uses a significant amount of energy even when performing no task, and several distinctive activity patterns have been observed.

M. Sato (✉) • W. Ishikawa • T. Suzuki • T. Matsumoto
Department of Electrical Engineering and Bioscience, Waseda University,
3-4-1 Kasumigaoka, Lions Garden Kamifukuoka #801, Fujimono-si, Saitama 356-006, Japan
e-mail: sato09@matsumoto.eb.waseda.ac.jp

T. Tsujii • K. Sakatani
Division of Optical Brain Engineering, Department of Neurological Surgery,
Nihon University School of Medicine, Tokyo, Japan

W.J. Welch et al. (eds.), *Oxygen Transport to Tissue XXXIV*, Advances in Experimental Medicine and Biology 765, DOI 10.1007/978-1-4614-4989-8_35,

One way of classifying patterns of brain activity is to detect the network structure of the activity using the hierarchical clustering algorithm [2]. Although such studies are often conducted with fMRI data, one of the nontrivial aspects is that the dimension of the fMRI data is typically tens of thousands.

We attempted to predict the degree of anxiety of individuals from the oxy- and deoxy-hemoglobin (Hb) levels acquired from two-channel portable near infrared spectroscopy (NIRS) data for the prefrontal cortex (PFC) in the resting states. fMRI measures mainly deoxy-Hb levels while NIRS measures both deoxy- and oxy-Hb, and our algorithm incorporates both oxy- and deoxy-Hb levels. A quantification of anxiety is needed for predicting the anxiety of individuals. We considered the state-trait anxiety inventory (STAI) index. An important aspect of this research project was the prediction method. It was formulated within a Bayesian machine learning framework [3] and implemented by Markov Chain Monte Carlo (MCMC).

2 Materials and Methods

2.1 Experimental Settings

The study population comprised 19 subjects (13 women; 6 men), aged 20–24 years. All the subjects were healthy, with no past history of psychiatric or neurological disorders. The subjects gave written informed consent on forms approved by the ethical committee of the Nihon University School of Medicine.

Each subject was seated in a comfortable chair in a dimmed room, and we measured the changes in oxy- and deoxy-Hb concentration using a two-channel NIRS system (PNIRS-10, Hamamatsu Photonics K.K., Japan). The NIRS probes were set symmetrically on the forehead; the positioning is similar to the midpoint between the electrode positions Fp1/Fp3 (left) and Fp2/Fp4 (right) of the international 10–20 system.

One trial consisted of the following steps: Step 1, relaxation period 30 s; Step 2, preparation period 30 s; Step 3, analysis period 3 min. Each subject performed four trials and the STAI questionnaires were completed before the trials commenced. Figure 35.1 shows the experimental protocol.

2.2 NIRS Data

The device was capable of acquiring NIRS data through two channels; each channel consisted of oxy- and deoxy-Hb levels. The acquired data consisted of four-dimensional values collected for 4 min per task. We used the latter 3-min data for analysis. The sampling frequency was 10 Hz; thus, there were 1,800 data points for each trial. Twenty features computed from the four-dimensional data are summarized in Table 35.1.

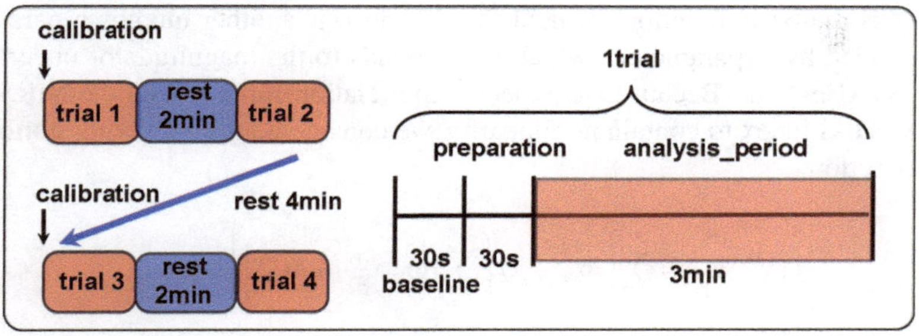

Fig. 35.1 Flow of tasks

Table 35.1 The features used in this study

No.	Mean	No.	Variance	No.	Covariance	No.	Correlation coefficient
1	Oxy(right PFC)	5	Oxy(r)	9	Oxy(r)/deoxy(r)	15	Oxy(r)/deoxy(r)
2	Deoxy(right PFC)	6	Deoxy(r)	10	Oxy(r)/oxy(l)	16	Oxy(r)/oxy(l)
3	Oxy(left PFC)	7	Oxy(l)	11	Oxy(r)/deoxy(l)	17	Oxy(r)/deoxy(l)
4	Deoxy(left PFC)	8	Deoxy(l)	12	Deoxy(r)/oxy(l)	18	Deoxy(r)/oxy(l)
–		–		13	Deoxy(r)/deoxy(l)	19	Deoxy(r)/deoxy(l)
–		–		14	Oxy(l)/deoxy(l)	20	Oxy(l)/deoxy(l)

The numbers in odd-numbered columns are feature numbers

2.3 *Prediction*

2.3.1 Prediction Flow

A leave-one-out prediction was conducted. Of the 76 (= 19 participants × 4 trials) data sets, one was reserved for testing, and the remaining 75 data sets and the STAI indices were used for training the machine After the parameters were learned, the reserved data were input to the machine for STAI index prediction.

2.3.2 Prediction Algorithm

Let $x_t^{(i)} := (x_{t1}^{(i)},\ldots,x_{tk}^{(i)}) \in R^K$, $t = 1,\ldots,T$, $i = 1,\ldots,N$, be the NIRS feature vector for the t-th trial of the i-th individual. For raw features, $K = 20$; however, the dimension will be later reduced to three in a principled manner. Let $y^{(i)}$, $i = 1,\ldots, N$ be the STAI state index calculated from the questionnaire ($N = 19$). We attempted to fit the data with the following learning model:

$$P(y^{(i)} \mid x_t^{(i)};\omega,\beta) = \sqrt{\frac{\beta}{2\pi}} \exp\left(-\frac{\beta}{2}(y^{(i)} - f(x_t^{(i)};\omega))\right)^2, \tag{35.1}$$

where f is the basis function for data fitting, and β is another unknown parameter (often called hyperparameter), which corresponds to the magnitude of uncertainty in the STAI indices. Because we expected the relationship between the NIRS data and the STAI index to contain nonlinearity, we considered the following **nonlinear** basis function:

$$f(x_t^{(i)};\omega) := \sum_{h=1}^{H}\left[\omega_{(K+1)h}\sigma\left(\sum_{k=1}^{K}[\omega_{kh}x_{tk}^{(i)}]+\omega_{0h}\right)\right]+\omega_{(K+1)0}, \tag{35.2}$$

where σ is a sigmoidal function that incorporates potential nonlinearities, and $\boldsymbol{\omega}=\{\omega_{kh}, \omega_{0h}, \omega_{(K+1)0}\}$ where $k=1,\ldots, K+1$, $h=1,\ldots, H$, are the unknown parameters associated with the basis function. All the unknown parameters, $\boldsymbol{\omega}$, needed to be learned from the available data $\{x_t^{(i)}, y^{(i)}\}$. This study formulated the prediction problem within a Bayesian framework, where a prior distribution is assumed about the unknown parameters that are incorporated into the data model given by (35.1).

The prior distribution for $\boldsymbol{\omega}_k$ was assumed to be specified by $P(\boldsymbol{\omega}_k|\alpha_k)=N(0,(1/\alpha_k)\boldsymbol{I})$ where $N(0,(1/\alpha_k)\boldsymbol{I})$ denotes the Gaussian distribution with a mean of 0 and a covariance matrix $(1/\alpha_k)\boldsymbol{I}$; α_k is another hyperparameter to be learned, and $\boldsymbol{I}$ denotes the H-dimensional identity matrix. This prior distribution indicates that the unknown parameters are often small in magnitude instead of large, and this fact often prevents overfitting. The prior distributions for α_k are assumed to follow the gamma distribution. In addition to the K parameter vectors $\{\boldsymbol{\omega}_k\}$, there were two parameters $\{\boldsymbol{\omega}_{0h}, \boldsymbol{\omega}_{(K+1)0}\}$ with which we associated another hyperparameter α_{K+1}. Let $\alpha = \{\alpha_k\}$, $x=\{x_t^{\{i\}}\}_{t,i}$, $y=\{y^{(i)}\}$. Assuming the data from each trial to be independent, the Bayes formula gives the posterior distribution:

$$P(\omega,\alpha \mid x,y,\beta) \propto \prod_{t,j} P(y^{(i)} \mid x_t^{(i)};\omega,\beta)P(\omega \mid \alpha)P(\alpha). \tag{35.3}$$

Equation (35.3) quantifies the plausibility of a particular value of the data for learning $(\boldsymbol{x}; \boldsymbol{y})$. Using (35.3), the prediction of the STAI index associated with the preserved data $y_{\text{preserved}}$ was performed. The calculation of (35.3) was nontrivial; therefore, we used Markov Chain Monte Carlo analysis to make approximations. This procedure was repeated 76 times while changing the data reserved for the test each time. Because the actual STAI index was available from the subject, the prediction error was computed.

We chose $H=8$ so that the number of unknown parameters was $(K+1)H+1=169$; this number was too large for the number of data sets used for training, i.e., 75. One way of automatically reducing the number of dimensions is to examine the posterior values of hyperparameter α_k in (35.3). A large α_k implies that the target quantity is concentrated around the origin; hence, this particular feature is relatively ineffective. Thus, by eliminating those features with large α_k values, one could reduce the number of features.

3 Results

The top left figure in Fig. 35.2 shows the posterior mean of the hyperparameters $\{\alpha_k\}$, $k=1,\ldots,20$. The top right figure in Fig. 35.2 shows the prediction errors summed over all the data sets corresponding to the full 20-dimensional feature vector, seven-dimensional feature vector consisting of 3, 15, 16, 17, 18, 19, and 20, five-dimensional feature vector consisting of 3, 15, 16, 19, and 20, three-dimensional feature vector consisting of 3, 15, and 16, and the one-dimensional feature that consists of only feature 3. The best performance appeared to be for the three-dimensional feature vector. The bottom left figure in Fig. 35.2 shows the prediction

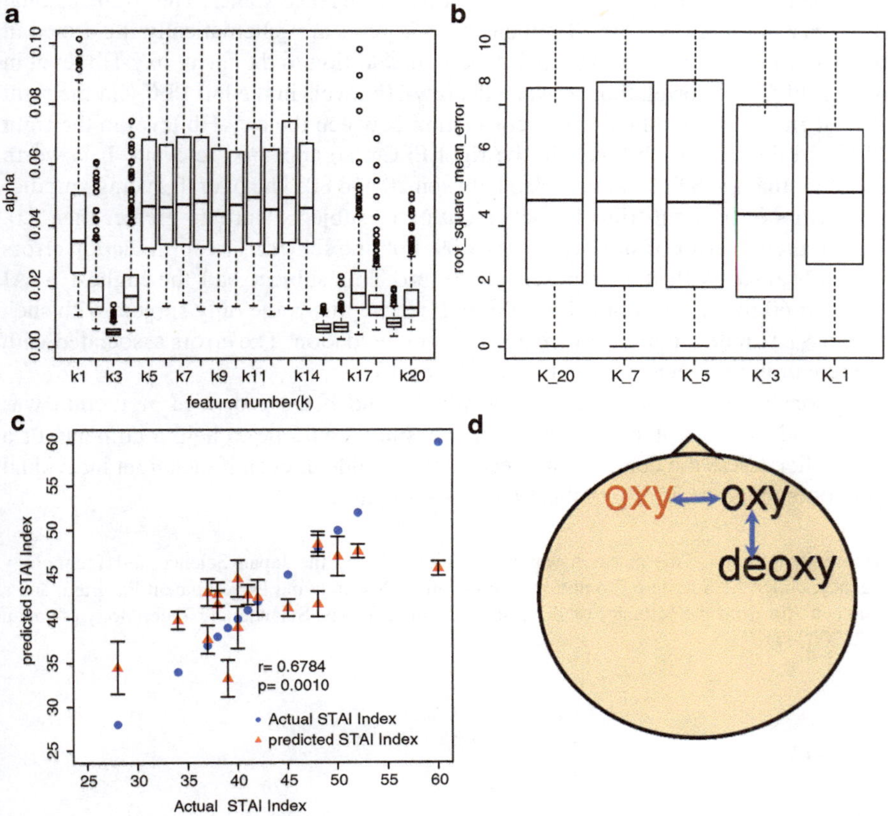

Fig. 35.2 *Top left* figure shows boxplots of the posterior mean of α_k. If α_k is large, feature x_k might be ineffective for prediction. *Top right* figure shows the root-mean-square errors with full 20-dimensional feature vector, along with 7-, 5-, 3-, and 1-dimensional feature vectors. Seven features were selected according to the smallest $7\alpha_k$'s. Other features were similarly selected. *Bottom left* shows the predicted and actual STAI indices. *Bottom right* shows a schematic of the three extracted features with the best performance

result associated with trial 2 of all the subjects with the three-dimensional feature. The horizontal axis denotes the actual STAI index and the vertical axis denotes the predicted value.

4 Discussion

The brain under stress is an active area of research. Recently, it has been demonstrated that there is left/right asymmetry in the PFC activity during tasks involving mental stress. These activity patterns were measured using a two-channel NIRS device and correlated with the systemic stress responses of the autonomic nervous system and the HPA axis system [4]. This study went beyond investigating such asymmetry by predicting the STAI index based on the NIRS values. The 3-dimensional features extracted from the 20-dimensional features are schematically illustrated in the bottom right figure in Fig. 35.2, where, in addition to the mean oxy-Hb level in the left PFC, the correlation between the oxy-Hb levels in the left PFC and the right PFC seems to be significant. The correlation between the oxy-Hb level in the right PFC and the deoxy-Hb levels in the right PFC also appeared relevant. It is worth recalling that the STAI index varied between 20 and 80. The overall average prediction errors of the four trials for seven out of 19 subjects were no greater than 5.0. The prediction error of subject 17 was 1.09. For one subject, larger prediction errors were observed with the proposed algorithm. This subject had the highest STAI index of 60 in the questionnaire session. Because this is the only subject with such a high STAI index, more data are needed for elucidation. The errors associated with the remaining subjects were between 5.0 and 10.0.

If considerably more data were available, and if the proposed prediction was functional, one possible application of this study could be to help a clinician or a researcher assess the degree of anxiety of an individual, even if the target individual did not respond truthfully to the STAI questionnaire.

Acknowledgments This research was partly supported by the Japan Science and Technology Agency, under the Strategic Promotion of Innovative Research and Development Program, and a Grant-in-Aid from the Ministry of Education, Culture, Sports, Science and Technology of Japan (B23300247).

References

1. Gusnard D, Akbudak E, Shulman G, Raichle M (2001) PNAS 98:4259–4264
2. Salvador R, Suckling J, Coleman MR, Pickard JD, Menon D, Bullmore E (2005) Cereb Cortex 15:1332–1342
3. Bishop C (2006) Pattern recognition and machine learning. Springer, Berlin
4. Tanida M, Katsuyama M, Sakatani K (2007) Brain Res 1184:210–216

Chapter 36
The Effect of Venous and Arterial Occlusion of the Arm on Changes in Tissue Hemodynamics, Oxygenation, and Ultra-Weak Photon Emission

Felix Scholkmann, Olaf Schraa, Roeland van Wijk, and Martin Wolf

Abstract Ultra-weak photon emission (UPE) is a general feature of living biological systems. To gain further insights into the origin of UPE and its physiological significance, the aim of the present study was to investigate the connection between hemodynamics (HD), oxygenation (OX), and UPE. Therefore, during venous and arterial occlusion (VO, AO), changes of UPE and surrogates of HD as well as OX were measured simultaneously using two photomultipliers and near-infrared spectroscopy, respectively. We showed that (1) changes in UPE correlate significantly nonlinearly with changes in oxyhemoglobin ($\Delta[O_2Hb]$), deoxyhemoglobin (Δ[HHb]), and hemoglobin difference ($\Delta[HbD]=\Delta[O_2Hb]-\Delta[HHb]$), indicating a complex association between UPE and tissue HD/OX; (2) UPE decreases significantly during AO but not during VO; (3) UPE increases significantly after AO; and (4) the view that ROS are the source of UPE is generally supported by the present study, although some findings remain unexplained in the context of the theory of ROS-mediated UPE generation. In conclusion, the present study revealed new insights into the interplay between HD, OX, and UPE and opens up new questions that have to be addressed by future studies.

Keywords Blood flow • Blood volume • Oxygenation • Ultra-weak photon emission • Reactive oxygen species • Oxidative stress

F. Scholkmann (✉) • M. Wolf
Division of Neonatology, Biomedical Optics Research Laboratory, University Hospital Zurich, Frauenklinikstr. 10, 8091 Zurich, Switzerland
e-mail: Felix.Scholkmann@usz.ch

O. Schraa • R. van Wijk
Meluna Research, 3821 AE Amersfoort, The Netherlands

W.J. Welch et al. (eds.), *Oxygen Transport to Tissue XXXIV*, Advances in Experimental Medicine and Biology 765, DOI 10.1007/978-1-4614-4989-8_36,

1 Introduction

Living biological systems spontaneously emit ultra-weak light (~10^2 photons/s cm^2 [1]) in the wavelength range of at least 200–800 nm from their surface. Previous experiments have shown that this ultra-weak photon emission (UPE) is associated with the relaxation of electronically excited states of molecules, e.g., lipids [2] and proteins [3]. The excited states are mainly consequences of oxidation reactions caused by reactive oxygen species (ROS). Reactions involving singlet molecular oxygen ($^1O_2$) [4] and triplet states of carbonyl [3] have been identified as causing UPE in particular. Early studies have demonstrated that UPE of isolated organs is highly oxygen dependent [5]. The decrease of UPE after application of antioxidative enzymes [6] supports the findings of ROS-based UPE generation.

Up to now, only a few studies [7–11] have investigated the effect of a temporally local shortage in blood supply (ischemia) and blood oxygenation (hypoxia) on UPE of intact organisms, some of these focusing on humans. To gain further insights into the interplay between UPE dynamics, changes in hemodynamics (HD) (i.e., blood flow/volume) and oxygenation (OX), the aim of the present study was to measure UPE, surrogates of HD and OX simultaneously for the first time on a human using two photomultipliers and near-infrared spectroscopy (NIRS).

2 Materials and Methods

2.1 NIRS and UPE Measurement

Surrogates of HD and OX were measured using a wireless NIRS device [12]. The NIRS device uses LEDs with two different wavelengths (760 and 870 nm) to determine absorption changes with a sampling frequency of 100 Hz. For the current study we selected four light paths (source-detector separation: 25 mm) from the device. The area of the probed tissue was 25×37.5 mm in total.

The UPE measurement was performed using a specially designed dark chamber with two compartments for recording on both human hands. On top of each chamber a single-photon counting photomultiplier tube (PMT) (Electron Tube 9235B, Electron Tubes Ltd., Ruislip, UK) was placed. The spectral sensitivity of the PMT was 200–650 nm. Subjects were dark-adapted and then inserted their left and right hand into the compartments of the dark chamber. The measurement was controlled using Labview (National Instruments, Austin, USA).

2.2 Measurement Procedure

Two healthy male adults participated in the study (ages: 28 and 44 year). Prior to the measurement, the hands of the subjects were covered for 45 min with black gloves

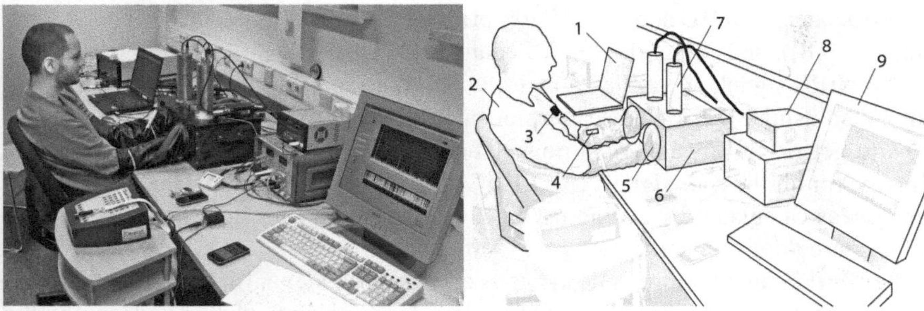

Fig. 36.1 Measurement setup. 1: near-infrared spectroscopy (NIRS) data recording, 2: subject, 3: occlusion cuff, 4: attached NIRS device, 5: light shielded opening of the dark chamber to insert the hand, 6: dark chamber, 7: PMT, 8: power supply and PMT controlling electronics, 9: PMT controlling and measurement

to reduce delayed luminescence [13]. The subjects then positioned their hands in the dark chamber. UPE of the palm side was measured. This side was selected for measurement because commonly it has stronger emission than the dorsal side of the hand [14, 15]. The NIRS optode was placed on the left forearm, in parallel to the *flexor digitorum profundus* muscle, ~7 cm below the crook of the arm. A blood pressure cuff was placed around the left upper arm. The experimental setup is shown in Fig. 36.1. The measurement protocol was as follows: (1) 0–6 min: baseline recording; (2) 6–11 min: venous cuff occlusion (VO) (at 40 mmHg); (3) 11–21 min: reperfusion; (4) 21–31 min: arterial occlusion (AO) (at 200 mmHg); (5) 31–40 min: reperfusion. Three measurements were made in total (i.e., one subject was measured twice).

2.3 *Signal Processing and Data Analysis*

All calculations were performed using Matlab (Mathworks, Natick, Massachusetts, USA). The UPE signal ($I_L(t)$) from the left hand was corrected by subtracting the dark count of the PMT ($I_D(t)=6$ counts/s (cps)): $I_L'(t)=I_L(t)-I_D(t)$. Afterwards, the sum of all values of $I_L'(t)$ for every minute of recording time was calculated and all obtained values were divided by 60, leading to the final signal $I_L''(t)=\{I_L'(t_i)\mid i=1,2,\ldots,40\}$ which contained the corrected UPE signal of the left hand, expressed in "cps."

From the absorption changes measured using NIRS, concentration changes in μM (relative to the first value obtained by the measurement) of oxyhemoglobin ($\Delta[O_2Hb]$) and deoxyhemoglobin ($\Delta[HHb]$) were calculated for every light path using the modified Lambert-Beer law [16]; the differential pathlength factor (DPF) was 4.48 (for 760 nm) and 3.81 (for 870 nm) [17]. Additionally, the following signals were calculated: total hemoglobin concentration ($\Delta[tHb]=\Delta[O_2Hb]+\Delta[HHb]$) and hemoglobin difference ($\Delta[HbD]=\Delta[O_2Hb]-\Delta[HHb]$). $\Delta[O_2Hb]$ and $\Delta[HHb]$

are associated with HD and OX, Δ[tHb] primarily with HD (especially blood volume changes [18]), and Δ[HbD] with OX changes [19].

All NIRS signals were then downsampled to a sampling frequency of 1 Hz and smoothed using robust local scatterplot smoothing (LOESS) with a window size of 30 s to get rid of physiological noise (such as heart pulsation and Mayer waves). Finally, for all the four signals (Δ[O_2Hb], Δ[HHb], Δ[tHb], and Δ[HbD]), the mean signal was calculated by averaging the signals from all four light paths having the same source-detector separation of 25 mm.

For further analysis, the NIRS signals and the UPE signal (of the left hand) from all three experiments were averaged leading to one signal each. Additionally, the mean values for 60-s long segments of the signals were calculated. To reduce the variance in the averaged UPE signal, the UPE signals were normalized so that that the average of each signal was equal to the average of all signals. The UPE signal of the right hand was not used for further analysis since it was recorded only as a control signal.

Two types of analysis were performed: (1) significance of NIRS and UPE changes depending on the 5 phases of the experiment, and (2) correlation between NIRS and UPE signals and their significance thereof. For the correlation analysis we empirically determined the mathematical function with the best fitting and we calculated the Pearson correlation between the fitting function and the NIRS and UPE signals. Unconstrained nonlinear minimization of the sum of squared residuals was used for the fitting. For calculating the p-values of the changes a Wilcoxon rank sum test was used. The p-values for the regression coefficients were computed using a Student's t distribution for a transformation of the correlation.

3 Results

Figure 36.2a shows the time courses of the UPE, Δ[O_2Hb], Δ[HHb], Δ[tHb], and Δ[HbD] signals obtained. It is evident that NIRS signals between experiments differ only slightly and are consistent between experiments. In contrast, for UPE it is evident that relatively larger and more complex fluctuations occur which may obscure small VO and/or AO related changes. The following significant ($p<0.05$) changes were observed: During the VO, Δ[HHb] and Δ[tHb] increase, Δ[HbD] decreases. After the VO, i.e., the first reperfusion phase, Δ[HHb], Δ[tHb], and Δ[HbD] return to baseline values. During the AO, Δ[O_2Hb] and Δ[HbD] decrease, Δ[HHb] and Δ[tHb] increase, UPE decreases above the baseline level. After the AO, i.e., the second reperfusion phase, Δ[O_2Hb], Δ[tHb], and Δ[HbD] show a transient increase, and Δ[HHb] a transient decrease. At the same time, UPE increases slowly.

Figure 36.2b shows the averaged signals that were used for the correlation analysis. It was found that the UPE and NIRS signals showed a correlation when an exponential function of the form $y=\alpha[\beta\exp(\gamma x)]^{-1}$ with three free parameters (α, β, γ) was used for correlation analysis. UPE correlates significantly ($p<0.001$) with Δ[O_2Hb], Δ[HHb], and Δ[HbD]: r(UPE/Δ[O_2Hb])=0.7182 ($p=1.8146\cdot10^{-7}$),

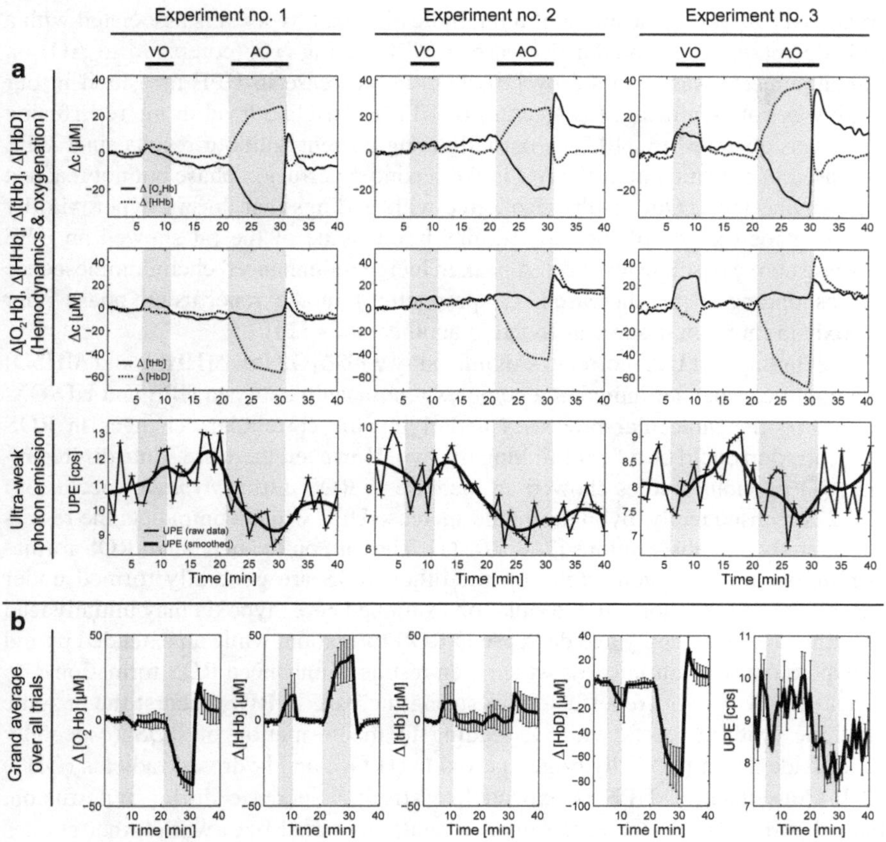

Fig. 36.2 Overview of the measured Δ[O_2Hb], Δ[HHb], Δ[tHb], Δ[HbD], and UPE signals before, during, and after VO and AO on the forearm. (**a**) Shows the signal from each experimental trial, (**b**) depicts the averaged signals and standard errors from all three measurements

r(UPE/Δ[HHb])=–0.7431 ($p=3.989\cdot10^{-8}$), and r(UPE/Δ[HbD])=0.7425, ($p=4.1513\cdot10^{-8}$). The correlation between UPE and Δ[tHb] was not significant: r(UPE/Δ[tHb])=–0.2229 ($p=0.4668$).

4 Discussion, Conclusion, and Outlook

The changes observed in Δ[O_2Hb], Δ[HHb], Δ[tHb], and Δ[HbD] during and after the occlusions are in agreement with previous studies (e.g., [19]).

The UPE changes agree, to some extent, with findings from other studies: (1) A decrease in UPE during AO was reported by [7, 10, 11], in agreement with our results. It is also in agreement with findings of [8] who found a decrease of UPE in brain tissue after cardiac arrest of a rat, indicating that hypoxia is associated with a UPE decrease. (2) A smaller decrease in UPE during VO (compared to AO) on human subjects was reported by [10, 11]; the decrease in UPE measured in our study was not significant. (3) A return of UPE to baseline level in the reperfusion phase was observed by [10, 11], partially in agreement with our results since UPE increased in our study after the AO in the second reperfusion phase but not reached the baseline level. Our results also agree with findings that during hyperoxia (by changing the oxygen of the inhaled air), brain tissue of the rat showed an UPE increase above baseline level [9]. A peak in lucigenin-enhanced chemiluminescence (corresponding to an increased $^1O_2$ production) in the reperfusion phase after hypoxia in rat brain tissue was found in another study [20].

The finding that UPE correlates significantly with $\Delta[O_2Hb]$, $\Delta[HHb]$, and $\Delta[HbD]$ in a nonlinear manner indicates a complex relationship between UPE and HD/OX. Regarding the molecular processes underlying the correlation, changes in ROS concentration could be a factor linking the two. However, there are some discrepancies: (1) previous studies showed an increase in ROS during *hypoxia* (occlusion) [21, 22]; consequently, hypoxia should increase UPE which contradicts the results of our study and observations [7, 8, 10, 11]. Though, on the one hand, ROS formation depends on oxygen while, on the other, ROS are generally formed under various stress conditions. In the situation examined here, hypoxia may initially lead to limited oxygen supply and decrease in ROS formation, while an extended period of hypoxia may result in stress which then results in enhanced ROS formation.

During *hyperoxia* (reperfusion), a strong increase in ROS occurs: the enzyme xanthine oxidase reacts with O_2 leading to the formation of ROS (especially superoxide anion ($O_2^{\bullet -}$), hydrogen peroxide (H_2O_2), and hydroxyl radicals ($OH^{\bullet}$)) [23]. Consequently, UPE is expected to strongly increase during reperfusion. Indeed, we found an increase (albeit a relatively small one) while other studies [10, 11] did not.

The exact origin of UPE measured from the hand is unknown. It has been estimated that about 60% of the UPE is based on internal sources. The rest may be attributable to oxidative processes on the skin surface [24].

Since the present work was only a pilot trial, there were certain limitations that need to be addressed in further studies: (1) small sample size and number of trials, (2) relatively short rest periods between the occlusions, (3) no direct measurement of OX (the usage of a multidistance NIRS algorithm is suggested), and (4) measurement of UPE and HD/OX at different muscle compartments of the arm. Further studies should extend the analysis of the relationship between UPE and HD/OX—especially since UPE could deliver some new information about hemodynamic and metabolic changes in tissue and the primary source of UPE.

In conclusion, we showed that (1) changes in UPE correlate significantly nonlinearly with $\Delta[O_2Hb]$, $\Delta[HHb]$, and $\Delta[HbD]$, indicating a complex association between UPE and tissue HD/OX; (2) UPE decreases significantly during AO but not during VO; (3) UPE increases significantly after AO; and (4) the view that ROS are the source of UPE is generally supported by the present study, although some findings remain unexplained.

Acknowledgments The authors thank Raphael Zimmerman and Andreas Metz for hardware and software support concerning the NIRS measurements, Rachel Folkes for proofreading the manuscript, as well as Dr. Lisa Holper, Dr. Daniel Fels, Dr. Michal Cifra, and Reto Kofmehl for stimulating discussions.

References

1. Van Wijk RV, Van Wijk EP (2005) An introduction to human biophoton emission. Forsch Komplementarmed Klass Naturheilkd 12(2):77–83
2. Boveris A, Cadinas E, Chance B (1980) Low level chemiluminescence of the lipoxygenase reaction. Photobiochem Photobiophys 1(3):175–182
3. Slawinski J (1988) Luminescence research and its relation to ultraweak cell radiation. Experientia 44(7):559–571
4. Hodgson EK, Fridovich I (1974) The production of superoxide radical during the decomposition of potassium peroxochromate(V). Biochemistry 13(18):3811–3815
5. Boveris A, Cadenas E, Reiter R et al (1980) Organ chemiluminescence: noninvasive assay for oxidative radical reactions. Proc Natl Acad Sci U S A 77(1):347–351
6. Sauermann G, Mei WP, Hoppe U et al (1999) Ultraweak photon emission of human skin in vivo: influence of topically applied antioxidants on human skin. Methods Enzymol 300:419–428
7. Edwards R, Ibson MC, Jessel-Kenyon J et al (1989) Light emission from the human body. Complement Ther Med 3(2):16–19
8. Kobayashi M, Takeda M, Ito K et al (1999) Two-dimensional photon counting imaging and spatiotemporal characterization of ultraweak photon emission from a rat's brain in vivo. J Neurosci Methods 93(2):163–168
9. Kobayashi M, Takeda M, Sato T et al (1999) In vivo imaging of spontaneous ultraweak photon emission from a rat's brain correlated with cerebral energy metabolism and oxidative stress. Neurosci Res 34(2):103–113
10. Van Wijk EP, Van Wijk RV (2005) Multi-site recording and spectral analysis of spontaneous photon emission from human body. Forsch Komplementarmed Klass Naturheilkd 12(2): 96–106
11. Yang J-M, Lee C, Yi S-H et al (2004) Biophoton emission and blood flow in the human hand. J Int Soc Life Inform Sci 22(2):344–348
12. Muehlemann T, Haensse D, Wolf M (2008) Wireless miniaturized in-vivo near infrared imaging. Opt Express 16(14):10323–10330
13. Yan Y, Popp FA, Sigrist S et al (2005) Further analysis of delayed luminescence of plants. J Photochem Photobiol B 78(3):235–244
14. Van Wijk R, Kobayashi M, Van Wijk EP (2006) Anatomic characterization of human ultraweak photon emission with a moveable photomultiplier and CCD imaging. J Photochem Photobiol B 83(1):69–76
15. Rastogi A, Pospisil P (2010) Ultra-weak photon emission as a non-invasive tool for monitoring of oxidative processes in the epidermal cells of human skin: comparative study on the dorsal and the palm side of the hand. Skin Res Technol 16(3):365–370
16. Delpy DT, Cope M, van der Zee P et al (1988) Estimation of optical pathlength through tissue from direct time of flight measurement. Phys Med Biol 33(12):1433–1442
17. Duncan A, Meek JH, Clemence M et al (1995) Optical pathlength measurements on adult head, calf and forearm and the head of the newborn infant using phase resolved optical spectroscopy. Phys Med Biol 40(2):295–304
18. Van Beekvelt MC, Colier WN, Wevers RA et al (2001) Performance of near-infrared spectroscopy in measuring local O2 consumption and blood flow in skeletal muscle. J Appl Physiol 90(2):511–519

19. Nioka S, Kime R, Sunar U et al (2006) A novel method to measure regional muscle blood flow continuously using NIRS kinetics information. Dyn Med 5:5
20. Schreiber SJ, Megow D, Raupach A et al (1995) Age-related changes of oxygen free radical production in the rat brain slice after hypoxia: on-line measurement using enhanced chemiluminescence. Brain Res 703(1–2):227–230
21. Guzy RD, Schumacker PT (2006) Oxygen sensing by mitochondria at complex III: the paradox of increased reactive oxygen species during hypoxia. Exp Physiol 91(5):807–819
22. Clanton TL (2007) Hypoxia-induced reactive oxygen species formation in skeletal muscle. J Appl Physiol 102(6):2379–2388
23. Jeroudi MO, Hartley CJ, Bolli R (1994) Myocardial reperfusion injury: role of oxygen radicals and potential therapy with antioxidants. Am J Cardiol 73(6):2B–7B
24. Nakamura K, Hiramatsu M (2005) Ultra-weak photon emission from human hand: influence of temperature and oxygen concentration on emission. J Photochem Photobiol B 80(2):156–160

Chapter 37
Metabolic Network Analysis of DB1 Melanoma Cells: How Much Energy Is Derived from Aerobic Glycolysis?

A.A. Shestov, A. Mancuso, D.B. Leeper, and J.D. Glickson

Abstract A network model has been developed for analysis of tumor glucose metabolism from ^{13}C MRS isotope exchange kinetic data. Data were obtained from DB1 melanoma cells grown on polystyrene microcarrier beads contained in a 20-mm diameter perfusion chamber in a 9.4 T Varian NMR spectrometer; the cells were perfused with 26 mM [1,6-$^{13}C_2$]glucose under normoxic conditions and 37° C and monitored by ^{13}C NMR spectroscopy for 6 h. The model consists of ~150 differential equations in the cumomer formalism describing glucose and lactate transport, glycolysis, TCA cycle, pyruvate cycling, the pentose shunt, lactate dehydrogenase, the malate-aspartate and glycerophosphate shuttles, and various anaplerotic pathways. The rate of oxygen consumption ($CMRO_2$) was measured polarographically by monitoring differences in pO_2. The model was validated by excellent agreement between model predicted and experimentally measured values of $CMRO_2$ and glutamate pool size. Assuming a P/O ratio of 2.5 for NADH and 1.5 for FADH2, ATP production was estimated as 46% glycolytic and 54% mitochondrial based on average values of $CMRO_2$ and glycolytic flux (two experiments).

Keywords Melanoma • Aerobic glycolysis

A.A. Shestov
CMRR, Department of Radiology, University of Minnesota Medical School, Minneapolis, MN, USA

A. Mancuso
Department of Radiology, University of Pennsylvania, Perelman School of Medicine, 626 Chatsworth Dr.Ambler, Philadelphia, PA 19002, USA

Abramson Comprehensive Cancer Center, Perelman School of Medicine, Philadelphia, PA, USA

D.B. Leeper
Department of Radiation Oncology, Thomas Jefferson University, Philadelphia, PA, USA

J.D. Glickson (✉)
Department of Radiology, University of Pennsylvania, Perelman School of Medicine, 626 Chatsworth Dr.Ambler, Philadelphia, PA 19002, USA
e-mail: glickson@mail.med.upenn.edu

W.J. Welch et al. (eds.), *Oxygen Transport to Tissue XXXIV*, Advances in Experimental Medicine and Biology 765, DOI 10.1007/978-1-4614-4989-8_37,

1 Introduction

Eighty-seven years ago Otto Warburg hypothesized (1) that the majority of tumors obtain most of their energy by aerobic glycolysis, and (2) that the reason why tumors rely so heavily on glycolysis despite its relatively low efficiency for ATP production was because their mitochondria were defective [1]. Defective mitochondrial function has been observed in some cancers [2, 3] but the vast majority do not possess this abnormality. Rather, upregulation of glycolysis is believed to be a response to the high-energy demand associated with rapid growth, the need of tumor cells to survive under hypoxic conditions, competition with host cells, and in order to facilitate tumor metastasis [4]. Aerobic glycolysis is observed in many rapidly proliferating cell types, both normal and malignant [5]. It can be induced with the expression of a single protein, pyruvate kinase type M2, which is known to be an important glycolytic regulator [6].

Here we describe the development of an accurate and reliable method based on ^{13}C NMR spectroscopy and metabolic network analysis to precisely determine metabolic flux through the glycolytic and oxidative pathways. Excellent agreement between predicted and experimental oxygen consumption rates and the glutamate pool size validate the model. This network model could play an important role in understanding cancer cell energetics and facilitate the development of therapeutics that interfere with normal metabolism to inhibit cell growth. In 1995 our laboratory [7] refined and validated a metabolic network model for analyzing fluxes through glycolytic and TCA cycle metabolism of the heart that was published in 1983 by Edwin Chance and John Williamson [8]. By including the malate-aspartate and glycerol-phosphate shuttles into the Chance and Williamson model we obtained excellent agreement between predicted and experimental oxygen consumption rates. Preliminary models of cancer energy metabolism have been reported [9–12], but they have not been comprehensive and have not been validated by comparing the calculated rate of oxygen consumption rate with the experimentally determined values.

2 Methods

2.1 *Experimental*

The origin and growth conditions for the DB-1 human melanoma cells have been previously described [13], as have the NMR bioreactor system and the growth of cells on polystyrene microcarrier beads [14].

2.2 *Bonded Cumomer Metabolic Network Model*

A three-compartment (extracellular, cytoplasm, and mitochondrion) dynamic metabolic model, adapted for perfusion experiments, was used to fit ^{13}C time courses to determine metabolic fluxes. The model was expressed mathematically using two types of mass balance equations: (1) mass balances for the total metabolite concentration in medium, cytosol, mitochondria, and (2) ^{13}C isotopomer mass balance for labeled metabolites. The metabolic network includes glycolysis, TCA cycle, pentose-phosphate pathway, α-ketoglutarate-glutamate and oxaloacetate-aspartate exchange, pyruvate carboxylase activity, anaplerosis at the succinyl-CoA level, pyruvate recycling through malic enzyme, and lactate dehydrogenase activity. Isotopomer balance equations were derived for every bonded cumomer of order 1, 2, and 3 of glutamate and glutamine. This resulted in a set of ~150 differential cumomer equations. Fine structure multiplets were completely described by each metabolite's bonded cumomers of order 1, 2, and 3. The use of bonded cumulative isotopomers or *bonded cumomers* leads to a reduced number of equations compared to a model including all possible isotopomers, while retaining all the NMR-measurable isotopomer information. The concept of cumomer was used in early studies by Muzykantov and Shestov [15] and the term cumomer was first proposed by Wiechert et al. [16]. There are connection matrices between the "bonded cumomer" $\boldsymbol{\pi}$ vector, which reflect sum of fractions of particular labeled isotopomers and the vector of singlets, doublets, triplets, and quartets of metabolite "fine structure" of ^{13}C NMR spectra. One can derive kinetic equations in the form of fine-structure spectroscopic-defined NMR data using connection equations.

For [1,6-$^{13}C_2$]glucose perfusion, the fitted time courses were: Glut-C4total, Glut C4D43, Glut-C3total, Glut-C2total, and Lac-C3total, for a total of five curves. The six following fluxes were determined: melanoma TCA cycle F_{TCA}, pyruvate carboxylase F_{PC}, exchange between glutamate and α-ketoglutarate F_X, glutamate-glutamine exchange F_{GG}, anaplerotic flux F_{ANA}, non-oxidative glycolytic flux CMR_{LAC}, and two other parameters—Michaelis-Menten lactate transport V_{max}^{LAC} and total cellular [Glut] concentration. Solving a system of differential equations (using the Runge–Kutta fourth order procedure for stiff systems) in terms of bonded cumomers yields time courses for all possible fine structure ^{13}C multiplets of glutamate, glutamine, and aspartate. Minimization was performed using Broyden-Fletcher-Goldfarb-Shanno or Simplex algorithms. The errors for the obtained values were estimated using Monte Carlo simulations with experimental noise levels. All numerical procedures were carried out in Matlab.

3 Results

A typical ^{13}C NMR spectrum for metabolically active DB-1 cells during perfusion with [1,6-$^{13}C_2$]glucose medium is shown in Fig. 37.1. The lower spectrum includes the labeled and natural abundance resonances. For the upper expanded spectrum, the natural abundance resonances have been removed by subtraction with a baseline spectrum. The resonances observed include those for lactate-3, glucose-6 (glucose-1 is not shown), the central three carbons of glutamate, and $-CH_2-$ groups of fatty acids [14]. A large amount of labeling was observed in glutamate-4, with lesser amounts in glutamate-2 and glutamate-3. The glutamate-3 resonance was a triplet

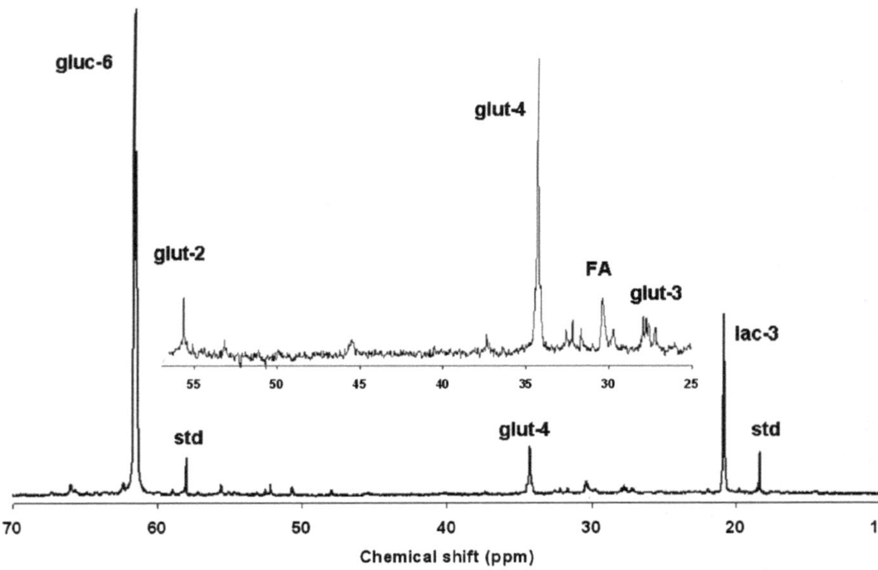

Fig. 37.1 ^{13}C NMR spectrum of living melanoma cells during perfusion with [1,6-$^{13}C2$]glucose

due to ^{13}C–^{13}C coupling with glutamate-4. Such coupling demonstrates complete TCA cycle function [17].

Figure 37.2 shows the experimental time courses for labeled glutamate during perfusion with [1,6-$^{13}C_2$]glucose medium (symbols). These values, along with the glucose and glutamine consumption rates and the lactate formation rate, were used to estimate the intracellular fluxes with the network model. For the calculations, equal fluxes through the malate-aspartate and glycerol-phosphate shuttles were assumed. The experiment was performed in duplicate and fluxes were calculated independently for the two experiments. The experimental and calculated cellular metabolic oxygen rate ($CMRO_2$) in mmol/L-cell/h and the glutamate pool size in mmol/L-cell for the two experiments (experimental/calculated) were Exp. #1; $CMRO_2$ 35/36, Glu 9.2/8.6; Exp. #2: $CMRO_2$ 32/30; Glu not measured/10.5. Therefore, there is excellent agreement between experimental and model calculated values of $CMRO_2$ and Glu, although the glutamate pool size was only measured in one experiment.

ATP production was also calculated assuming a P/O ratio of 2.5 for NADH and 1.5 for $FADH_2$ [18]. The average ATP production rates for the two experiments were CMR_{atp}(mitochondrial) = 174 mmol/L-cell/h and CMR_{atp}(glycolytic) = 150 mmol/L-cell/h, which indicates that 54% of the energy was produced by oxidative phosphorylation and 46% was from glycolysis.

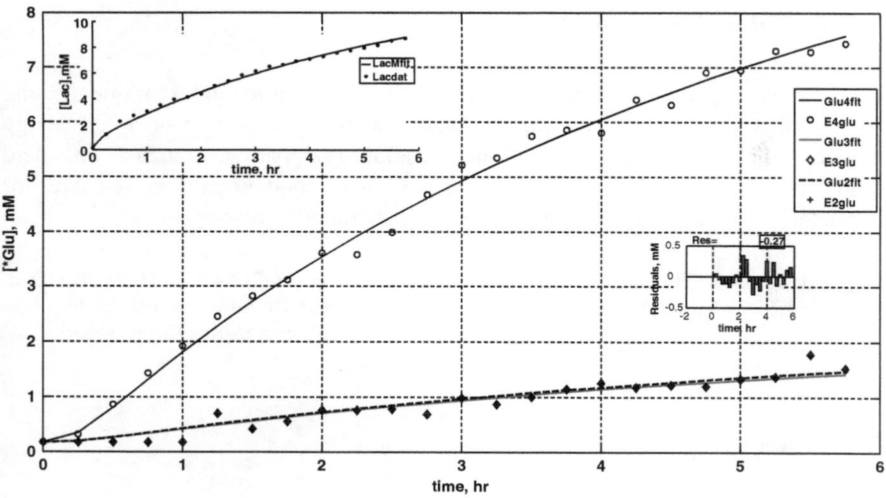

Fig. 37.2 [1,6-$^{13}C_2$]Glc perfusion. First experiment. Total Glu C4,C3,C2 labeling time courses. Inset: total LacC3 labeling time course. *Continuous lines* represent the best fits

4 Discussion

The excellent agreement between the predicted and experimental $CMRO_2$ values demonstrates that the model accurately estimates fluxes for oxidative metabolism. Consistent with the Waburg Hypothesis, we found that glycolysis accounts for approximately half of total cellular energy. The cells were well oxygenated but still produced a large amount of lactate.

The labeling patterns in glutamate observed for this cell line are similar to those observed in other cancer cells [17]. The high level of labeling in glutamate-4 relative to glutamate-2 and 3 indicates that a significant amount of label leaves the TCA cycle in a form other than CO_2 [17]. In highly aerobic tissues such as the beating heart, labeling in glutamate-4 from the intermediate [3-^{13}C]pyruvate precedes that for carbons 2 and 3, but eventually all 3 central carbons are equally labeled [7, 8, 19]. The labeling pattern observed in melanoma cells is believed to be the result of a high level of anaplerotic exchange that is associated with biosynthesis [20]. To date, this is only a hypothesis, which needs to be investigated experimentally. A high ratio of glutamate-4 to glutamate-2 and 3 can also result from a high rate of glutamine consumption, but this was not observed with DB-1 melanomas. These cells have a low glutamine consumption that is only ~1.2% of the glucose consumption rate.

5 Conclusions

We have developed and examined the accuracy of a versatile model for quantitating the contributions of glycolysis and oxidative phosphorylation to cancer cell energy production. Further refinements will include addition of pathways such as fatty acid synthesis that are important for growth. The network model could prove useful for characterizing cancers in vivo and monitoring therapeutic response.

Acknowledgments This research was supported by NIH grants 2U24-CA083105-10 (JDG), P41RR08079 (AAS), and R01NS3867 (AAS). The authors thank Dr. Edwin M. Chance for inspiration and helpful discussion and David S. Nelson for assistance in preparing this manuscript.

References

1. Warburg O, Posener K, Negelein E (1924) Ueber den Stoffwechsel der Tumoren. Biochmeische Zeitschrift 152:319–344
2. Chatterjee A, Mambo E, Sidransky D (2006) Mitochondrial DNA mutations in human cancer. Oncogene 25:4663–4674
3. Hoberman HD (1975) Is there a role for mitochondrial genes in carcinogenesis. Cancer Res 35:3332–3335
4. Gatenby RA, Gillies RJ (2004) Why do cancers have high aerobic glycolysis? Nat Rev Cancer 4:891–899
5. Wang T, Marquardt C, Foker J (1976) Aerobic glycolysis during lymphocyte-proliferation. Nature 261:702–705
6. Christofk HR, Vander Heiden MG, Harris MH et al (2008) The M2 splice isoform of pyruvate kinase is important for cancer metabolism and tumour growth. Nature 452:230–233
7. Chatham JC, Forder JR, Glickson JD et al (1995) Calculation of absolute metabolic flux and the elucidation of the pathways of glutamate labeling in perfused rat-heart by C-13 NMR-spectroscopy and nonlinear least-squares analysis. J Biol Chem 270:7999–8008
8. Chance EM, Seeholtzer SH, Kobayashi K et al (1983) Mathematical analysis of isotope labeling in the citric acid cycle with applications to 13C NMR studies in perfused hearts. J Biol Chem 258:13785–13794
9. Yang C, Richardson AD, Osterman A et al (2008) Profiling of central metabolism in human cancer cells by two-dimensional NMR, GC-MS analysis, and isotopomer modeling. Metabolomics 4:13–29
10. Miccheli A, Tomassini A, Puccetti C et al (2006) Metabolic profiling by C-13-NMR spectroscopy: 1,2-C-13(2) glucose reveals a heterogeneous metabolism in human leukemia T cells. Biochimie 88:437–448
11. Paredes C, Sanfeliu A, Cardenas F et al (1998) Estimation of the intracellular fluxes for a hybridoma cell line by material balances. Enzyme Microb Technol 23:187–198
12. Portais JC, Voisin P, Merle M et al (1996) Glucose and glutamine metabolism in C6 glioma cells studied by carbon 13 NMR. Biochimie 78:155–164
13. Hill LL, Korngold R, Jaworsky C et al (1991) Growth and metastasis of fresh human-melanoma tissue in mice with severe combined immunodeficiency. Cancer Res 51:4937–4941
14. Mancuso A, Beardsley NJ, Wehrli S et al (2004) Real-time detection of C-13 NMR labeling kinetics in perfused EMT6 mouse mammary tumor cells and beta HC9 mouse insulinomas. Biotechnol Bioeng 87:835–848
15. Muzykantov VS, Shestov AA (1986) Kinetic-equations for the redistribution of isotopic molecules due to reversible dissociation—homoexchange of methane. React Kinet Catal Lett 32:307–312

16. Wiechert W, Mollney M, Isermann N et al (1999) Bidirectional reaction steps in metabolic networks: III. Explicit solution and analysis of isotopomer labeling systems. Biotechnol Bioeng 66:69–85
17. Mancuso A, Zhu AZ, Beardsley NJ et al (2005) Artificial tumor model suitable for monitoring P-31 and C-13 NMR spectroscopic changes during chemotherapy-induced apoptosis in human glioma cells. Magn Reson Med 54:67–78
18. Berg JM, Tymoczko JL, Stryer L (2002) Biochemistry. W.H. Freeman, New York
19. Weiss RG, Chacko VP, Glickson JD et al (1989) Comparative C-13 and P-31 nmr assessment of altered metabolism during graded reductions in coronary flow in intact rat hearts. Proc Natl Acad Sci USA 86:6426–6430
20. Malloy CR, Sherry AD, Jeffrey FMH (1987) Carbon flux through citric acid cycle pathways in perfused heart by 13C NMR spectroscopy. FEBS Lett 212:58–62

Chapter 38 Muscle Oxygen Saturation Heterogeneity Among Leg Muscles During Ramp Exercise

Shun Takagi, Ryotaro Kime, Masatsugu Niwayama, Norio Murase, and Toshihito Katsumura

Abstract We examined whether O_2 saturation in several leg muscles changes as exercise intensity increases. Twelve healthy young males performed 20 W/min ramp bicycle exercise until exhaustion. Pulmonary O_2 uptake (VO_2) was monitored continuously during the experiments to determine peak oxygen uptake. Muscle O_2 saturation (SmO_2) was also monitored continuously at the belly of the vastus lateralis (VL), rectus femoris, vastus medialis, biceps femoris, gastrocnemius lateralis, gastrocnemius medialis, and tibialis anterior by near-infrared spatial resolved spectroscopy. Although the VL muscle mainly contributes during cycling exercise, deoxygenation was enhanced not only in the VL muscle but also in the other thigh muscles and lower leg muscles with increased exercise intensity. Furthermore, SmO_2 response during ramp cycling exercise differed considerably between leg muscles.

Keywords Muscle oxygen saturation • Heterogeneity • Oxygen uptake • Cycling exercise • Near-infrared spatial resolved spectroscopy

1 Introduction

Near-infrared spectroscopy (NIRS) has been widely used in measuring muscle oxygenation. Muscle oxygen saturation (SmO_2), which is measured by NIRS, is an indicator of the balance between local O_2 delivery and O_2 consumption. SmO_2 in the

S. Takagi (✉) • R. Kime • N. Murase • T. Katsumura
Department of Sports Medicine for Health Promotion, Tokyo Medical University, 6-1-1 Shinjuku, Shunjuku-ku, Tokyo 160-8402, Japan
e-mail: stakagi@tokyo-med.ac.jp

M. Niwayama
Department of Electrical and Electronic Engineering, Shizuoka University, Shizuoka 432-8561, Japan

W.J. Welch et al. (eds.), *Oxygen Transport to Tissue XXXIV*, Advances in Experimental Medicine and Biology 765, DOI 10.1007/978-1-4614-4989-8_38,

vastus lateralis (VL) muscle decreased during incremental cycling [1], and the decline of SmO_2 was likely attributed to higher O_2 consumption than O_2 delivery.

During cycling exercise, anterior thigh muscles, especially the vastus lateralis muscle (VL), mainly contribute. Many studies utilizing NIRS only measure from a single site (usually the distal site or midpoint of VL muscle), which is viewed as the representative response of the whole leg muscle. However, it has been reported that the O_2 balance is distributed heterogeneously between and within skeletal muscles [1, 2]. Previous studies reported that other thigh muscles and lower leg muscles may also contribute during cycling exercise [3, 4]. Therefore, SmO_2 responses in the other thigh muscles and lower leg muscles may differ from the responses in VL muscle. Furthermore, available data about SmO_2 response within a single muscle with increased exercise intensity is still limited.

The main purpose of this study was to examine dynamic O_2 balance responses in several leg muscles and within a single muscle during cycling exercise.

2 Methods

2.1 *Subjects*

Twelve healthy young males (age: 24±2 years; height: 174.1±7.3 cm; weight 66.0±7.0 kg; mean±SD) participated in the study, which was approved by the Tokyo Medical University Local Research Ethics Committee, Japan. All volunteers were non-obese, nonsmokers, and were physically active but none had been involved in any type of endurance training program for at least 12 months prior to the study. All volunteers were informed of the purpose and nature of the study, after which their written, informed consent was given.

2.2 *Experimental Design*

Twelve healthy young males performed 20 W/min ramp bicycle exercise (after a 3-min warm up at 0 W) until exhaustion (Strength Ergo 8, Fukuda-Denshi, Tokyo, Japan). Pulmonary O_2 uptake (VO_2) was monitored continuously during the experiments to determine VO_2peak, and carbon dioxide production (VCO_2) was assessed breath-by-breath with an online metabolic system (AE310S, Minato Medical Science, Osaka, Japan). Pedal frequency of 60 rpm was maintained by keeping time with a metronome.

Muscle O_2 saturation (SmO_2) was monitored at the distal site of the VL (VLd), rectus femoris (RF), vastus medialis (VM), biceps femoris (BF), gastrocnemius lateralis (GL), gastrocnemius medialis (GM), and tibialis anterior (TA) in the left leg by multichannel near-infrared spatial resolved spectroscopy (NIR_{SRS}). VLd was defined as 10–14 cm above from the patella (30% of the length between the patella and the greater trochanter). The SmO_2 values were defined as the SmO_2 averaged over the last 10 s at rest, warm up, 20, 40, 60, 80, and 100%VO_2peak. Coefficients of variation (CV) of SmO_2 values of thigh muscles (VLd, RF, VM, and BF), lower leg muscles (GL, GM, and TA), and whole leg muscles (VLd, RF, VM, BF, GL, GM, and TA) were calculated as an index of SmO_2 heterogeneity between muscles for 10 s. To examine regional difference of SmO_2 in a single muscle, an additional probe was placed on a proximal point (VLp) of 30 % of the length between the patella and the greater trochanter, from the VLd muscle. A regional difference within the VL muscle was defined as SmO_2 at VLp–SmO_2 at VLd.

2.3 *Multichannel* NIR_{SRS} *System*

We used a two wavelength (770 and 830 nm) light-emitting diode NIR_{SRS} (Astem Co., Japan). The probe consisted of one light source and two photodiode detectors, and the optode distances were 20 and 30 mm, respectively. The SmO_2 was derived from the relative absorption coefficients obtained from the slope of light attenuation over a distance measured at two focal points from the source of light emission. The relative absorption coefficients were converted to relative concentrations of oxygenated and deoxygenated-Hb. Therefore, this device can essentially calculate tissue oxygen saturation. A previous study has reported that fat layer thickness affects SmO_2 [5]. In contrast, Niwayama et al. [6] recently reported that SmO_2 can be quantified by the correction of fat layer thickness effects, and the specifications of the NIR_{SRS} were fully described. In this study, we measured fat layer thickness at each measurement site in the muscles to correct these effects using an ultrasound device (LogiQ3, GE-Yokokawa Medical Systems, Japan) by placing an ultrasound probe on the same sites as the NIR_{SRS} probes had been placed.

2.4 *Statistics*

Normal distribution of the dependent variable was documented using the Kolmogorov-Smirnoff test. Consequently, data are given as means ± standard deviation (SD). The statistical differences among values were tested using one-way ANOVA. When a significant difference was found, multiple comparisons were conducted using Dunnett's test to compare between values at baseline and each exercise intensity. For all statistical analyses, significance was accepted at $p < 0.05$.

3 Results

In all measurement sites, SmO_2 was significantly decreased as exercise intensity increased. However, SmO_2 response was considerably different between and within exercising leg muscles. VLd and RF muscle were significantly decreased at 80 and

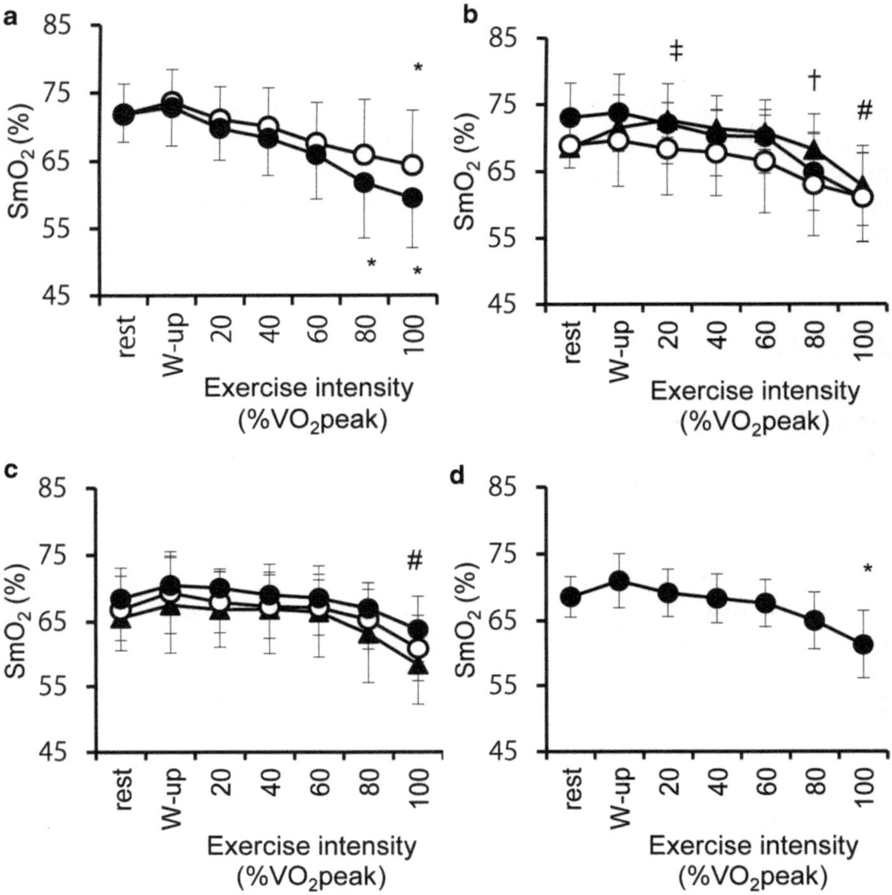

Fig. 38.1 SmO_2 in VLd and VLp muscles (**a**), RF, VM, and BF muscles (**b**), GL, GM, and TA muscles (**c**), and mean SmO_2 in whole leg muscles (VLd, RF, VM, BF, GL, GM, and TA) (**d**) during ramp cycling exercise. The *closed circles* show SmO_2 at VLd and the *open circles* show SmO_2 at VLp (**a**). The *closed circles* show SmO_2 at RF; the *open circles* show SmO_2 at VM; the *closed triangles* show SmO_2 at BF (**b**). The *closed circles* show SmO_2 at GL; the *open circles* show SmO_2 at GM; the *closed triangles* show SmO_2 at TA (**c**). Significant difference from rest (*: $p<0.05$). Significant difference from rest at the three muscles shown in the figure (#: $p<0.05$). Significant difference from rest at RF muscle (†: $p<0.05$). Tendency to be higher than rest at BF muscle (‡: $p=0.07$)

100%VO_2peak from rest ($p<0.05$). SmO_2 in VLp, VM, GL, GM, and TA were significantly decreased at peak exercise ($p<0.05$). SmO_2 in BF tended to be increased at 20%VO_2peak ($p=0.07$), then, significantly decreased with increased exercise intensity ($p<0.05$) (Fig. 38.1a–c). Consequently, the decline in the mean SmO_2 of whole leg muscles (VLd, RF, VM, BF, GL, GM, and TA) seemed to begin at around 30 % of VO_2peak, and a significant decrease was observed at exhaustion from rest

($p<0.05$) (Fig. 38.1d). However, CV of SmO_2 between thigh muscles, lower leg muscles, and whole leg muscles showed no significant change (thigh muscles: $p=0.18$, lower leg muscles: $p=0.48$, whole leg muscles: $p=0.28$). Conversely, regional difference in a single muscle significantly increased at 80 and 100%VO_2peak ($p<0.05$). SmO_2 of VLd was significantly lower than VLp at 80%VO_2peak (VLd: 61.8±8.2 %, VLp: 65.9±8.1 %, $p<0.05$) and peak exercise (VLd: 60.7±7.4 %, VLp: 64.3±8.1, $p<0.05$) while no significant difference was found at rest (VLd: 71.9±4.0 %, VLp: 71.9±4.5, $p=0.99$) (Fig. 38.1a).

4 Discussion

The results of the present study provided three major findings. First, SmO_2 of not only anterior thigh muscles, especially the VL muscle, but also lower leg muscles significantly decreased as exercise intensity increased. Second, SmO_2 response in the other thigh muscles and lower leg muscles is quite different from SmO_2 in the VLd muscles. Finally, regional difference of SmO_2 within a single muscle changed as exercise intensity increased. These results are attributed to the fact that blood flow and metabolic demand differ considerably between leg muscles, as well as within a synergetic muscle group, and heterogeneous metabolic functions have been observed even within a single muscle, such as in the VL muscle, during incremental cycling exercise.

Deoxygenation patterns were considerably different between muscles during ramp cycling exercise. Muscle perfusion and muscle VO_2 were presumably affected by differences in muscle fiber composition, microvascular structure, and motor unit recruitment pattern [7, 8]. Moreover, SmO_2 response was likely influenced by the anatomical differences and action of the muscles (e.g., VL is one of the knee extensor muscles; RF is one of the knee extensor/hip flexor muscles) during ramp cycling. In contrast, SmO_2 heterogeneity among thigh muscles, among lower leg muscles, and among whole leg muscles was relatively maintained. These results may be attributed predominantly to compensation by the other thigh muscles (RF, VM, and BF) and lower leg muscles (GL, GM, TA), although previous studies reported that the VL muscle mainly contributes during cycling exercise. There is a possibility that the subjects' training statuses [9] and/or shoe-pedal interface [10] affected SmO_2 responses and their heterogeneity between muscles. In addition, because of the low number of subjects, a significant change of heterogeneity among muscles may not have been detected.

Exercise intensity is an important factor in determining regional difference of SmO_2 within the VL muscle. Kime et al. [1] demonstrated that SmO_2 at both distal and proximal sites of the VL significantly decreased and regional difference seemed to increase as exercise intensity increased during incremental cycling. In the present study, SmO_2 at the distal site was significantly lower than at the proximal site at peak exercise, while no difference was observed at rest. These results may be partly explained by higher muscle activity at the distal site than at the proximal site as exercise intensity increases [11]. Therefore, the disparities in the O_2 balance may be partly caused by recruitment of additional muscle fibers, especially at the distal site.

In conclusion, during ramp cycling exercise, SmO_2 response in the other thigh muscle and lower leg muscles was different from SmO_2 in the VL muscles. SmO_2 heterogeneity between muscles did not change, even though deoxygenation in all muscles was enhanced as exercise intensity increased. Furthermore, regional difference of SmO_2 within the VL muscle increased during ramp cycling exercise.

Acknowledgments The authors are grateful for revision of this manuscript by Andrea Hope. We also thank Mikiko Yonemitsu and Ayaka Kime for their helpful technical assistance. This study was supported in part by Grant-in-Aid for Young Scientists from the Japan Society for Promotion of Science 22700702 to R. Kime.

References

1. Kime R, Osada T, Shiroishi K et al (2006) Muscle oxygenation heterogeneity in a single muscle at rest and during bicycle exercise. Jpn J Phys Fitness Sports Med 55(suppl):S19–S22
2. Kalliokoski KK, Oikonen V, Takala TO et al (2001) Enhanced oxygen extraction and reduced flow heterogeneity in exercising muscle in endurance-trained men. Am J Physiol Endocrinol Metab 280:E1015–E1021
3. Komi PV, Kaneko M, Aura O (1989) EMG activity of the leg extensor muscles with special reference to mechanical efficiency in concentric and eccentric exercise. Int J Sports Med 8(suppl 1):22–29
4. Praagman M (2001) A method for biomechanical modeling of energy consumption. J Clin Bimech 15(suppl 1):S53–S54
5. Komiyama T, Quaresima V, Shigematsu H et al (2001) Comparison of two spatially resolved near-infrared photometers in the detection of tissue oxygen saturation: poor reliability at very low oxygen saturation. Cli Sci (Lond) 101(6):715–718
6. Niwayama M, Sone S, Murata H et al (2007) Errors in muscle oxygenation measurement using spatially-resolved NIRS and its correction (Abstract in English). J Jpn Coll Angiol 47(1):17–20
7. Murrant CL, Sarelius IH (2000) Coupling of muscle metabolism and muscle blood flow in capillary units during contraction. Acta Physiol Scand 168(4):531–541
8. Poole DC, Mathieu-Costello O (1996) Relationship between fiber capillarization and mitochondrial volume density in control and trained rat soleus and plantaris muscles. Microcirculation 3(2):175–186
9. Takahashi T, Yamamoto T, Ono T et al (1998) Neuromuscular, metabolic, and kinetic adaptations for skilled pedaling performance in cyclist. Med Sci Sport Exerc 30:442–449
10. Ericson M (1986) On the biomechanics of cycling. A study of joint and muscle load during exercise on the bicycle ergometer. Scand J Rehabil Med 16(suppl):1–43
11. Malek MH, Coburn JW, Weir JP et al (2006) The effects of innervation zone on electromyographic amplitude and mean power frequency during incremental cycle ergometry. J Neurosci Methods 155:126–133

Chapter 39
PET Imaging of the Impact of Extracellular pH and MAP Kinases on the *p*-Glycoprotein (Pgp) Activity

Oliver Thews, Wolfgang Dillenburg, Frank Rösch, and Marco Fellner

Abstract The functional activity of *p*-glycoprotein (Pgp) can be increased in vitro by an extracellular acidosis via activation of MAP kinases (p38, ERK1/2). In order to study these effects in vivo a new ^{68}Ga-labeled PET tracer was developed which serves as a substrate of the Pgp and therefore indirectly mirrors the Pgp activity. For in vivo studies, experimental tumors were imaged under acidic conditions (inspiratory hypoxia, injection of lactic acid) and during inhibition of MAP kinases in a μ-PET system. In vitro, [^{68}Ga]MFL6.MZ showed an accumulation within the cells of about 20% which was increased to 30% by Pgp inhibition. In solid tumors a marked tracer uptake was observed showing spatial heterogeneity. When the tumors were acidified, the PET tracer accumulation was reduced by 20–30%. Changing the inspiratory O_2-fraction to 8% led dynamically to a decrease in pH and in parallel to a reduced tracer concentration. Inhibition of the p38 pathway reduced the Pgp transport rate. The new ^{68}Ga-labeled tracer is suitable for PET imaging of the tissue Pgp activity. In vivo imaging reveals that an acidosis activates the Pgp markedly, a mechanism in which the p38-MAPK pathway seems to play an important role.

Keywords *p*-Glycoprotein • Acidosis • MAP kinases • PET • ^{68}Ga

---

This study forms part of the doctoral thesis of Wolfgang Dillenburg.

O. Thews (✉)
Julius-Bernstein Institute of Physiology, University of Halle-Wittenberg, Magdeburger Str. 6, 06112 Halle (Saale), Germany
e-mail: oliver.thews@medizin.uni-halle.de

W. Dillenburg
Institute of Physiology and Pathophysiology, University Medicine Mainz, Mainz, Germany

F. Rösch • M. Fellner
Institute of Nuclear Chemistry, University of Mainz, Mainz, Germany

W.J. Welch et al. (eds.), *Oxygen Transport to Tissue XXXIV*, Advances in Experimental Medicine and Biology 765, DOI 10.1007/978-1-4614-4989-8_39,

1 Introduction

Solid tumors show several pronounced differences as compared to normal tissues with respect to physiological characteristics at the cellular and tissue level. As a consequence of insufficient tumor perfusion [1] there is inadequate oxygen delivery with forced anaerobic metabolism resulting in an increased lactic acid formation [2] and extracellular acidosis [1]. Numerous studies have demonstrated that the abnormal physiological microenvironment reduces the cytotoxicity of chemotherapeutic drugs [3].

On the other hand, reduced chemosensitivity may result from drug transporters which actively pump amphiphilic xenobiotics out of the cell [4]. Modulation of these drug transporters (e.g., by inhibitors) can modify the cytotoxic efficacy of chemotherapy [5]. The best studied member of the ABC-transporter family is the *p*-glycoprotein (Pgp) responsible for a multidrug-resistant phenotype of many human tumors. Besides constitutional differences in the Pgp expression of tumor entities or cell lines, the expression as well as the functional activity of the transporter has been shown to be regulated. In vitro studies [6, 7] demonstrated that lowering the extracellular pH to 6.5 functionally increases the Pgp activity and by this reduces the cytotoxicity of chemotherapeutics which are a substrate of the *p*-glycoprotein. In vivo it was shown that lowering the tumor pH by forcing anaerobic glycolysis decreases the cytotoxicity of daunorubicin [6] which could be attributed to an acidosis-induced activation of the Pgp. However, in these experiments the Pgp transport rate itself could not be measured in vivo. The study also revealed that MAP kinases (p38, ERK1/2) are a signaling pathways responding to extracellular pH and that activate Pgp [6]. Inhibition of p38 or ERK1/2 reduced the Pgp activity in vitro.

From these results, there is strong evidence that extracellular acidosis functionally increases the Pgp activity via a p38-mediated pathway in vivo [6]. With functional imaging such as PET it might be possible to measure the functional Pgp activity noninvasively and to confirm the impact of extracellular acidosis on drug transport as well as to identify the role of MAP kinases in the signaling pathway in vivo.

2 Material and Methods

2.1 PET Tracer and Tumor Model

For functional Pgp imaging, six tracers (MFL1.MZ-MFL6.MZ) were synthesized which consist of hexadentate Schiff-bases labeled with the positron emitter ^{68}Ga ($T_{1/2}$=67.7 min, β^+ branching=89 %). ^{68}Ga was eluted from a ^{68}Ge/^{68}Ga generator [8] with 10 mL HCl and immobilized on an acidic cation exchanger. Labeling was performed with 20 μL of each ligand solution (1 mg/mL in EtOH, 30 nmol). Radiochemical labeling yield and complex formation were determined by thin layer chromatography (TLC). For injection the tracer pH was adjusted to 7.4 with NaOH and diluted with isotonic saline. The tracers themselves are substrates of the Pgp and are actively pumped out of the cell. The intratumoral concentration reflects therefore inversely the Pgp activity.

In all experiments the subline AT1 of the R3327 Dunning prostate carcinoma was used which functionally expresses the *p*-glycoprotein [7]. Cells were grown in RPMI 1640 medium (+10 % fetal calf serum) and passaged once per week. Solid carcinomas of this cell line were heterotopically induced by injection of AT1 cells (0.4 mL approximately 10^4 cells/μL) subcutaneously into the dorsum of the hind foot. Tumors were used when they reached a volume of between 1.0 and 2.0 mL approximately 10–14 days after tumor cell inoculation.

2.2 Acidosis Model and Tumor Treatment

In order to dynamically change the extracellular pH of the tumors, the inspiratory gas mixture was changed from pure oxygen (100 % O_2) starting 10 min prior to the PET tracer injection to a hypoxic gas containing 92 % N_2+8 % O_2 at 15 min after tracer injection (i.e., during the μ-PET measurement). It was continued until the end of the μ-PET measurement at 60 min. In parallel, the extracellular pH was measured with steel-shafted pH glass electrodes (type MI-418B, Microelectrodes Inc, Bedford NH, USA) with an outer diameter of 800 μm inserted into the center of the tumor. Before and after the PET imaging, the electrode was calibrated and pH measurements were corrected for signal shift.

In a second series the extracellular tumor space was acidified by small amounts of lactic acid injected directly intratumorally. Therefore, 50 μL of a 0.222 mM solution of lactic acid were injected in the tumor tissue at a depth of 2–3 mm about 5–10 min prior to the tracer application. The same amount of a 0.222 mM sodium lactate solution was applied in the contralateral tumor and this tumor served as intra-individual control.

In order to inhibit different MAP kinase pathways specific inhibitors (SB203580 and U0126 for inhibition of p38 and ERK1/2, respectively) were injected intratumorally. Therefore, inhibitors were dissolved in DMSO at a concentration of 1 mmol/L and tumors were treated with a single injection of a small amount (20 μL) of these inhibitors 5–10 min prior to the PET measurements. The tumor on the contralateral hind foot was treated with 20 μL DMSO alone and served as intra-individual control.

2.3 μ-PET Imaging

The μ-PET imaging was performed on a microPET Focus 120 small animal PET (Siemens/Concorde, Knoxville, USA). During measurements the anaesthetized (pentobarbital 40 mg/kg, i.p.) rats were placed in supine position and breathed room air spontaneously through a tracheal tube. After a 15 min transmission scan with an external ^{57}Co source, dynamic PET studies were acquired in 2D mode. The radiotracer was administered as a bolus injection (0.4–0.7 mL, mean activity 44.0 ± 1.5 MBq) via a catheter placed in the left jugular vein. Time activity curves were obtained with varying time frames (1–5 min) for a total measuring interval of 60 min. Volumes-of-interests (VOIs) were defined for tumor and reference tissue in the field of view (testis). Ratios of tumor to reference tissue were calculated from integral image between 10 and 60 min after tracer injection.

3 Results

In order to measure the cellular uptake of the ^{68}Ga-Schiff base complexes, AT1 cells were incubated with the tracer in the presence or absence of the Pgp inhibitor verapamil (VPL) for 30 min after which the suspension was centrifuged and the activity in the cells and in the supernatant was determined. Figure 39.1a shows the fraction of ^{68}Ga-complexes detected inside the cells for all tracers. Tracers MFL2.MZ, MFL5.MZ, and MFL6.MZ were found to be substrates of Pgp since the active efflux could be inhibited by VPL resulting in a statistically significant higher intracellular concentration (Fig. 39.1a; e.g., 178 ± 5 % for [^{68}Ga]MFL2.MZ and 144 ± 3 % for [^{68}Ga]MFL6.MZ). From these in vitro experiments it became evident that [^{68}Ga]MFL2.MZ and [^{68}Ga]MFL6.MZ showed the best compromise of good passive uptake into the cells and pronounced Pgp-mediated efflux inhibitable by VPL. Subsequently, both these tracers were used in PET experiments with tumor-bearing rats. Figure 39.1b shows the averaged activity in tumor and testes indicating that [^{68}Ga]MFL6.MZ is strongly accumulated in the tumors resulting in a three-times

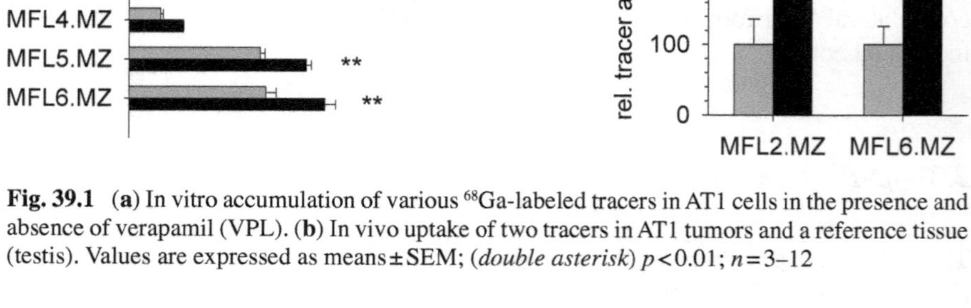

Fig. 39.1 (**a**) In vitro accumulation of various ^{68}Ga-labeled tracers in AT1 cells in the presence and absence of verapamil (VPL). (**b**) In vivo uptake of two tracers in AT1 tumors and a reference tissue (testis). Values are expressed as means ± SEM; (*double asterisk*) $p < 0.01$; $n = 3$–12

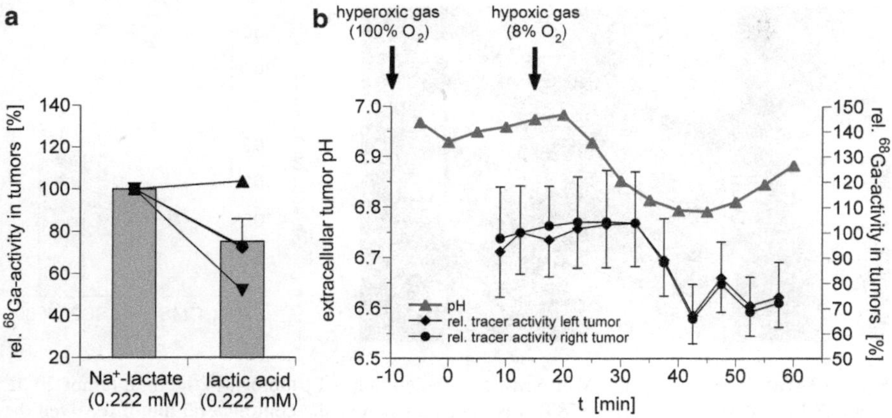

Fig. 39.2 (**a**) Intratumoral [^{68}Ga]MFL6.MZ concentration in tumors treated with lactic acid (50 µL, 0.222 mM) or with an equivalent volume of Na^+-lactate. *Each line* represents one individual animal with two tumors on each hind foot. *Columns* represent means ± SEM of all experiments. (**b**) Extracellular pH and [^{68}Ga]MFL6.MZ concentration during change of the inspiratory O_2-fraction

higher concentration as compared to the reference tissue whereas [^{68}Ga]MFL2.MZ is enriched by only 74 %. Therefore, [^{68}Ga]MFL6.MZ was chosen for further experiments.

As a weak indication of a pH dependency of Pgp-mediated transport the extracellular pH and the tracer concentration were analyzed in tumors of different volume. A clear inverse correlation was found between the pH and tumor volume ($r=-0.941$; data not shown) as well as between tracer concentration (inversely indicating the Pgp activity) and tumor volume ($r=-0.550$; data not shown). Direct injection of lactic acid into one tumor (and Na^+-lactate in the contralateral tumor) showed a markedly reduced tracer concentration in most of experiments (Fig. 39.2a) indicating a higher Pgp transport rate. Figure 39.2b illustrates an experiment where the pH was dynamically reduced during the imaging period by inspiratory hypoxia. Changing the inspiratory gas mixture from 100 to 8 % O_2 15 min after tracer injection reduced the pH from 6.95 to values around 6.8 (Fig. 39.2b, triangles). With a slight delay of 10 min the intratumoral tracer concentration decreased by 25–30 % in both tumors.

Figure 39.3a illustrates an example where in the right tumor the p38 inhibitor SB203580 was injected whereas the left tumor received only DMSO. The right tumor shows a much higher tracer concentration indicating a reduced Pgp activity. This effect was seen in all animals (Fig. 39.3b). On average, the SB203580-treated tumors showed a more than 70 % higher tracer concentration than the contralateral control tumors. Inhibition of the ERK1/2 pathway by U0126 had only a minor effect on the Pgp activity (increase of the tracer concentration by only 29 %; data not shown).

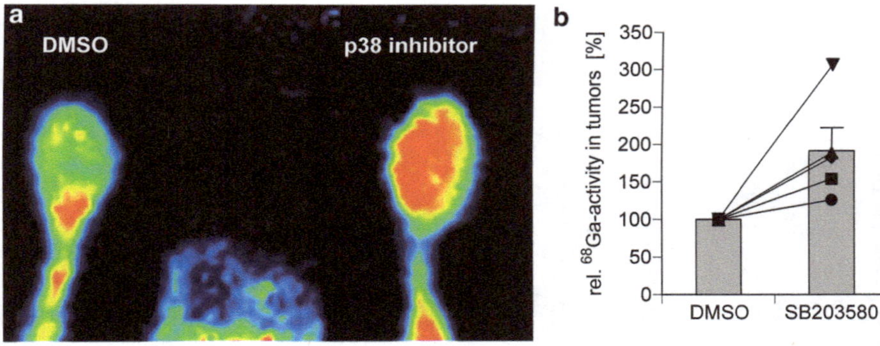

Fig. 39.3 (**a**) Example of [^{68}Ga]MFL6.MZ accumulation in AT1 tumors. In the right tumor 20 μL of a p38 MAPK inhibitor (SB203580) was injected whereas the contralateral tumor received the same volume of the vehicle (DMSO). (**b**) Relative [^{68}Ga]MFL6.MZ concentration in all tumors treated with the p38 inhibitor or with DMSO. *Each line* represents one individual animal with tumors on each hind foot. *Columns* represent means ± SEM of all experiments

4 Discussion

A suitable PET tracer for measuring the Pgp transport rate should fulfill two features: (1) it should enter the cell easily (by passive diffusion) and (2) the tracer should be a substrate of the Pgp and this transport should be inhibitable by Pgp inhibitors. All tracers show high cellular uptake but with marked concentration differences. Taking into account the volume fraction of the cells (~0.3 %) the concentration ratio between cells and medium was approximately 16:1 for most of the tracers but ~90:1 for [^{68}Ga]MFL3.MZ which is the most lipophilic compound. Since the intracellular concentration depends on the Pgp activity, the ratio of cellular tracer concentration in the presence and absence of verapamil could be used as an indicator of the Pgp-mediated efflux [9]. This ratio was ~1.2–1.8 for most tracers (Fig. 39.1a) but for the lipophylic tracer [^{68}Ga]MFL3.MZ Pgp inhibition had practically no impact. The uptake measurements in vivo correspond well with the in vitro data as indicated by a higher tumor concentration of the tracer [^{68}Ga]MFL6.MZ compared to [^{68}Ga]MFL3.MZ.

With this tracer μ-PET images were analyzed after acidifying the tumor tissue by direct injection of lactic acid. Since the interstitial fluid pressure is elevated in many human malignancies, the injection of even a small fluid volume of lactic acid into the tissue may somehow disturb tumor microcirculation. For this reason it cannot be excluded that the direct injection may reduce tumor perfusion locally and by this limit the tracer transport. For this reason, the contralateral tumor of each animal served as an intra-individual control by injecting the same volume of an equimolar Na^+-lactate solution leading to a comparable increase of the interstitial hydrostatic pressure. The comparison of the tracer concentration in both tumors therefore reflects solely the impact of the lower pH on Pgp activity.

The second method for acidifying the extracellular space was to force anaerobic glycolysis leading to an increased endogenous formation of lactic acid. Breathing a hypoxic gas mixture with 8 % oxygen is suitable to markedly lower the tumor-pO_2 [10] and to reduce the tumor pH [6]. In the present study hypoxia led to decrease in pH followed by an increase in Pgp activity resulting in a lower tracer accumulation (Fig. 39.2a). The increase of pH 30 min after the onset of the hypoxic breathing remains unclear at the moment. Perhaps changes in the blood pressure or a hypoxia-induced vasodilation led to an increase in tumor perfusion and by this to a better oxygen supply.

Previous in vitro data revealed that MAP kinases (p38, ERK1/2) play a relevant role for signal transduction in the acidosis-induced Pgp activation [6]. The present study clearly shows that MAP kinases are also important for the Pgp transport activity in vivo. Inhibition of p38 by SB203580 increased the intratumoral tracer concentration indicating a reduced Pgp transport rate (Fig. 39.3). A problem of the experimental design might be the direct intratumoral injection of drugs which could impair tumor perfusion. However, since in the contralateral tumor the same volume of DMSO was injected, the specific effect of SB203580 can be derived. In all tumors the activity in the SB203580 treated tumor was markedly higher (on average by the factor of 2). In the present study inhibition of ERK1/2 on the Pgp activity had only a minor impact on the Pgp activity which is in good accordance with previous indirect results on chemosensitivity of tumors [6].

The new ^{68}Ga-labeled tracer is suitable for PET imaging of the tissue Pgp activity. With this tracer it becomes possible to identify patients with multidrug-resistant tumors pre-therapeutically. In vivo imaging reveals that tumor acidosis activates the Pgp markedly, a mechanism in which the p38-MAPK pathway seems to play an important role. From these results new strategies for overcoming multidrug resistance (e.g., increasing tumor pH, inhibition of p38) may be developed.

Acknowledgments This study was supported by Deutsche Krebshilfe (grants 106774/109136).

References

1. Vaupel P, Kallinowski F, Okunieff P (1989) Blood flow, oxygen and nutrient supply, and metabolic microenvironment of human tumors: a review. Cancer Res 49:6449–6465
2. Mueller-Klieser W, Vaupel P, Streffer C (1998) Energy status of malignant tumors in patients and experimental animals. In: Molls M, Vaupel P (eds) Blood perfusion and microenvironment of human tumors. Springer, Berlin, pp 193–207
3. Sanna K, Rofstad EK (1994) Hypoxia-induced resistance to doxorubicin and methotrexate in human melanoma cell lines in vitro. Int J Cancer 58:258–262
4. Schinkel AH, Jonker JW (2003) Mammalian drug efflux transporters of the ATP binding cassette (ABC) family: an overview. Adv Drug Deliv Rev 55:3–29
5. Fojo T, Bates S (2003) Strategies for reversing drug resistance. Oncogene 22:7512–7523
6. Sauvant C, Nowak M, Wirth C et al (2008) Acidosis induces multi-drug resistance in rat prostate cancer cells (AT1) in vitro and in vivo by increasing the activity of the p-glycoprotein via activation of p38. Int J Cancer 123:2532–2542

7. Thews O, Gassner B, Kelleher DK et al (2006) Impact of extracellular acidity on the activity of p-glycoprotein and the cytotoxicity of chemotherapeutic drugs. Neoplasia 8:143–152
8. Zhernosekov KP, Filosofov DV, Baum RP et al (2007) Processing of generator-produced ^{68}Ga for medical application. J Nucl Med 48:1741–1748
9. Kunikane H, Zalupski MM, Ramachandran C et al (1997) Flow cytometric analysis of p-glycoprotein expression and drug efflux in human soft tissue and bone sarcomas. Cytometry 30:197–203
10. Thews O, Wolloscheck T, Dillenburg W et al (2004) Microenvironmental adaptation of experimental tumours to chronic vs acute hypoxia. Br J Cancer 91:1181–1189

Chapter 40
Meconium and Transitional Stools May Cause Interference with Near-Infrared Spectroscopy Measurements of Intestinal Oxygen Saturation in Preterm Infants

Alecia Thompson, Paul Benni, Sara Seyhan, and Richard Ehrenkranz

Abstract In a pilot study, we found that the NIRS measurements were highly variable and dramatically decreased or disappeared sporadically while monitoring gastro-intestinal (GI) StO_2 on preterm neonates. We hypothesized that the light absorption characteristics of different neonatal and infant stool types may be interfering with NIRS GI measurements. Methods: Meconium, transitional, and mature stool samples collected from preterm neonates and older infants (ranging from a few days to weeks in age) were analyzed by a bench-top spectrometer to measure light absorbance in the near infrared range (650–950 nm). Results: Some meconium and transitional stool samples were noted to have a more marked increase in light absorbance as wavelength decreases towards 650 nm, when compared to more mature stools. When compared to the light absorbance spectra of deoxy-hemoglobin (Hb) and oxy-hemoglobin (HbO_2), there is a high potential that Hb and HbO_2 may be erroneously calculated, resulting in falsely low GI StO_2. Discussion: Meconium and green colored transitional stools demonstrate a light absorbance spectra that may interfere with NIRS GI measurements, possibly due to their higher biliverdin content, which exhibits a broad spectral absorption peak at 660 nm. Caution is warranted in interpreting GI StO_2 NIRS results in neonates still passing meconium and transitional stools until NIRS-based tissue oximeters to measure GI StO_2 can compensate for the presence of these stool types.

Keywords Meconium • Stools • Near-infrared spectroscopy • StO_2 • Intestine

A. Thompson
Division of Neonatology, Children's Hospital at Montefiore, Albert Einstein College of Medicine, Bronx, NY, USA

P. Benni (✉) • S. Seyhan
CAS Medical Systems, Inc, Branford, CT, USA
e-mail: pbenni@casmed.com

R. Ehrenkranz
Division of Newborn Medicine, Yale New Haven Hospital, New Haven, CT, USA

W.J. Welch et al. (eds.), *Oxygen Transport to Tissue XXXIV*, Advances in Experimental Medicine and Biology 765, DOI 10.1007/978-1-4614-4989-8_40,

1 Introduction

Light in the near infrared range (650–950 nm) is capable of penetrating biologic tissue and is absorbed in an oxygen-dependent manner, by chromophores such as hemoglobin, myoglobin, and cytochrome aa3 amongst others. Near-infrared spectroscopy (NIRS) monitors, such as tissue oximeters and cerebral oximeters, measure regional tissue oxygen saturation by measuring the difference in intensity between a transmitted and received light at specific wavelengths generally by using an algorithm based on the modified Beer-Lambert Law [1]. Calculation of tissue oxygen saturation (StO_2) can be made by measuring the ratio of the regional concentration of oxyhemoglobin (HbO_2) and deoxyhemoglobin (Hb) from the following expression [$HbO_2/(HbO_2 + Hb)$], provided that the ratio of HbO_2 and Hb can be accurately determined.

NIRS has been successfully used as a monitor of brain oxygenation in neonates [2, 3] and there is increasing research and clinical interest in using NIRS to monitor gastro-intestinal tract (GI) StO_2, [4] particularly for premature neonates with concern for necrotizing enterocolitis. With informed consent, we initiated a study using NIRS GI StO_2 in preterm neonatal patients of varying gestational and postnatal ages using a prototype NIRS tissue oximeter (FORE-SIGHT®, CASMED, Branford, CT, USA) using a conventional NIRS algorithm used for cerebral StO_2 measurements [1]. During the study, we found that the GI StO_2 measurements were highly variable and dramatically decreased or disappeared sporadically during the same evaluation period, in the same patient. Other emerging research publications reveal a highly variable and unusually low tissue oxygen saturation of the neonatal GI tract as well [5, 6]. A study using fetal pulse oximetry revealed that pulse oximetry sensors over skin stained by a thin coating of meconium resulted in artificially low SpO_2 values [7]. Because pulse oximetry and NIRS monitors both use near-infrared light (650–950 nm) to determine arterial and tissue oxygen saturation, respectively, we hypothesized that the presence of meconium under the NIRS sensor resulted in the erroneous GI tissue oxygen saturation measurements of neonates. In a pilot investigation, we compared the light absorption characteristics of different neonatal and infant stool types. From the obtained spectroscopic data, we hoped we could identity possible chromophores in meconium that may interfere with NIRS measurement.

2 Methods

Meconium, the earliest newborn stool containing a mix of sloughed intestinal cells, mucous, blood, and bile, was obtained and a portion was analyzed by a bench-top spectrometer (EPP2000, StellarNet, Inc., Tampa, FL USA). In addition, further meconium samples from neonates up to a few days old, transitional stool (post meconium clearance) samples from neonates a few days to weeks old, and mature stool samples of infants several weeks old were collected to be analyzed by the

EPP2000 bench-top spectrometer to measure light absorbance. After normalizing the bench-top spectrometer with a saline-filled cuvette, the different stool samples were mixed with saline and placed inside the cuvette for measurement. The subject's age along with type and color of the stool sample were recorded. The objective of the bench-top spectrometer was to measure any difference in the light absorbance between the stool samples and to delineate the spectral patterns of each stool type. The light-scattering properties of stools can affect the results, so the light absorbance results were normalized at 850 nm to compare differences.

3 Results

The light absorbance of eight stool samples of varying types (Table 40.1) were collected and compared against each other, as shown in Fig. 40.1. For 2 of 3 different samples (#2 & #3), meconium is noted to have a more marked increase in light absorption as the wavelength decreases towards 650 nm, when compared to more mature stools. Interestingly, one of the transitional stool samples (#5b) also demonstrated marked increase in light absorbance as wavelength decreases towards 650 nm. Stool samples #5a and #5b are from the same larger stool sample from one subject which had a yellow portion which was sampled for #5a and a green portion which was sampled for #5b. Sample #8 was interesting in that it contained high bilirubin levels with the appearance of light green color as the subject had hyperbilirubinemia, but did not exhibit marked absorption increase at 650 nm. In Fig. 40.2, samples 2, 5a, 5b, and 8 were selectively plotted along with Hb and HbO_2 on a second light absorption plot to visually analyze the similarities and differences. It can be seen in Fig. 40.2 that samples 2 and 5b mimic the light absorption of Hb.

Table 40.1 Stool samples were from eight different neonate or infant subjects

Sample/subject	Age (days)	Stool color/type
1	5	Green-brown meconium
2	2	Forest green meconium
3	Few days	Green-brown meconium
4	3	Brownish-green ET
5A	36 (ex-28 week)	Yellow-light brown LT
5B	36 (ex-28week)	Green LT
6	78	Dark yellow-brown OS
7	30	Yellow LT
8	28	Light green, hyperbilirubinemia LT

The age of the subjects is reported in days. The stool color and type [meconium, early-transitional (ET), late-transitional (LT), or older stools (OS)] are reported. Note that since subject 5 had a stool sample that consisted of a yellow portion (5a) and a green portion (5b), these two stool portions had very different light absorbance characteristics (as shown in the figures)

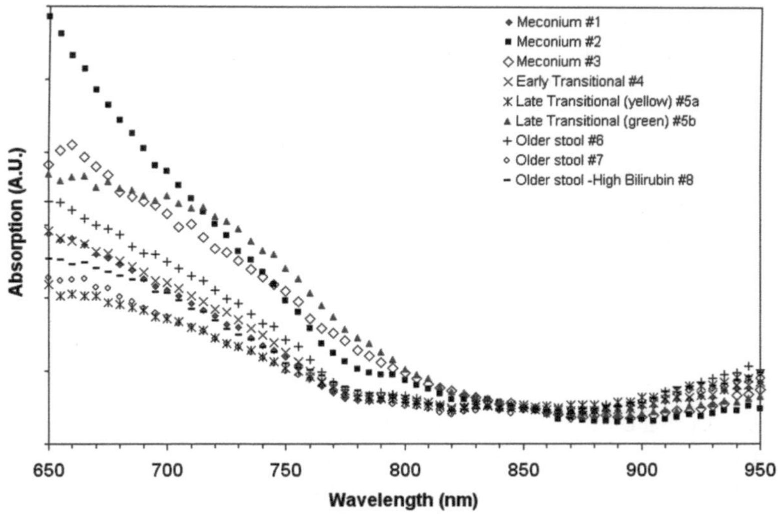

Fig. 40.1 Light absorption spectra of all stool samples. The individual stool absorption spectra were normalized at 850 nm to allow for mutual comparison. Note that the magnitude of light absorption tends to increase with decreasing wavelength to 650 nm, but to various rates

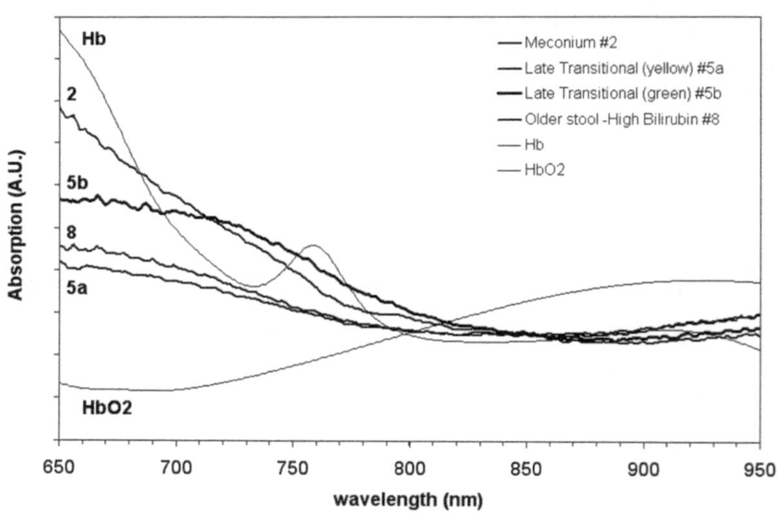

Fig. 40.2 Light absorption spectra of select stool samples from Fig. 40.1, including a meconium sample (#2), a late transitional sample from the same subject with yellow stool portion (#5a) and green portion (#5b) plotted separately to demonstrate the differences in their light absorption spectra, along with a late transitional stool sample from a hyperbilirubinemia subject (#8). Light absorption spectra of Hb and HbO_2 are included for reference

4 Discussion

Meconium is a composite of desquamated intestinal lining, mucous, blood, and bile. In term infants meconium is passed in the first few days of life. Stool appearance and composition typically changes after that interval, once enteral feeds and digestion occur. However, enteral feedings may be delayed in the first few days of life in preterm infants due to unstable medical condition. Even if fed, preterm patients have impaired GI function and motility. Hence, meconium may be present for a longer duration in the GI tract of preterm infants compared to term patients of equivalent postnatal age.

These preliminary results show that some meconium samples and a green colored transitional stool sample exhibited marked increase in light absorption as the wavelength decreases towards 650 nm. This is the same light spectral band that is utilized in NIRS-based monitors that measure tissue or arterial oxygen saturation by measuring the ratio of relative oxyhemoglobin and deoxyhemoglobin. Because the light absorbance of meconium is similar to that of Hb in pattern where the relative magnitude increases towards 650 nm (Fig. 40.2), the presence of meconium along with Hb and HbO_2 may corrupt the calculation of GI StO_2 resulting in lower than expected values if meconium is not accounted for in the NIRS algorithm. This error could occur because the unaccounted light absorbance of meconium would be computed as a higher relative concentration of Hb and/or a lower relative concentration of HbO_2 which in any of these combinations would compute a lower GI StO_2 than expected.

Assuming that the presence of meconium and some other types of stools under the abdominally placed NIRS sensor interfere with GI StO_2 measurement, variability of StO_2 measurement may also result, which could explain the observations of mesenteric oxygen saturation variability [5, 6]. This inconsistency is possibly because stools in the GI tract are moving, as well as the intestine changing position relative to the abdominal wall and the NIRS sensor secured to the skin. Placing the NIRS sensor at different positions on the abdomen over different parts of the GI tract may also result in very different GI NIRS measurements.

Because a green colored transitional stool exhibited the same marked increase in light absorption as the wavelength decreases towards 650 nm like meconium, a particular molecule contained in neonatal stools in varying concentrations may be the interfering chromophore of interest. Biliverdin, a molecule produced by the breakdown of hemoglobin, is a green tetrapyrrolic bile pigment. Biliverdin naturally reduces to bilirubin, a yellow pigmented molecule. Bile in the neonatal stools contains both biliverdin and bilirubin, among other substances. Separating a transitional stool sample by color (#5a yellow & #5b green) revealed very different light absorbance characteristics. The green portion may contain a high level of biliverdin while the yellow portion may contain a high level of bilirubin. Bilirubin exhibits a broad light absorbance peak at 455 nm, and a low light absorbance greater than 530 nm [8], which could partially explain the flatter light absorbance spectra of the yellow stool #5a. Biliverdin, on the other hand, exhibits two light absorbance peaks, one at

350–400 nm and the second a broad peak centered at 660 nm which steadily declines with increasing wavelengths, then levels off at greater than 800 nm [9, 10]. Whether or not biliverdin is the responsible chromophore, it appears that both meconium and some transitional stools potentially interfere with NIRS GI StO_2 measurement on neonates.

In conclusion, the spectrometer testing of meconium and some transitional stool samples from preterm infants of varying postnatal ages potentially demonstrate that some neonatal stools can interfere with NIRS StO_2 measurement, resulting in falsely low values. Meconium and green colored transitional stools affect NIR absorption of light possibly due to its higher biliverdin content. Caution is warranted in interpreting GI StO_2 NIRS results in neonates still passing meconium and transitional stools until NIRS-based tissue oximeters to measure GI StO_2 can compensate for the presence of these stool types.

References

1. Benni PB, Chen B, Dykes FD et al (2005) Validation of the CAS neonatal NIRS system by monitoring vv-ECMO patients: preliminary results. Adv Exp Med Biol 566:195–201
2. Rais-Bahrami K, Rivera O, Short BL (2006) Validation of a noninvasive neonatal optical cerebral oximeter in veno-venous ECMO patients with a cephalad catheter. J Perinatol 26(10):628–635
3. Fenik JC, Rais-Bahrami K (2009) Neonatal cerebral oximetry monitoring during ECMO cannulation. J Perinatol 29(5):376–381
4. Dave V, Brion LP, Campbell DE et al (2009) Splanchnic tissue oxygenation, but not brain tissue oxygenation, increases after feeds in stable preterm neonates tolerating full bolus orogastric feeding. J Perinatol 29(3):213–218
5. Cortez J, Gupta M, Amaram A et al (2011) Noninvasive evaluation of splanchnic tissue oxygenation using near-infrared spectroscopy in preterm neonates. J Matern Fetal Neonatal Med 24:574–582
6. McNeill S, Gatenby JC, McElroy S et al (2011) Normal cerebral, renal and abdominal regional oxygen saturations using near-infrared spectroscopy in preterm infants. J Perinatol 31:51–57
7. Johnson N, Johnson VA, Bannister J et al (1990) The effect of meconium on neonatal and fetal reflectance pulse oximetry. J Perinat Med 18(5):351–355
8. Agati G, Fusi F (1990) New trends in photobiology recent advances in bilirubin photophysics. J Photochem Photobiol 7:1–14
9. Wagnière G, Blauer G (1976) Calculations of optical properties of biliverdin in various conformation. J Am Chem Soc 98(24):7806–7810
10. Avila L, Huang H, Rodríguez JC et al (2000) Oxygen activation by axial ligand mutants of mitochondrial cytochrome b5: oxidation of heme to Verdoheme and Biliverdin. J Am Chem Soc 122(31):7618–7619; ja001137s supplementary material

Chapter 41
Acute Effects of Physical Exercise on Prefrontal Cortex Activity in Older Adults: A Functional Near-Infrared Spectroscopy Study

Takeo Tsujii, Kazutoshi Komatsu, and Kaoru Sakatani

Abstract We examined the acute effect of physical exercise on prefrontal cortex activity in older adults using functional near-infrared spectroscopy (NIRS). Fourteen older adults visited our laboratory twice: once for exercise and once for the control condition. On each visit, subjects performed working memory tasks before and after moderate intensity exercise with a cycling ergo-meter. We measured the NIRS response at the prefrontal cortex during the working memory task. We found that physical exercise improved behavioral performance of the working memory task compared with the control condition. Moreover, NIRS analysis showed that physical exercise enhanced the prefrontal cortex activity, especially in the left hemisphere, during the working memory task. These findings suggest that the moderate intensity exercise enhanced the prefrontal cortex activity associated with working memory performance in older adults.

Keywords NIRS (near-infrared spectroscopy) • Physical exercise • Older adults • Working memory • Prefrontal cortex

1 Introduction

It is often claimed that physical exercise may ameliorate or protect against age-related cognitive decline [1, 2]. For example, a neuroimaging study on chronic exercise effect indicated that aerobic exercise training for 6 months increased the volume of grey and white matter in the prefrontal cortex [3]. Similarly, acute effects of exercise on the performance of a higher-order cognitive task (flanker task) have also

T. Tsujii (✉) • K. Komatsu • K. Sakatani
Department of Neurological Surgery, Division of Optical Brain Engineering, Nihon University School of Medicine, 30-1 Oyaguchi-Kamicho, Itabashi-ku, Tokyo 173-8610, Japan
e-mail: tsujiitakeo@gmail.com

W.J. Welch et al. (eds.), *Oxygen Transport to Tissue XXXIV*, Advances in Experimental Medicine and Biology 765, DOI 10.1007/978-1-4614-4989-8_41,

been demonstrated in older subjects [2]. It is, however, still unclear whether the acute exercise could enhance the prefrontal cortex activity during higher-order cognitive tasks in older adults.

The aim of this study was to examine the acute effect of physical exercise on prefrontal cortex activity in older adults using functional near-infrared spectroscopy (NIRS), an emergent imaging technique for investigating cortical hemodynamic response. Since oxygenated hemoglobin (oxy-Hb) and deoxygenated hemoglobin (deoxy-Hb) have different absorption spectra in the infrared range, changes in oxy-Hb and deoxy-Hb can be calculated by detecting infrared light at different wavelengths on the skull. In general, enhanced oxy-Hb and reduced deoxy-Hb are associated with regional cortical activation. NIRS is noninvasive, is robust against body movement, and has been validated as a suitable technique for investigating neural mechanisms in psychological experiments.

Recently, an NIRS study found that moderate exercise enhanced prefrontal cortex activity associated with improved performance of higher-order cognitive tasks in young adults [4]. However, it is still unknown about the exercise effect on the prefrontal cortex activity in older adults. In the present study, 14 older adults visited our laboratory twice: once for the exercise condition and once for the control condition. On each visit, subjects performed working memory tasks before and after moderate intensity aerobic exercise with a cycling ergo-meter. We measured the NIRS response at the prefrontal cortex during the working memory task, whose procedure is known to activate the prefrontal cortex [5–7]. We compared NIRS responses between the exercise and control conditions.

2 Methods

2.1 *Procedures*

Fourteen subjects participated in this study (mean age = 65.9 ± 1.0 years, 9 female and 7 male) without history of cerebrovascular disease. The study was conducted in accordance with the principles of the Declaration of Helsinki, and all protocols were approved by the Ethics Committee of Nihon University School of Medicine. Written informed consent was obtained from all subjects prior to enrolment in the study.

The experiment was run individually, consisting of cardiopulmonary exercise (CPX) session and NIRS-recording session. The CPX session was conducted to determine the individual's exercise intensity for the subsequent NIRS recording session. Exercise was performed with a cycling ergo-meter (Strength Ergo 240, Mitsubishi electric engineering, Tokyo, Japan). Because our subjects were elderly persons, we indirectly estimated the maximum oxygen uptake (VO2max) from a reference value of Japanese subjects [8].

In the NIRS recording session, subjects visited our laboratory twice: once for the exercise and once for the control condition. In the exercise condition, subjects performed moderate intensity exercise (about 40% of VO2max) with the cycling ergo-meter for 10 min. Immediately before (pretest) or 10 min after (posttest) the exercise session, we measured NIRS response at the prefrontal cortex while subjects performed the working memory task for about 5 min. In the control condition, subjects took a rest during the exercise session.

In the working memory task, two test blocks were sandwiched between three baseline blocks. Subjects were required to remember four digits on the test blocks and one on the baseline blocks. Each block was presented for about 50 s, involving 5 trials. Each trial began with a central fixation cross for 500 ms, followed by the presentation of a set of one or four digits to be learned for 3 s. The delay period was inserted for 5 s, which was followed by the probe digit until response. Subjects were required to indicate whether they thought the probe digit was contained within the learned stimulus set by pressing one of two mouse buttons as quickly and as accurately as possible.

2.2 NIRS Recordings

During the working memory task, relative changes in oxy- and deoxy-Hb concentration values were measured in the prefrontal region using a two-channel portable-type NIRS system (PNIRS-10, Hamamatsu Photonics K.K., Hamamatsu, Japan). This device uses near-infrared light at three wavelengths, ranging around 735 ± 15, 810 ± 18, and 850 ± 20 nm. The oxy-Hb and deoxy-Hb values were estimated from the detected changes of the near-infrared light using the modified Beer-Lambert Law [9]. Because the individual optical path length is unknown, the hemoglobin concentration value is not an absolute but a relative value; this value is expressed as a change from baseline concentration (a.u., arbitrary units). The sampling frequency was 10.2 Hz (97.8 ms/data).

The flexible NIRS probe (P-Probe-1M, Hamamatsu Photonics K.K., Hamamatsu, Japan) consisted of a LED emitter and a detector, separated by a distance of 30 mm. The emitters were always placed more medially relative to the detectors on the subject's forehead. The emitters were placed at the Fp1 position for left hemisphere (LH) measurements and at the Fp2 position for the right hemisphere measurements (international 10–20 system).

3 Results

3.1 Behavioral Performance

The reaction time (RT) during the working memory task is summarized in Fig. 41.1. The significant acute exercise effect was found in the test block, which is a more demanding task than the base block. After the exercise, subjects responded significantly faster compared with control condition ($p < 0.05$) in posttest. On the

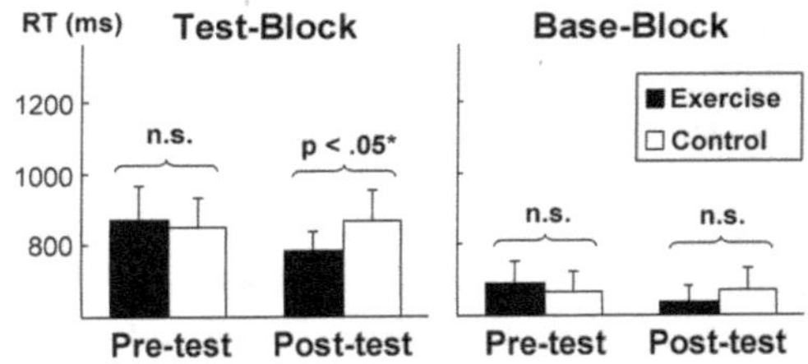

Fig. 41.1 Mean reaction times (RTs) for the test block and base block in the working memory task. The pretest was performed immediately before the exercise (or control) session while the posttest was performed 10 min after the exercise session

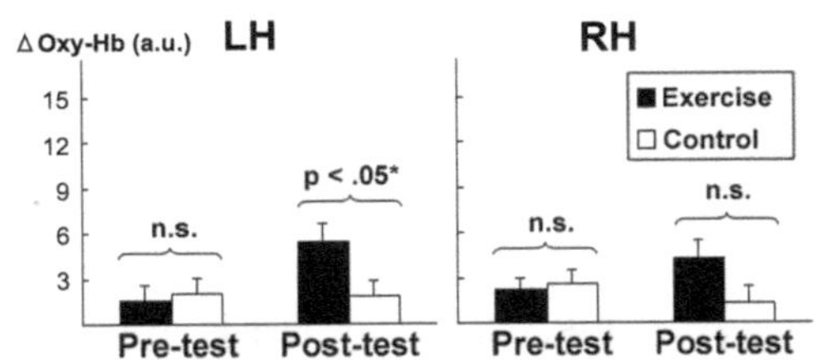

Fig. 41.2 Mean Δoxy-Hb concentration during the working memory task at the left-hemisphere (LH) and right-hemisphere (RH) prefrontal cortex. The pretest was performed immediately before the exercise (or control) session while the posttest was performed 10 min after the exercise session

other hand, there was no significant difference between the exercise and control condition in pretest ($p=0.35$). In the base block, the exercise effect was not significant both in pretest ($p=0.16$) and posttest ($p=0.16$).

The STAI score was also statistically tested. We found that subjects felt more relaxed after exercise (posttest; STAI = 31.07) compared with before exercise (pretest; STAI = 36.57) in the exercise condition ($p<0.05$). On the other hand, there was no significant difference between pretest (STAI = 35.21) and posttest (STAI = 33.14) in the control condition ($p=0.25$). These findings suggest that acute exercise made the older subjects more relaxed and improved their performance in the working memory task.

3.2 NIRS Response

For the evaluation of the prefrontal cortex activation, we analyzed the Δoxy-Hb during the time window from 0 s (trigger-point: start time of the test block) to 50 s. Baseline epochs were set for a pre-trigger of 10 s in the baseline block. A summary of mean Δoxy-Hb values for each condition is shown in Fig. 41.2. The posttest analysis showed that exercise significantly more activated the left prefrontal cortex compared with the control condition ($p<0.05$), whereas there was no significant difference between the exercise and control condition for the right prefrontal cortex activation ($p=0.11$). The pretest analysis showed that there was no significant difference between the exercise and control condition both for the right hemisphere ($p=0.40$) and the left hemisphere prefrontal activation ($p=0.41$).

4 Discussion

The present study examined the acute effect of physical exercise on prefrontal cortex activity in older adults using NIRS. Older subjects visited our laboratory twice: once for the exercise condition and once for the control condition. On each visit, subjects performed working memory tasks before and after exercise (or control) sessions. We measured the Δoxy-Hb change at the prefrontal cortex during the working memory task. Behavioral analysis found that physical exercise improved behavioral performance of the working memory task compared with the control condition. Older subjects can respond more quickly after the exercise. This is consistent with several behavioral studies which reported that physical exercise may ameliorate or protect against age-related cognitive decline [1, 2]. STAI analysis also showed that subjects felt more relaxed after aerobic exercise.

NIRS analysis showed that physical exercise enhanced the prefrontal cortex activation during the working memory task compared with the control condition. These findings suggest that the moderate intensity exercise enhanced the prefrontal cortex activity associated with working memory performance. This is in line with a previous NIRS study which demonstrated the acute exercise effect on the prefrontal cortex activity in young adults [4]. The present findings indicated the hemispheric asymmetry of the exercise effect in the prefrontal cortex activity. Acute physical exercise enhanced activity of the left hemisphere more than the right hemisphere. One possible reason may be that the present working memory task used verbal materials. Numerous studies suggested that the verbal working memory performance was more associated with the left prefrontal cortex activity [10]. In conclusion, the present study could successfully demonstrate, for the first time, the acute exercise effect on the prefrontal cortex activity in older subjects using NIRS.

Acknowledgments This study was supported by funds of the Japan Science and Technology Agency (JST), under Strategic Promotion of Innovative Research and Development Program and by the fund for Grant-in Aid for Scientific Research (B) and Young Scientist (B23300247) from the Ministry of Education, Culture, Sports, Science, and Technology (MEXT) in Japan.

References

1. Kramer AF, Erickson KI, Colcombe SJ (2006) Exercise, cognition, and the aging brain. J Appl Physiol 101:1237–1242
2. Kamijo K, Hayashi Y, Sakai T et al (2009) Acute effects of aerobic exercise on cognitive function in older adults. J Gerontol 64B:356–363

3. Colcombe S, Erickson KI, Scalf PE et al (2006) Aerobic exercise training increases brain volume in aging humans. J Gerontol 61:1166–1170
4. Yanagisawa H, Dan I, Tsuzuki D et al (2010) Acute moderate exercise elicits increased dorsolateral prefrontal activation and improves cognitive performance with Stroop test. Neuroimage 50:1702–1710
5. Tsujii T, Yamamoto E, Ohira T et al (2007) Effects of sedative and non-sedative H1 antagonists on cognitive tasks: behavioral and near infrared spectroscopy (NIRS) examinations. Psychopharmacology 194:83–91
6. Tsujii T, Yamamoto E, Masuda S et al (2009) Longitudinal study of spatial working memory development in young children. Neuroreport 20:759–763
7. Tsujii T, Yamamoto E, Ohira T et al (2010) Antihistamine effects on prefrontal cortex activity during working memory process in preschool children: a near-infrared spectroscopy (NIRS) study. Neurosci Res 67:80–85
8. Murayama M (1922) Japanese standard values of respiratory and circulation in physical exercise (in Japanese). Jpn Circl J 56(suppl V):1514–1523
9. Delpy DT, Cope M, van der Zee P et al (1988) Estimation of optical pathlength through tissue from direct time of flight measurement. Phys Med Biol 33:1433–1442
10. Brahmbhatt SB, McAuley T, Barch DM (2008) Functional developmental similarities and differences in the neural correlates of verbal and nonverbal working memory tasks. Neuropsychologia 46:1020–1031

Chapter 42
Blood Flow and Oxygenation Status of Prostate Cancers

Peter Vaupel and Debra K. Kelleher

Abstract Hypoxia is a characteristic of many solid tumors, can lead to the development of an aggressive phenotype and acquired treatment resistance, and is an independent, adverse prognostic indicator. In this literature review, we show that hypoxia is also a typical feature in prostate cancer (PC), the most commonly diagnosed cancer among men in most western countries. Data on blood flow (a major determinant of oxygenation status in malignancies) and on the oxygenation status (as assessed by O_2-sensitive electrodes) are presented. Where possible, data on prostate cancers are compared to normal prostate (NP) tissue and benign prostate hyperplasia (BPH). The average blood flow rate in NP is 0.21 vs. 0.28 mL/g/min in BPH. Blood flow in PC is approximately three times higher than in NP (mean flow: 0.64 mL/g/min) and shows pronounced intra- and inter-tumor variability. Despite relatively high flow rates in PC, the overall mean pO_2 in cancers is 6 mmHg compared to 26 mmHg in NP. As was the case with blood flow, tissue oxygenation was extremely heterogeneous with no clear dependency on a series of tumor (Gleason score, clinical size, androgen deprivation) and patient characteristics (serum PSA levels, age).

Keywords Prostate • Blood flow • Oxygenation

P. Vaupel (✉)
Department of Radiotherapy and Radiooncology, Klinikum rechts der Isar,
Technische Universität München, Ismaninger Strasse 22, 81675 Munich, Germany
e-mail: vaupel@uni-mainz.de

D.K. Kelleher
Institute of Functional and Clinical Anatomy, University Medical Center,
Johannes-Joachim-Becher-Weg 13, 55128 Mainz, Germany

W.J. Welch et al. (eds.), *Oxygen Transport to Tissue XXXIV*, Advances in Experimental Medicine and Biology 765, DOI 10.1007/978-1-4614-4989-8_42,

1 Introduction

Tumor hypoxia is known to play a crucial role in the development of acquired treatment resistance via direct and/or indirect mechanisms (e.g., [1, 2]). Additionally, intermittent (acute) tumor hypoxia seems to be a major driving force for malignant progression by promoting locoregional invasion of cancer cells and metastatic spread to distant sites (e.g., [3–7]). As a result, hypoxia has been shown to act as an independent, adverse prognostic factor [8, 9].

Due to this seminal role of tumor hypoxia in malignant progression and acquired treatment resistance, knowledge of the oxygenation status of malignant solid tumors in terms of the O_2 partial pressure distributions and detection of hypoxia is indispensable in the clinical setting. Data on the oxygenation status of PCs have been compiled in this review together with blood flow data, which is the major determinant of the delivery of oxygen and anticancer agents [3]. For comparison, flow data for normal prostate (NP) and benign prostate hyperplasia (BPH) are also presented. Information on PCs is of utmost interest since this malignancy is the most commonly diagnosed tumor in males in the United States and Central Europe and the second leading cause of cancer deaths in males. Prostate cancer represents 26% of new cancer cases in Central Europe and 29% in the United States in 2011 [10]. In the USA, the incidence in African American men is approximately 1.6 times higher than in white men [10]. With the aging of the population and improved screening, the incidence is still rising.

2 Blood Flow in Normal Prostate, Benign Prostate Hyperplasia, and Prostate Carcinomas

Current knowledge of blood flow in NP is summarized in Table 42.1. The overall mean flow value is 0.21 mL/g/min (or mL/mL/min, depending on the measurement method used) [11–17]. Data analysis shows that NP blood flow decreases with increasing prostate volume [18] and patient age from 20 to 80 years [11].

Table 42.1 Blood flow of normal prostate (*n* = number of tumors investigated)

Blood flow rate (mL/g/min)	[range]	Method	*n*	References
0.16 ± 0.08	[0.13–0.45]	$H_2{}^{15}O$-PET	9	[11]
	[0.08–0.31]	Thermoclearance	30	[12]
0.12 ± 0.07	[0.04–0.32]	CT imaging	22	[13]
0.14 ± 0.03		CT imaging	9	[14]
0.13		Quantitative MRI	21	[15]
0.32 ± 0.36		DCE-MRI	22	[16]
0.34	[0.01–2.81]	DCE-MRI	32	[17]

Table 42.2 Blood flow of benign prostate hyperplasia (*n* = number of tumors investigated)

Blood flow rate (mL/g/min)	[range]	Method	*n*	References
0.18 ± 0.05	[0.11–0.24]	$H_2{}^{15}O$-PET	6	[11]
0.32 ± 0.08	[0.23–0.38]	$H_2{}^{15}O$-PET	3	[40]
0.30	[0.12–0.60]	Thermoclearance	10	[41]

Table 42.3 Blood flow of prostate cancers (*n* = number of tumors investigated)

Blood flow rate (mL/g/min)	[range]	Method	*n*	References
0.42 ± 0.22		H_2-clearance	19	[18]
0.29 ± 0.08	[0.16–0.40]	$H_2{}^{15}O$-PET	11	[11]
0.55 ± 0.29	[0.28–0.99]	$H_2{}^{15}O$-PET	9	[40]
0.10 ± 0.08	[0.03–0.20]	Thermoclearance	10	[42]
0.37 ± 0.12		Functional CT-imaging	9	[14]
0.34		Quantitative MRI	21	[15]
0.66 ± 0.43	[0.30–0.81]	DCE-MRI	22	[16]
0.97	[0.02–7.61]	DCE-MRI	35	[17]
1.16 ± 1.19		DCE-MRI	42	[43]

Blood flow in BPH is slightly higher than in NP (0.28 vs. 0.21 mL/g/min; Table 42.2). These differences in flow are independent of the detection method used [11].

PC blood flow data are summarized in Table 42.3. The overall mean flow (0.64 mL/g/min) is approximately three times higher than in NP. In PCs, flow is very heterogeneous. The intra- and inter-tumor variability indicates the importance of measuring flow in individual tumors and the necessity to sample as much of the tumor volume as possible. These higher perfusion levels may be due to rigorous angiogenesis activity in aggressive tumor cell clusters, a notion supported by the finding that blood flow in PCs increases with decreasing differentiation (i.e., increasing Gleason scores/histopathological grades [19–21]), enlarging PC volume and increasing volume of the vessel-containing stromal compartment [18]. Upon therapeutic androgen deprivation therapy ("hormonal castration"), PC blood flow decreased within 1 month of starting treatment [22]. Reductions in levels of vascular endothelial growth factor (VEGF) are the most likely explanation for this flow decline, since studies have shown that androgen withdrawal results in rapid and profound decreases in VEGF expression and concomitant reductions in tumor angiogenesis [22].

Due to the marked flow differences described above, blood flow enabled clinical discrimination of PC and NP [17]. Although blood volume, interstitial volume, mean vascular volume, and microvascular permeability of PCs are also higher than in NP, the latter parameters did not reliably correlate with the different histologies [16, 17, 23].

3 Oxygenation Status of Prostate Cancers

In 1999, hypoxic regions were first described in PCs by direct assessment of pO_2 values [24, 25]. Clinically relevant levels of hypoxia are detectable in 30–90% of locally advanced PCs, as repeatedly documented [26, 27]. As in other solid tumors, PC oxygenation is quite heterogeneous, with pronounced intra- and inter-tumor variability (Table 42.4). The overall median pO_2 value is 6 mmHg and is thus distinctly lower than the median pO_2 value of NP (26 mmHg, [24]). The mean fractional hypoxic volume is 18% in PCs compared to 9% in head and neck cancers (assessed by ^{18}FMISO-PET, [26]). This is in accordance with the average median pO_2 values (6 vs. 10 mmHg) and the fraction of pO_2 values ≤5 mmHg (48 vs. 30% as measured with O_2-sensitive electrodes [25]). The obvious discrepancy between relatively high blood flow rates and poor oxygenation status might be explained by a high shunt perfusion in tumor cell clusters with an increased angiogenic activity (provided a high respiratory capacity of prostate cancer cells can be excluded).

Patient and tumor characteristics as well as different pathophysiological conditions have no consistent impact on the oxygenation status: Controversial data on the role of patient age ([24, 28] vs. [29]), clinical stage ([24, 28] vs. [29, 30]) and androgen withdrawal ([30] vs. [31]) have been reported. No association of Gleason scores and serum PSA levels, and oxygenation have been found [29, 30]. Positive correlations have been described for PC oxygenation status and VEGF expression [32]. Contradictory findings on the impact of androgen deprivation have been ascribed to an anti-VEGF effect leading to a "pseudonormalisation" of tumor vasculature (i.e., increased vascular efficiency) and a subsequent reduction in tumor hypoxia [30] on the one hand, and to vascular rarefaction on the other [22, 31]. Nevertheless, hypoxia has been clearly documented as a poor prognostic indicator in PCs [30, 33].

4 Intrinsic Hypoxia Markers in Prostate Cancers

As a consequence of severe hypoxia, prostate cancer cells overexpress HIF-1α [34, 35] and VEGF, most probably in response to transcriptional regulation by HIF [32, 36, 37]. NP generally expresses no or only very low levels of these hypoxic biomarkers [35, 36]. In addition to these markers, GLUT-1 is also overexpressed in

Table 42.4 Pretherapeutic oxygenation status of prostate cancers (n = number of tumors investigated, HF 5 = fraction of pO_2 values ≤5 mmHg)

Median pO_2				
(mmHg)	[range]	HF5 (%)	n	References
10	[2–38]	34	10	[24]
10			55	[28]
20	[0.5–45]		13	[32]
2			57	[33]
4		60	55	[29]
6		40	248	[30]

hypoxic PCs [38]. Direct (by pO_2 measurements) or indirect detection of hypoxia (via increased expression of HIF-1α, VEGF, and GLUT-1) may thus help to identify patients at high risk, and thus poor prognosis [39].

References

1. Vaupel P (2009) Physiological mechanisms of treatment resistance. In: Molls M, Vaupel P, Nieder C et al (eds) The impact of tumor biology on cancer treatment and multidisciplinary strategies. Springer, Berlin, Heidelberg, pp 273–290
2. Vaupel P (2008) Hypoxia and aggressive tumor phenotype: implications for therapy and prognosis. Oncologist 13(suppl 3):21–36
3. Vaupel P (2004) Tumor microenvironmental physiology and its implications for radiation oncology. Semin Radiat Oncol 14:198–206
4. Vaupel P (2004) The role of hypoxia-induced factors in tumor progression. Oncologist 9:10–17
5. Vaupel P, Mayer A (2007) Hypoxia in cancer: significance and impact on clinical outcome. Cancer Metastasis Rev 26:225–239
6. Vaupel P, Mayer A, Hoeckel M (2004) Tumor hypoxia and malignant progression. Methods Enzymol 381:335–354
7. Hoeckel M, Vaupel P (2001) Tumor hypoxia: definitions and current clinical, biologic, and molecular aspects. J Natl Cancer Inst 93:266–276
8. Hoeckel M, Knoop C, Schlenger K et al (1993) Intra-tumoral pO_2 predicts survival in advanced cancer of the uterine cervix. Radiother Oncol 26:45–50
9. Hoeckel M, Schlenger K, Aral B et al (1996) Association between tumor hypoxia and malignant progression in advanced cancer of the uterine cervix. Cancer Res 56:4509–4515
10. Siegel R, Ward E, Brawley O et al (2011) Cancer statistics 2011. CA Cancer J Clin 61:212–236
11. Inaba T (1992) Quantitative measurements of prostatic blood flow and blood volume by positron emission tomography. J Urol 148:1457–1460
12. Bolmsjö M, Sturesson C, Wagrell L et al (1998) Optimizing transurethral microwave thermotherapy: a model for studying power, blood flow, temperature variations and tissue destruction. Br J Urol 81:811–816
13. Harvey CJ, Blomley MJK, Dawson P et al (2001) Functional CT imaging of the acute hyperemic response to radiation therapy of the prostate gland: early experience. J Comput Assist Tomogr 25:43–49
14. Hendersen E, Milosevic MF, Haider MA et al (2003) Functional CT imaging of prostate cancer. Phys Med Biol 48:3085–3100
15. Kershaw LE, Logue JP, Hutchinson CE et al (2008) Late tissue effects following radiotherapy and neoadjuvant hormone therapy of the prostate measured with quantitative magnetic resonance imaging. Radiother Oncol 88:127–134
16. Buckley DL, Roberts C, Parker GJM et al (2004) Prostate cancer: evaluation of vascular characteristics with dynamic contrast-enhanced T1-weighted MR imaging—initial experience. Radiology 233:709–715
17. Franiel T, Lüdemann L, Rudolph B et al (2009) Prostate MR imaging: tissue characterization with pharmacokinetic volume and blood flow parameters and correlation with histologic parameters. Radiology 252:101–108
18. Toma H, Nakamura R, Onitsuka S et al (1988) Effect of endocrine treatment on prostatic blood flow in patients with prostatic adenocarcinoma. J Urol 140:91–95
19. Franiel T, Lüdemann L, Lutz MS et al (2008) Evaluation of normal prostate tissue, chronic prostatitis, and prostate cancer by quantitative perfusion analysis using a dynamic contrast-enhanced inversion-prepared dual-contrast gradient echo sequence. Invest Radiol 43:481–487

20. Ives EP, Burke MA, Edmonds PR et al (2005) Quantitative computed tomography perfusion of prostate cancer: correlation with whole-mount pathology. Clin Prostate Cancer 4:109–112
21. Mitterberger M, Aigner F, Pinggera GM et al (2010) Contrast-enhanced colour Doppler-targeted prostate biopsy: correlation of a subjective blood-flow rating scale with the histopathological outcome of the biopsy. BJU Int 106:1315–1318
22. Alonzi R, Padhani AR, Taylor NJ et al (2011) Antivascular effects of neoadjuvant androgen deprivation for prostate cancer: an in vivo human study using susceptibility and relaxivity dynamic MRI. Int J Radiat Oncol Biol Phys 80:721–727
23. Franiel T, Hamm B, Hricak H (2011) Dynamic contrast-enhanced magnetic resonance imaging and pharmacokinetic models in prostate cancer. Eur Radiol 21:616–626
24. Movsas B, Chapman JD, Horwitz EM et al (1999) Hypoxic regions exist in human prostate carcinoma. Urology 53:11–18
25. Vaupel P, Hoeckel M, Mayer A (2007) Detection and characterization of tumor hypoxia using pO_2 histography. Antioxid Redox Signal 9:1221–1235
26. Rasey JS, Koh WJ, Evans ML et al (1996) Quantifying regional hypoxia in human tumors with positron emission tomography of [^{18}F]fluoromisonidazole: a pretherapy study of 37 patients. Int J Radiat Oncol Biol Phys 36:417–428
27. Chan N, Milosevic M, Bristow RG (2007) Tumor hypoxia, DNA repair and prostate cancer progression: new targets and new therapies. Future Oncol 3:329–341
28. Movsas B, Chapman JD, Greenberg RE et al (2000) Increasing levels of hypoxia in prostate carcinoma correlate significantly with increasing clinical stage and patient age. Cancer 89:2018–2024
29. Parker C, Milosevic M, Toi A et al (2004) Polarographic electrode study of tumor oxygenation in clinically localized prostate cancer. Int J Radiat Oncol Biol Phys 58:750–757
30. Milosevic M, Chung P, Parker C et al (2007) Androgen withdrawal in patients reduces prostate cancer hypoxia: implications for disease progression and radiation response. Cancer Res 67:6022–6025
31. Anastasiadis AG, Stisser BC, Ghafar MA et al (2002) Tumor hypoxia and the progression of prostate cancer. Curr Urol Rep 3:222–228
32. Cvetkovic D, Movsas B, Dicker AP et al (2001) Increased hypoxia correlates with increased expression of the angiogenesis marker vascular endothelial growth factor in human prostate cancer. Urology 57:821–825
33. Movsas B, Chapman JD, Hanlon AL et al (2002) Hypoxic prostate/muscle pO_2 ratio predicts for biochemical failure in patients with prostate cancer: preliminary findings. Urology 60:634–639
34. Zhong H, de Marzo AM, Laughner E et al (1999) Overexpression of hypoxia-inducible factor 1α in common human cancers and their metastases. Cancer Res 59:5830–5835
35. Du ZX, Fujiyama C, Chen YX et al (2003) Expression of hypoxia-inducible factor 1α in human normal, benign, and malignant prostate tissue. Chin Med J 116:1936–1939
36. Green MML, Hiley CT, Shanks JH et al (2007) Expression of vascular endothelial growth factor (VEGF) in locally invasive prostate cancer is prognostic for radiotherapy outcome. Int J Radiat Oncol Biol Phys 67:84–90
37. Ferrer FA, Miller LJ, Andrawis RI et al (1997) Vascular endothelial growth factor (VEGF) expression in human prostate cancer: in situ and in vivo expression of VEGF by human prostate cancer cells. J Urol 157:2329–2333
38. Jans J, van Dijk JH, van Scheiven S et al (2010) Expression and localization of hypoxia proteins in prostate cancer: prognostic implications after radical prostatectomy. Urology 75:786–792
39. Vergis R, Corbishley CM, Norman AR et al (2008) Intrinsic markers of tumour hypoxia and angiogenesis in localized prostate cancer and outcome of radical treatment: a retrospective analysis of two randomized radiotherapy trials and one surgical cohort study. Lancet Oncol 9:342–351
40. Muramoto P et al (2002) $H_2{}^{15}O$ positron emission tomography validation of semiquantitative prostate blood flow determined by double-echo dynamic MRI: a preliminary study. J Comput Assist Tomogr 26:510–514

41. Venn SN, Hughes SW, Montgomery BSI et al (1996) Heating characteristics of a 434 MHz transurethral system for the treatment of BPH and interstitial thermometry. Int J Hyperthmia 12:271–278
42. Franiel T, Lüdemann L, Taupitz M et al (2009) Pharmacokinetic MRI of the prostate: parameters for differentiating low-grade and high-grade prostate cancer. Fortschr Röntgenstr 181:536–542
43. van Vulpen M, Raaymakers BW, de Leeuw AAC et al (2002) Prostate perfusion in patients with locally advanced prostate carcinoma treated with different hyperthermia techniques. J Urol 168:1597–1602

Chapter 43
Targeted Delivery of VEGF to Treat Myocardial Infarction

Bin Wang, Rabe'e Cheheltani, Jenna Rosano, Deborah L. Crabbe, and Mohammad F. Kiani

Abstract Noninvasive injection of pro-angiogenic compounds such as vascular endothelial growth factor (VEGF) has shown promising results in regenerating cardiac microvasculature. However, these results have failed to translate into successful clinical trials in part due to the short half-life of VEGF in circulation. Increasing the dose of VEGF may increase its availability to the target tissue, but harmful side-effects remain a concern. Encapsulating and selectively targeting VEGF to the MI border zone may circumvent these problems. Anti-P-selectin conjugated immunoliposomes containing VEGF were developed to target the infarct border zone in a rat MI model. Targeted VEGF therapy significantly improves vascularization and cardiac function after an infarction.

Keywords Targeted drug delivery • Pro-angiogenic compounds • Vasculature • Cardiac function • Myocardial infarction

1 Introduction

Myocardial infarction (MI) involves loss of blood flow in the infarcted region of the heart. Therapeutic angiogenesis, a method for noninvasively reestablishing reperfusion of infarcted myocardium by systemic injection of vascular growth factors, has shown promising results in animal studies. However, it has failed to demonstrate significant therapeutic potential in clinical trials. A likely reason may be in part due

---

B. Wang
Temple University, 1947 North 12th Street, Philadelphia, PA 19140, USA

Widener University, Chester, PA, USA

R. Cheheltani • J. Rosano • D.L. Crabbe • M.F. Kiani (✉)
Temple University, 1947 North 12th Street, Philadelphia, PA 19140, USA
e-mail: mkiani@temple.edu

W.J. Welch et al. (eds.), *Oxygen Transport to Tissue XXXIV*, Advances in Experimental Medicine and Biology 765, DOI 10.1007/978-1-4614-4989-8_43,

to the short circulation time of vascular endothelial growth factor (VEGF) and inadequate dose to the target tissue.

Recent studies have shown that MI-induced inflammatory response leads to increased expression of several cell-adhesion molecules (e.g., P-selectin and ICAM-1) in the vasculature of the infarcted region [1]. These cell-adhesion molecules can serve as targeting moieties for the recruitment of leukocytes and other circulating cells (e.g., stem cells) to the infarcted region [2]; these cell-adhesion molecules can also be used to selectively deliver drugs to the diseased area [3].

Previously, we have shown that the cell-adhesion molecule, P-selectin, is up-regulated in the infarct border zone, and can be used to selectively deliver drug carrying particles to cardiac tissue post-MI [1]. Here we have developed anti-P-selectin-coated immunoliposomes to selectively deliver VEGF to the infarcted zone in a rat MI model using a time course that may be more clinically relevant. Our hypothesis is that selective delivery of VEGF using these immunoliposomes to the infarcted tissue results in significant improvements in cardiac function and vascular structure after MI.

2 Materials and Methods

2.1 Rodent Model of Myocardial Infarction

An MI model in 6-week-old male Sprague–Dawley rats (150–200 g) (Harlan Laboratories, Indianapolis, Indiana) was used in these studies [4]. In brief, following induction of anesthesia, an anterior transmural MI was induced by ligation of the left anterior descending coronary artery. The MI was detected by ECG [5]. The protocol was approved by the institution's Animal Care and Use Committee.

2.2 Preparation of VEGF Immunoliposomes Conjugated to Anti-P-Selectin

Liposomes (200 nm in diameter) will be generated according to our established protocol [1, 6, 7]. Immunoliposomes will be prepared by conjugated anti-P-selectin antibody to the surface of liposomes [6]. VEGF165A will be encapsulated into the hydrophilic interior of our immunoliposomes [6]. Drug encapsulation efficiency, the release profile, and stability of immunoliposomes will be tested in blood serum or under storage condition [6].

2.3 Targeting Immunoliposome to Infarct Tissue

Previously, we have shown that the maximal up-regulation time for P-selectin is at 1–4 h after MI, and we have shown targeted delivery of VEGF-encapsulated immunoliposomes immediately after MI significantly improved the heart function 4 weeks post-MI [6]. To study the efficacy of our proposed therapy in a more clinically relevant time course, we injected VEGF-encapsulated immunoliposomes (VEGF dose: 0.12 μg/kg, $n=6$) via tail vein at 4 h post-MI. For systemic therapy, $rhVEGF_{165}$ was administrated by IV infusion (0.05 μg/kg/min) for 200 min in rats. This IV infusion of rhVEGF165 was repeated on days 4 and 7 post-MI (total dose of 30 μg/kg animal weight over the 3 infusions). Previously we have studied the effects of systemic delivery of VEGF on cardiac structure function [6]; these experiments were not repeated here due to their high cost. Serial echocardiographic measurements were obtained over a period of 28 days after surgery. Four weeks after MI, 0.1 mL of DiOC7 was injected via tail vein and allowed to circulate in the blood stream for 1 min [8]. Then the hearts were collected and immediately quick frozen using dry ice, and stored at −80°C. Hearts were sectioned (10 μm thickness) at −20°C using a cryostat (Lecia CM3050 S). Sections were mounted on poly-L-lysine-coated glass slides for later staining and imaging.

2.4 *Transthoracic Echocardiographic Studies*

Upon recovery from surgery, an echocardiogram was performed following a standard protocol to assess regional wall motion, and systolic and diastolic function. Regional wall motion assessment was used to give a noninvasive evaluation of left ventricular function and internal chamber dimensions. The left ventricular wall displacement (LVWD), the left ventricular end-diastolic diameter (LVEDD), and fractional shortening (FS) were measured over a 4-week period.

2.5 *Immunohistochemistry*

The changes in vascular perfusion and density were determined using immunohistochemistry. As described previously [4], $DiOC_7$ was used to stain perfused vessels, and CD31 staining was used to identify anatomical vessels. The distance between cells and their nearest perfused vessel can be used as an index of effective oxygen diffusion distance in tissue [4, 9, 10].

2.6 *Statistical Analysis*

All results are expressed as the mean value ± SEM. Differences between groups were analyzed using a two-way analysis of variance (ANOVA). The intraclass correlation was used to assess reproducibility between measurements made by two separate observers. In all cases, statistical significance was defined as $p<0.05$.

3 Results

3.1 Changes in Cardiac Function After MI

Serial changes in cardiac function over 4 weeks following the induction of MI were determined by echocardiography. An M-mode trace of LV heart function is shown in Fig. 43.1. MI heart lost the LV-free wall motion (Fig. 43.1b) compared to normal heart (Fig. 43.1a). However, this loss was moderated by targeted VEGF therapy (Fig. 43.1c) as compared with the untreated MI group (Fig. 43.1b). The LVWD reduction due to myocyte necrosis was moderated by targeted VEGF therapy as compared to untreated rats starting at 1 week after MI and persisting up to 4 weeks after MI (Fig. 43.2). LVEDD dilation due to MI was improved (LVEDD reduction) by targeted VEGF therapy starting at 1 week and persisting up to 4 weeks after MI as compared to the controls (Fig. 43.3). FS was also improved by targeted VEGF therapy starting at 2 weeks and persisting up to 4 weeks after MI compared to the controls (Fig. 43.4). At the end time point of 4 weeks after MI, animals treated with targeted VEGF clearly showed significant improvements in cardiac function compared to controls.

3.2 Changes in Vascular Density and Perfusion After MI

Changes in cardiac function were accompanied by improvements in vasculature. As shown in Fig. 43.5, DiOC7 stains perfused vessels (green) and CD31 stains anatomical vessels (brown). Compared to untreated MI, at 4 weeks post-MI the numbers of both anatomical and perfused vessels in the infarcted areas were significantly increased (20, 75 %, respectively) after targeted VEGF treatments (Fig. 43.6). In addition, the fraction of anatomical vessels that were perfused was also significantly ($p<0.01$) increased after targeted VEGF therapy. Using less than 1 % of typical systemic doses, targeted VEGF therapy reached the same therapeutic outcome as systemic VEGF therapy.

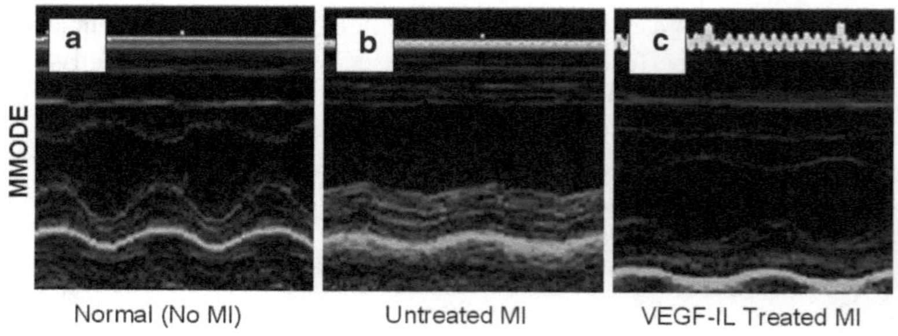

Fig. 43.1 Targeted VEGF therapy results in significant improvement in left ventricular wall motion 4 weeks post-MI (**a–c**)

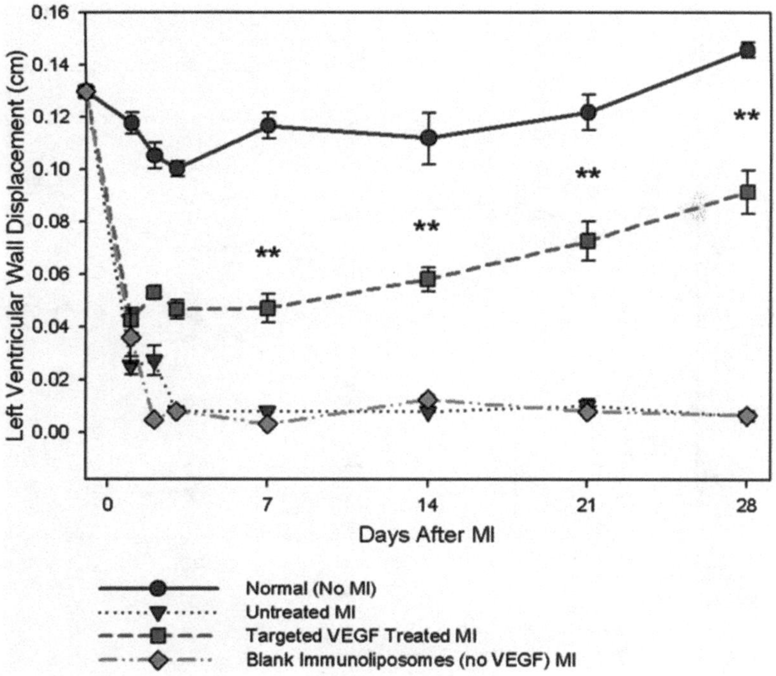

Fig. 43.2 Targeted VEGF therapy results in significant improvements in left ventricular wall displacement over 4 weeks

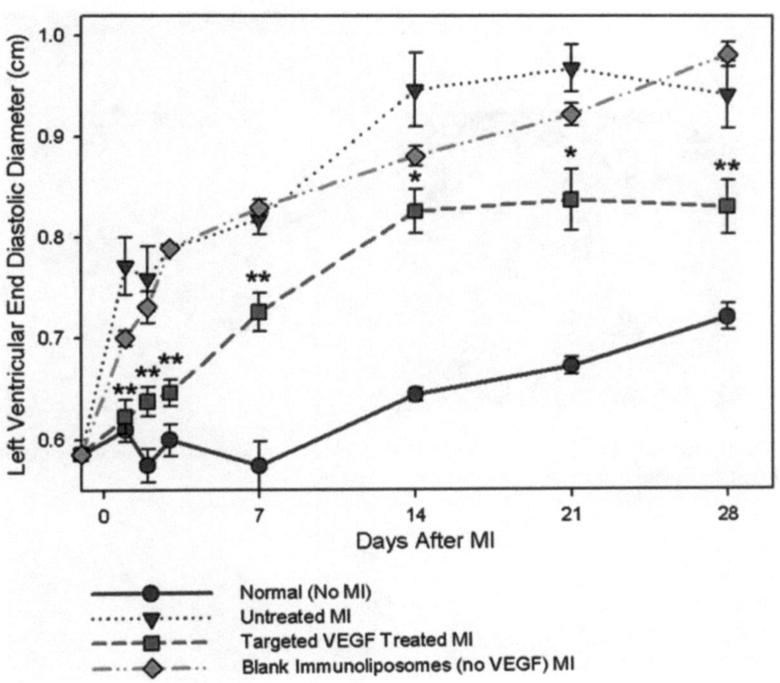

Fig. 43.3 Targeted VEGF therapy results in significant improvements in left ventricular end diastolic dimension over 4 weeks

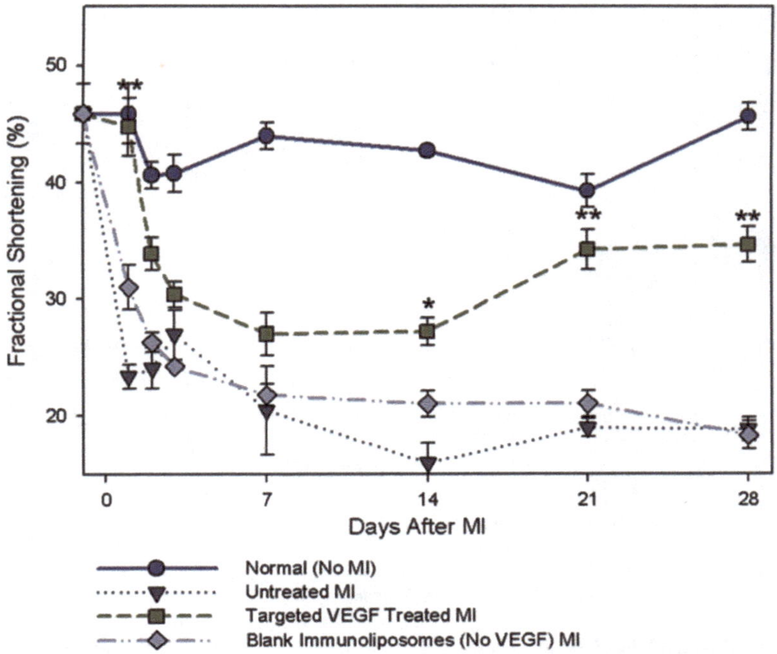

Fig. 43.4 Targeted VEGF therapy results in significant improvements in fractional shortening over 4 weeks

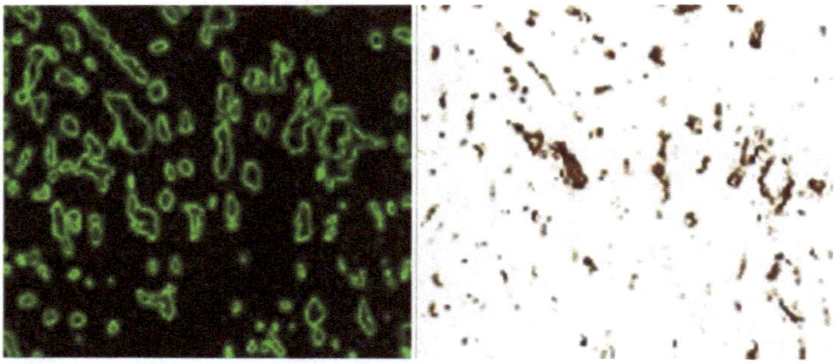

Fig. 43.5 DiOC7 staining (*left panel*) and CD31 (*right panel*) were used to quantify the number of perfused and anatomical vessels, respectively

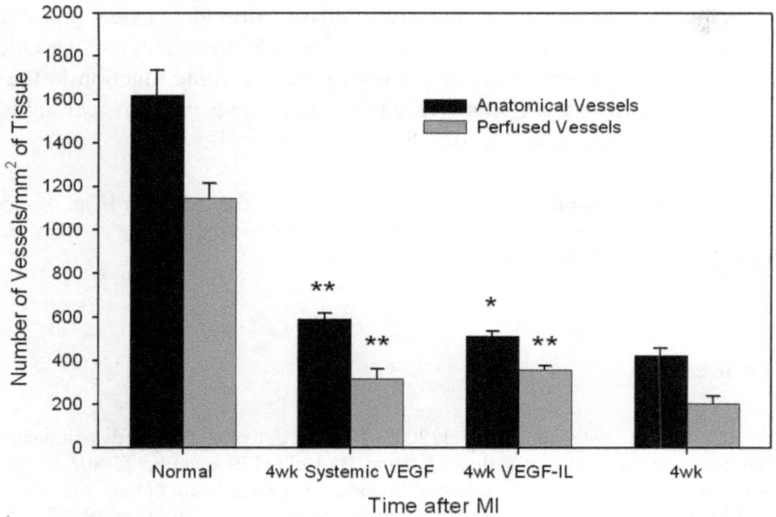

Fig. 43.6 Targeted VEGF therapy results in significant improvements in vessel density 4 weeks post-MI, the same therapeutic outcome as systemic VEGF therapy

4 Discussion

In this study, we have shown that targeted delivery of a single dose of VEGF to post-MI tissue via anti-P-selectin conjugated immunoliposomes results in increasing vascular perfusion, which decreases oxygen diffusion distance, and improves cardiac function 4 weeks after MI.

Previous studies have shown that in order to have significant therapeutic outcomes using systemic infusion of VEGF, a large dose (total dose of 1–30 μg/kg animal weight) has to be used over several days after MI. However, this treatment can only improve vascular density modestly and does not necessarily lead to significant improvements in cardiac function [11, 12]. This may be in part due to its very short half-life (several minutes) in plasma [13]. To increase the half-life of VEGF in the circulatory system, immunoliposome formulations may be used. Immunoliposome can encapsulate VEGF in its inner hydrophilic core, which can protect VEGF from immediate degradation during circulation in the blood. In our studies, using anti-P-selectin conjugated immunoliposomes to targeted deliver a small dose of VEGF (0.12 μg/kg, <1 % of typical systemic doses) to the infarcted tissue not only resulted in similar improvements in vascular density as a large dose of systemic VEGF delivery, but also significantly improved cardiac function up to 4 weeks post-MI, the result that a large dose of systemic VEGF delivery did not reach.

In conclusion, this novel targeted drug delivery provides a new and exciting therapeutic option. This unique system will be able to regenerate vasculature, enhance myocardial regeneration, and further restore cardiac function in the ischemic myocardium, which will improve survival in patients with myocardial infarction and ultimately help in eradicating heart attack mortality.

Acknowledgments This work was supported by grants from The American Heart Association, and The National Heart, Lung and Blood Institute. The human $VEGF_{165A}$ was generously provided by Genentech, Inc., San Francisco, CA.

References

1. Scott RC, Wang B, Nallamothu R et al (2007) Targeted delivery of antibody conjugated liposomal drug carriers to rat myocardial infarction. Biotechnol Bioeng 96:795–802
2. Misao Y, Takemura G, Arai M et al (2006) Importance of recruitment of bone marrow-derived CXCR4+ cells in post-infarct cardiac repair mediated by G-CSF. Cardiovasc Res 71:455–465
3. Scott RC, Crabbe D, Krynska B et al (2008) Aiming for the heart: targeted delivery of drugs to diseased cardiac tissue. Expert Opin Drug Deliv 5:459–470
4. Wang B, Ansari R, Sun Y et al (2005) The scar neovasculature after myocardial infarction in rats. Am J Physiol Heart Circ Physiol 289:H108–H113
5. Sun Y, Weber KT (1996) Angiotensin converting enzyme and myofibroblasts during tissue repair in the rat heart. J Mol Cell Cardiol 28:851–858
6. Scott RC, Rosano JM, Ivanov Z et al (2009) Targeting VEGF-encapsulated immunoliposomes to MI heart improves vascularity and cardiac function. FASEB J 23:3361–3367
7. Pattillo CB, Sari-Sarraf F, Nallamothu R et al (2005) Targeting of the antivascular drug combretastatin to irradiated tumors results in tumor growth delay. Pharm Res 22:1117–1120
8. Burch EE, Shinde PV, Camphausen RT et al (2002) The N-terminal peptide of PSGL-1 can mediate adhesion to trauma-activated endothelium via P-selectin in vivo. Blood 100:531–538
9. Wang B, Scott RC, Pattillo CB et al (2007) Microvascular transport model predicts oxygenation changes in the infarcted heart after treatment. Am J Physiol Heart Circ Physiol 293:H3732–H3739
10. Wang B, Scott RC, Pattillo CB et al (2008) Modeling oxygenation and selective delivery of drug carriers post-myocardial infarction. Adv Exp Med Biol 614:333–343
11. Lopez JJ, Laham RJ, Stamler A et al (1998) VEGF administration in chronic myocardial ischemia in pigs. Cardiovasc Res 40:272–281
12. Freedman SB, Isner JM (2001) Therapeutic angiogenesis for ischemic cardiovascular disease. J Mol Cell Cardiol 33:379–393
13. Rudge JS, Holash J, Hylton D et al (2007) VEGF Trap complex formation measures production rates of VEGF, providing a biomarker for predicting efficacious angiogenic blockade. Proc Natl Acad Sci U S A 104:18363–18370

Chapter 44
Magnetic Nanoparticles and Thermally Responsive Polymer for Targeted Hyperthermia and Sustained Anti-Cancer Drug Delivery

Sarah Y. Wang, Michelle C. Liu, and Kyung A. Kang

Abstract A novel cancer treatment method is being designed using a combination of iron oxide (Fe_3O_4) nanoparticles (IONPs) and Pluronic F-127 (PF127). IONPs have been used for heating tumors via an alternating electromagnetic (AEM) field. PF127 is a polymer possessing thermo-reversible and concentration-dependent gelation properties in aqueous solutions. PF127, as a gel, is an attractive drug delivery vehicle due to its zero-order drug release property. The combination of IONPs and PF127 would allow both short-term, tumor-specific, hyperthermic treatment, and long-term sustained drug delivery. As a preliminary study, the gelling and heating properties of IONPs/PF127 mixtures were investigated: 18% (w/w) PF127 was found to be ideal for our purpose because it gels at 28.0°C, i.e., it would be injectable at room temperature (20–25°C) and forms gel upon injection into the body (37°C). IONPs in PF127 showed little effect on gelation temperatures. The heating performance of IONPs in PF127 slightly, but linearly decreased with PF127. In the IONP concentration range of 0.01–0.05% (w/v) mixed with PF127 at 18% (w/w), the heating performance increased linearly with the increase in IONP concentration.

Keywords Nanoparticles • Anti-cancer drug delivery

S.Y. Wang • M.C. Liu
DuPont Manual High School, Louisville, KY, USA

Chemical Engineering Department, University of Louisville, Louisville, KY 40292, USA

K.A. Kang (✉)
Chemical Engineering Department, University of Louisville, Louisville, KY 40292, USA
e-mail: kyung.kang@louisville.edu

W.J. Welch et al. (eds.), *Oxygen Transport to Tissue XXXIV*, Advances in Experimental Medicine and Biology 765, DOI 10.1007/978-1-4614-4989-8_44,

1 Introduction

1.1 Nanoparticle-Mediated Hyperthermia

Traditional cancer treatment has been mainly by chemo- and/or radiation therapy. These treatments, however, inflict many harmful side-effects due to their systemic (noncancer-specific) nature. Low-heat hyperthermia is a cancer treatment in which the treatment site is exposed to temperatures between 41 and 45°C, with minimal damage to normal tissue. Tumors are more vulnerable to heat than healthy tissue due to their disorganized vascular structure. Recent research suggests that hyperthermia treatments can be localized to tumors using IONPs 0. When they are applied to the tumor and heated by a noninvasively applied alternating electromagnetic (AEM) field, tumor cells can be effectively treated without damaging surrounding normal tissue. Due to the non-toxicity of IONPs and the minimally invasive nature of nanoparticle-mediated hyperthermia, this method has great potential to be an ideal tumor treatment.

1.2 Pluronic F-127

Poly(ethylene oxide)-poly(propylene oxide)-poly(ethlyene oxde)-Poloxamer 407, (Pluronic F-127; PF127) has received considerable attention for being an injectable and controlled drug delivery vehicle in recent years [1, 2]. PF127 is a biocompatible polymer that can deliver both water-soluble and water-insoluble drugs due to its unique structure of hydrophobic core and hydrophilic shell [3]. Most anti-cancer drugs are hydrophobic and have low solubility in water. The drug can be encapsulated in the hydrophobic core. The hydrophilic chains would still allow it to be solubilized in water and also significantly decrease systemic toxicity [4].

PF127 is also thermo-reversible, i.e., it has a sol–gel transition at a particular temperature [5–7]. As the concentration increases, the gelation temperature decreases. Therefore, the gelation of PF127 can be manipulated to a desired temperature by adjusting its concentration [8]. Additionally, as a gel, PF127 displays a zero-order (constant) drug release behavior [9–15].

1.3 Hyperthermia in Conjunction with Sustained Drug Delivery

Our study aims to develop a minimally invasive, dual cancer therapy by combining (1) IONPs with AEM field generator for tumor-specific heating, and (2) PF127 for sustained drug release of anti-cancer agents. Figure 44.1 shows the preliminary design for the proposed system. The concentration of PF127 can be specifically fit so that, at room temperatures (20–25°C), the PF127/IONP mixture containing the cancer drug is a liquid and, at body temperature, it forms a gel [16].

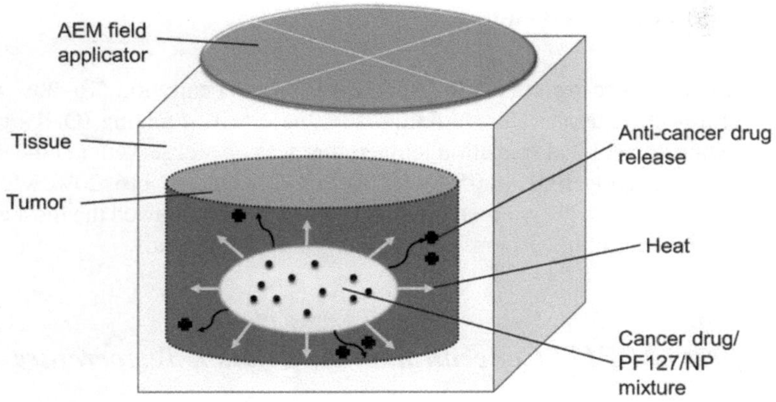

Fig. 44.1 Preliminary design for the dual tumor therapy using PF127/IONP mixture containing anti-cancer drugs for hyperthermic tumor treatment and sustained drug delivery

Once the tumor is detected and located, the liquid mixture can be injected into the tumor via a syringe. Body temperature will then turn the mixture into gel. An AEM field will be noninvasively applied, which will initiate NP-mediated, low-heat hyperthermia. Any remaining tumor cells can be treated with the long-term, sustained drug delivery by PF127.

2 Materials, Methods, and Instruments

PF127 in prill form was kindly donated by BASF Global (New Milford, CT). IONPs (30 nm, concentration 5 mg/mL) were purchased from Ocean NanoTech (Springdale, AR). To heat the NPs, an AEM field generator with a solenoid probe (Taylor-Winfield, Inc.; Brookfield, OH) was used.

Aqueous PF127 solutions at various concentrations were prepared using the cold method [16]. PF127 in prill form was dissolved in cold D.I. water by mixing at 4°C and then placing it in a refrigerator (4°C) overnight. One milliliter samples of the solutions in centrifuge tubes were arranged on a test tube rack in a water bath. The temperature of the water bath was raised by 1°C every 3 min. After each 3-min interval, the samples in the centrifuge tubes were inverted. The temperature at which the mixture displayed no movement was recorded as the gelation temperature.

For heating PF127/IONP by AEM field, a solenoid probe was used. The probe configuration was a 3 turn 3-cm outer diameter made of 5 (W) × 5 (H) mm coil with 3-mm spacing between adjacent turns [17]. Samples were positioned in the center of an insulated solenoid probe (2 × 3 cm). They were exposed to the AEM field for 5 min at 20% power (~57 A) at a frequency of 287 KHz. The sample temperatures were measured using a digital thermometer (Thermo Fisher Scientific Inc.; Waltham, MA) before and after applying AEM field. This process was used for the IONPs in D.I. water with 0.01 ~ 0.05% (w/v) and in PF127 solutions.

3 Results and Discussion

The studies performed were: (1) the effect of PF127 concentration (16–20% w/w) on sol–gel transition temperature. Additionally, the effect of adding IONPs to the PF127 mixtures on sol–gel transition temperature was also observed; (2) the effect of IONP concentration (0.01–0.05% w/v) in PF127 solutions (16–20% w/w) on heating performance. In previous studies [18], 30 nm IONPs showed the most effective heating and, therefore, 30 nm IONPs were used for this study.

3.1 Effect of PF127 Concentration on Gelation Temperature

Figure 44.2 shows the gelation temperatures for varying PF127 in water concentrations. In general, as PF127 concentration increases, gelation temperature decreases. Sixteen percent PF127 showed gelation at 36.7°C, very close to body temperature. This PF127 concentration is not appropriate for our purpose since it may not gel completely after injection into the body. The gelation temperature of the 18% was 28.0°C and is ideal as it can be injected as a liquid at room temperature and its gelation is ensured upon injection into the body. The gelation temperature for 20% PF127 was 24.3°C, very close to room temperature, and would be too viscous to inject into the body.

3.2 Effect of IONP Concentration on Gelation Temperature

Since IONPs will eventually be mixed with PF127 containing anti-cancer drugs for a dual cancer treatment, the effect of adding IONPs to the PF127 mixtures on the sol–gel transition temperature was studied. Figure 44.3 shows gelation temperatures

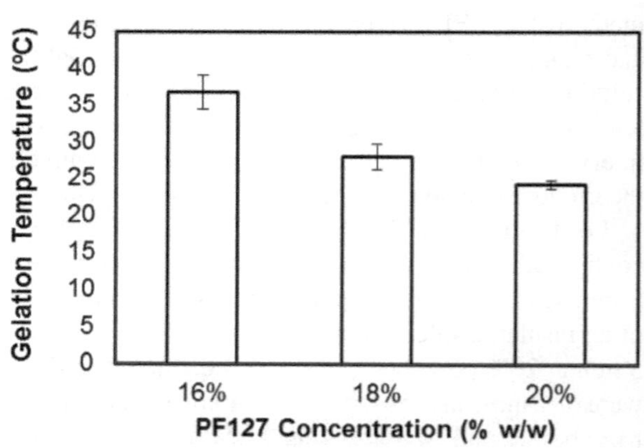

Fig. 44.2 Comparison of gelation temperatures of aqueous solutions of PF127 with varying concentrations of PF127 (16–18% w/w)

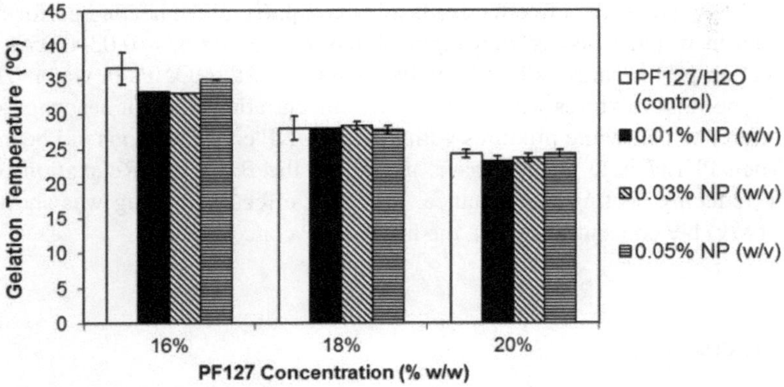

Fig. 44.3 Gelation temperatures of IONP/PF127 mixtures with varying PF127 (16–20% w/w) and NP (0.01–0.05% w/v) concentrations

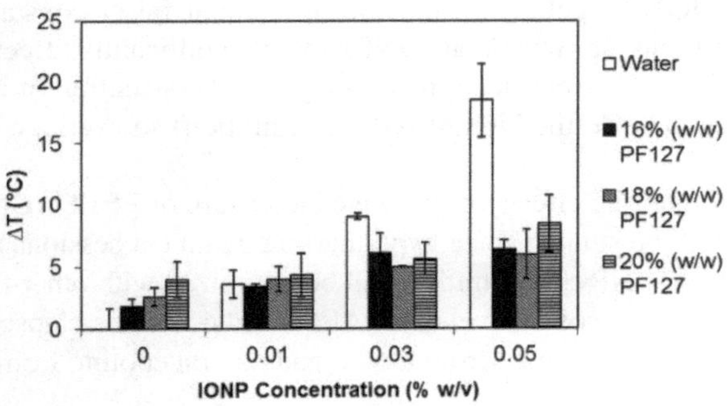

Fig. 44.4 Heating performance of IONP/PF127 mixture with changes in IONP and PF127 concentrations

for PF127/IONP mixtures with varying NP (0.01–0.05% w/v) and PF127 (16–20% w/w) concentrations. For all PF127 concentrations, gelation temperatures remained fairly constant with a change in IONP concentration, indicating that adding IONP to PF127 solutions does not significantly affect gelation temperature.

3.3 *Effect of PF127 and IONP Concentration on Heating Performance*

Figure 44.4 shows the heating performances of IONPs at various concentrations in aqueous solution of PF127 (16, 18, 20% w/w) after 5 min of AEM exposure. Adding PF127 appears to decrease IONP heating performance. Increasing NP concentration increased the temperature. PF127/IONP mixtures with lower IONP concentrations

(0–0.01% w/v) showed more effective heating compared to the heating performance of IONPs in water, whereas with higher IONP concentrations (0.03–0.05% w/v), their heating performance is less effective than that of the IONPs in water. This is probably because mixtures with lower IONP concentrations did not heat enough for the mixtures to gel, while mixtures with higher IONP concentrations did become a gel. When PF127 is in its gel phase, it reduces the Brownian Relaxation of the IONPs, reducing heating performance. The most effective heating was shown by 0.05% (w/v) NP concentration i.e., the highest NP concentration.

4 Conclusions

According to our study results, gelation temperature decreases with an increase in PF127 concentration. 18% PF127 was found to be appropriate for the study because its gelation temperature is 28.0°C, i.e., between room and body temperatures. With the addition of IONPs, gelation temperatures remain fairly constant for various PF127 concentrations, showing that IONPs do not significantly affect gelation temperature. For heating performance, increasing IONP concentration increases heating performance. 0.05% (the highest NP concentration) showed the most effective heating.

In the near future, the effects of prolonged exposure of PF127/IONP mixtures to the AEM field will be studied since hyperthermia treatment sessions may last much longer than 5 min. IONP concentration will be optimized with other factors, such as probe configuration, the power level of the AEM field generator, elapsed time of treatment sessions, distance of probe from tumor, and natural cooling from the body.

Acknowledgments Special thanks go to Ms. Jianting Wang and Dr. Robert Lupitskyy for their help. The authors also thank BASF Global for their kind donation of PF127.

References

1. Miyazaki S, Ohkawa Y, Takada M et al (1992) Anti-tumor effect of Pluronic F-127 containing Mitomycin C on Saroma 180 ascites tumor. Chem Pharm Bull 40:2224–2226
2. Hatefi A, Amsden B (2002) Review: biodegradable injectable in situ forming drug delivery systems. Release 80:9–28
3. Escobar-Chávez JJ, López-Cervantes M, Naïk A et al (2006) Applications of thermo-reversible pluronic F-127 gels in pharmaceutical formulations. J Pharm Sci 9:339–358
4. Rapoport N (2007) Physical stimuli-responsive polymeric micelles for anti-cancer drug delivery. Prog Polym Sci 32:962–990
5. Jeong B, Kim SW, Bae YH (2002) Thermosensitive sol-gel reversible hydrogels. Adv Drug Del Rev 54:37–51
6. Cabana A, Ait-Kadi A, Juhász J (1997) Study of the gelation process of polyethylene oxide-polypropylene oxide-polyethylene oxide copolymer (Poloxamer 407) aqueous solutions. J Colloid Interface Sci 190:307–312

7. Chung HJ, Go DH, Bae JW et al (2005) Synthesis and characterization of Pluronic® grafted chitosan copolymer as a novel injectable biomaterial. Curr Appl Phys 5:485–488
8. Lenaerts V, Triqueneaux C, Quartern M et al (1987) Temperature-dependent rheological behavior of Pluronic F-127 aqueous solutions. Int J Pharm 39:121–127
9. Anderson BC, Pandit NK, Mallapragada SK (2001) Understanding drug release from poly (ethylene oxide)-b-poly(propylene oxide)-b-poly(ethylene oxide) gels. J Control Release 70:157–167
10. Jeong B, Bae YH, Kim SW (2000) Drug release from biodegradable injectable thermosensitive hydrogel of PEF-PLGA-PEG triblock copolymers. J Control Release 63:155–163
11. Zhang L, Parsons DL, Navarre C et al (2002) Development and in-vitro evaluation of sustained release Poloxamer 407 (P407) gel formulations of ceftiofur. J Control Release 85:73–81
12. Moore T, Croy S, Mallapragada S et al (2000) Experimental investigation and mathematical modeling of Pluronic® F127 gel dissolution: drug release in stirred systems. J Control Release 67:191–202
13. Kaowumpai W, Koolpiruck D, Viravaidya K (2007) Development of a 3D mathematical model for a doxorubicin controlled release system using pluronic gel for breast cancer treatment. Proceedings of the World Academy of Science, Engineering and Technology December: 287–292
14. Xu X, Lee P (1993) Programmable drug delivery from an erodible association polymer system. Pharm Res 10:1144–1152
15. Hong YJ, Lee HY, Kim J (2009) Preparations and temperature-dependent release properties of Pluronic F-127 containing microcapsules prepared by a double emulsion technique. J Ind Eng Chem 15:758–762
16. El-Kamel AH (2002) In vitro and in vivo evaluation of Pluronic F127-based ocular delivery system fortimolol maleate. Int J Pharm 241:47–55
17. Sanapala KK, Hewparakrama K, Kang KA (2009) Effect of AEM energy applicator configuration of magnetic nanoparticles mediated hyperthermia for breast cancer. Adv Exp Med Biol Oxygen Trans Tissue 32:143–147
18. Jin H, Kang KA (2007) Application of novel metal nanoparticles as optical/thermal agents in optical mammography and hyperthermia treatment for breast cancer. Adv Exp Med Biol Oxygen Trans Tissue 28:45–52

Chapter 45
NIR Fluorophore-Hollow Gold Nanosphere Complex for Cancer Enzyme-Triggered Detection and Hyperthermia

Jianting Wang, Damon Wheeler, Jin Z. Zhang, Samuel Achilefu, and Kyung A. Kang

Abstract Hollow gold nanospheres (HGN) may be delicately tuned to absorb near infrared light (NIR) by tailoring the diameter-to-shell ratio. This unique property can be utilized for enhancing the contrast for the NIR and X-ray/CT imaging, and also noninvasive and local, photothermal hyperthermia by conjugating cancer-targeting molecules on the particle surface. In addition, when an NIR fluorophore is placed on the surface of the NIR-tuned HGNs, the fluorescence can be significantly quenched due to the emitted light absorption by the HGNs. Combining the NIR fluorescence quenching property of HGNs and the enzyme secreting nature of cancer, we have developed a novel enzyme-triggered NIR contrast agent for cancer detection with high specificity. NIR fluorophore Cypate (Indocyanine Green based) was conjugated to HGN via a short spacer for fluorescence quenching. The spacer contains an enzyme-substrate-motif (G-G-R) that can be cleaved by urokinase-type plasminogen activator (uPA, a breast cancer enzyme). The nano-complex normally does not emit fluorescence but, in the presence of uPA, the fluorescence was restored, providing high specificity. The enzyme-specific emission allows us to characterize the nature of the cancer (e.g., invasive, metastatic, etc.). Once the cancer is detected, the same HGNs can be used to deliver heat to the cancer site for cancer-specific hyperthermia.

Keywords NIR fluorophore • Hyperthermia

J. Wang • K.A. Kang (✉)
Department of Chemical Engineering, University of Louisville, Louisville, KY 40292, USA
e-mail: kyung.kang@louisville.edu

D. Wheeler • J.Z. Zhang
Department of Chemistry and Biochemistry, University of California, Santa Cruz, CA, USA

S. Achilefu
Department of Radiology, Washington University School of Medicine, St. Louis, MO, USA

W.J. Welch et al. (eds.), *Oxygen Transport to Tissue XXXIV*, Advances in Experimental Medicine and Biology 765, DOI 10.1007/978-1-4614-4989-8_45,

1 Introduction

Nanoparticles, including metal nanoparticles, are being extensively studied for biomedical application because they can interact with biomolecules at a molecular level and have a longer circulation time in the body [1, 2]. Gold nanoparticle (GNP) is especially good for biomedical application because it is nontoxic and its surface properties accommodate easy conjugation of biomolecules. GNPs also generate a strong electro-magnetic (plasmon) field upon receiving light, which is particularly beneficial for optical imaging.

Solid, spherical GNPs are known to generate surface plasmon field resonance at around 520 nm [3]. Other gold nanostructures such as shells, thin films, nanorods, and triangles also generate strong surface plasmon field by the visible light, and the absorption peak strongly depends on the dimension of the nanostructure [4]. The Zhang group developed hollow gold nanospheres (HGNs) with various plasmon resonance peaks in the entire range of the visible and NIR wavelengths by varying the diameter-to-wall thickness ratio [5–7].

HGNs tuned for NIR feature several advantages. Their strong absorption in NIR makes the HGNs an effective absorption contrast agent for NIR imaging. For example, the extinction coefficient of hemoglobin, one of the strongest bio-chromophore in NIR, is in the range of 3×10^{2}/cm M [8]. From our measurement and calculation, the extinction coefficient of the HGNs that we currently use is in the range of $5–12\times10^{10}$/cm M, which is approximately eight orders of magnitude higher than that of hemoglobin. Because of this high absorption in NIR they were used for NIR hyperthermia very effectively [9, 10]. GNPs are also strong scatterers for X-ray and therefore HGNs can be an effective X-ray/CT contrast agent [10]. These multiple functionalities allow HGNs to be excellent candidates for both cancer detection (NIR fluorescence/X-ray/NIR imaging) and treatment (photothermal hyperthermia) [9, 11, 12].

According to our theoretical analysis on fluorescence quenching by GNP [13, 14], when the emission wavelength of a fluorophore is close to the plasmon resonance wavelength of the GNP, the fluorescence is strongly quenched. Therefore, HGNs with plasmon resonance peak in the NIR region can be a highly effective quencher for NIR fluorophores.

Using this unique property of the NIR tuned HGNs, we tested the conditional emission of Cypate fluorescence (Fig. 45.1), using a short spacer containing –G-G-R motif, which is to be cleaved by urokinase type tissue plasminogen activator (uPA) [13, 15]. The HGNs studied were: (1) HGN_{654}: The particle has an outer diameter of 41 nm with a hollow center diameter of 28 nm (28/41 nm). Its plasmon resonance peak is at 654 nm; (2) HGN_{786}: The hollow diameter and outer diameter are 43 and 53 nm, respectively, and the plasmon resonance peak is at 786 nm.

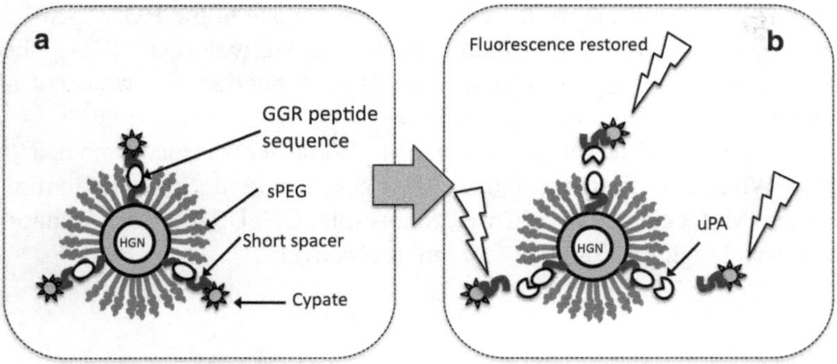

Fig. 45.1 Schematic diagram of the enzyme-triggered fluorescence contrast agent: (**a**) Cypate conjugated HGN, via a short spacer, emits little fluorescence (**b**) In the presence of the cancer secreting enzyme, the spacer is cleaved and the fluorescence is restored

2 Materials and Methods

HGN_{654} (3.32×10^{-6} μM) and HGN_{786} (2.49×10^{-6} μM) stabilized with citric acid in water were produced by the Zhang group at the University of California, Santa Cruz. The coating material of HGNs, (1-Mercapto-11-undecyl) tri(ethylene glycol) [HS-$(CH_2)_{11}$-$(CH_2CH_2O)_3$; MW, 380.58; sPEG] was from ProChimia Surfaces (Ul. Zacisze, Poland). The spacer (SP) was SH-$(CH_2)_2$-Gly-Gly-Arg-Gly-Gly-Gly-NH_2 containing the substrate motif of uPA, G-G-R, and it was synthesized and conjugated to Cypate (SP-Cy) by the solid phase peptide synthesis technique as described by Wang et al. [15] uPA was purchased from Innovative Research (Novi, MI, USA).

Conjugation of sPEG and SP-Cy onto GNPs was done via the thiol-gold reaction. According to Duchesne et al. [16], a tightly packed, mixed monolayer of sPEG and peptide on the GNP surface corresponds to 3.6 molecule/nm^2-GNP surface area. Here, the amount of SP-Cy and sPEG was at 60 molecule/nm^2-HGN surface area (excess by 15 times). SP-Cy and sPEG were mixed at a molar ratio of 1:4 and diluted in ethanol at the same volume of the HGN solution to be added to result in a 1:1 ethanol/water mixture for the reaction. The HGN solution was added to the SP/sPEG mixture and the mixture was stirred for 4 h at room temperature. After the reaction, the solution was placed in a Slide-A-Lyzer Dialysis Cassette (MW Cut-Off of 20 kD, Thermo Scientific; Rockford, IL, USA) and the cassette was placed in 2 L of deionized (DI) water in the dark with stirring overnight. The dialyzed sample was then centrifuged using an Eppendorf 5415 R centrifuge (Eppendorf AG; Hamburg, Germany) at 13,000 RPM for 60 min. The pellet, HGN-SP-Cy, was re-dispersed in 1 mL of DI water and sonicated for 5 min using a Sonic Dismembrator (Fisher Science; Chicago, IL,USA).

For HGN_{654}, the concentration of HGN_{654} and Cypate in the HGN_{654}-SP-Cy was quantified by the absorption at 654 and 780 nm, respectively. For HGN_{786}, because the absorption of HGN_{786} and Cypate significantly overlap, we were not able to quantify them by the spectroscopic method.

The fluorescence of the sample in 0.001 M PBS buffer was measured in a 96-well Uniplate (Whatman; Florham Park, NJ, USA) using a Spectra Gemini XPS fluorometer (Molecular Devices Corp.; Sunnyvale, CA, USA) at the excitation and emission wavelengths of 780 and 830 nm, respectively.

3 Results and Discussion

3.1 *Fluorescence Quenching and Restoration of HGN_{654}-SP-Cy*

The fluorescence of HGN_{654}-SP-Cy was measured and compared to the free SP-Cy at the same SP-Cy concentration (5 μM) (Fig. 45.2). At this Cypate concentration the HGN_{654} concentration was 3.5×10^{-3} nM. For HGN_{654}-SP-Cy, the Cy fluorescence was quenched by $83 \pm 11\%$.

When uPA was added (1,030 unit/mL, at a sufficient amount) to the HGN_{654}-SP-Cy sample, the fluorescence was restored up to $80 \pm 1\%$ of the free Cypate within 10 min. The restoration rate is high compared to our previous study using solid spherical GNPs [13].

3.2 *Fluorescence Quenching and Restoration of HGN_{786}-SP-Cy*

HGN_{786} has strong absorption at NIR, and its absorption peak (786 nm) is very close to the emission wavelength of Cypate (830 nm). HGN_{786} and Cypate in the final product (HGN_{786}-SP-Cy) could not be determined because the HGN absorption

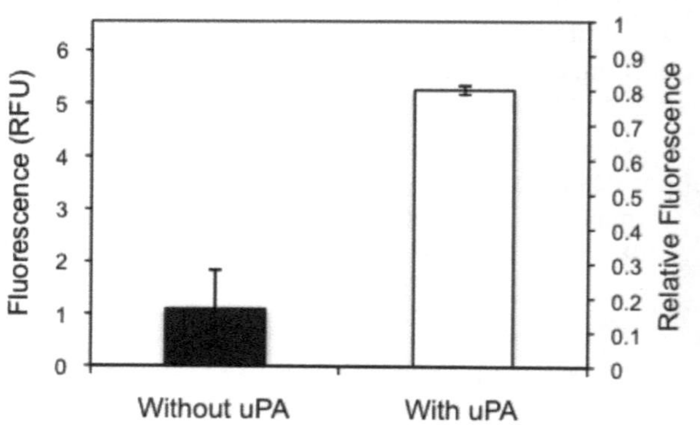

Fig. 45.2 Relative fluorescence levels of HGN_{654}-SP-Cy before and after adding uPA (30 min after uPA addition), compared to the Cypate fluorescence without HGN (experiment conditions: Cypate conc., 5 μM; HGN conc., 3.5×10^{-3} nM; Ex/Em 780/830 nm)

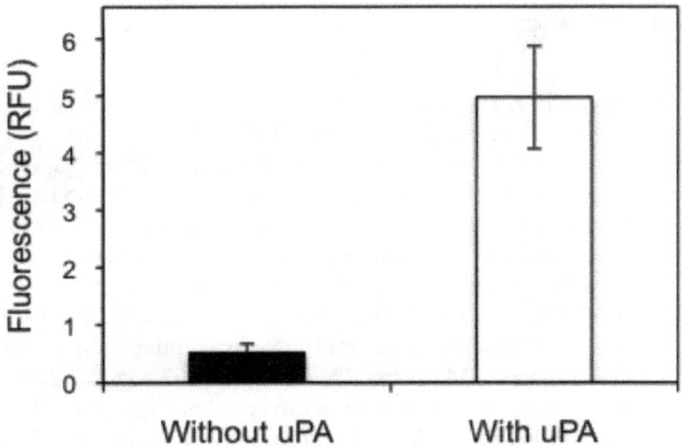

Fig. 45.3 Fluorescence of HGN_{786}-SP-Cy before and after adding uPA (30 min after uPA addition), with Ex/Em 780/830 nm (relative fluorescence data were not available due to the absorption overlap of GNP and Cypate)

peak considerably overlaps with that of Cypate. Therefore, only the absolute fluorescence before and after adding uPA were obtained (Fig. 45.3), and the relative fluorescence values of HGN_{786}-SP-Cy compared to SP-Cy were not available.

The fluorescence of HGN_{786}-SP-Cy without uPA was 0.52±0.15. RFU, at the lower limit of the instrument, suggested that the Cypate fluorescence was almost or completely quenched by HGN_{786}. When uPA was added, the fluorescence increased to 5.0±0.9 RFU. Assuming that both HGN_{654} and HGN_{786} have a similar restoration rate, the initial fluorescence of HGN_{786}-SP-Cy is 53% lower than that of HGN_{654}-SP-Cy, and HGN_{786} appears to quench more effectively than HGN_{654} does, due to its strong absorption in NIR.

4 Conclusions

HGNs, especially HGN_{786}, demonstrated an excellent ability for quenching Cypate fluorescence when Cypate is conjugated on them via a short spacer. The conditional fluorescence emission in the presence of the cancer secreting enzyme was well confirmed. The conditional emission property is especially important because it adds an imaging contrast ability to the nano-entity that already possesses multifunctions of NIR absorption contrast, X-ray scattering contrast, and photo-thermal property. In the near future, this nano-entity will be optimized to be used for minimally invasive cancer detection in two different image modalities (optically and by CT), characterization and hyperthermic treatment.

Acknowledgments The authors thank the US Army (DoD) Breast Cancer Program (BC074387) for their financial support. The authors are also grateful to the US National Science Foundation for the financial support for the HGN development.

References

1. Davis ME, Chen Z, Shin DM (2008) Nanoparticle therapeutics: an emerging treatment modality for cancer. Nat Rev Drug Discov 7:771–782
2. Shi J, Votruba AR, Farokhzad OC, Langer R (2010) Nanotechnology in drug delivery and tissue engineering: from discovery to applications. Nano Lett 10:3223–3230
3. Eustics S, El-Sayed MA (2006) Why gold nanoparticles are more precious than pretty gold: noble metal surface plasmon resonance and its enhancement of the radiative and non-radiative properties of nanocrystals of different shapes. Chem Soc Rev 35:209–217
4. Hu M, Chen J, Li Z-Y, Au L, Hartland GV, Li X et al (2006) Gold nanostructures: engineering their plasmonic properties for biomedical applications. Chem Soc Rev 35:1084–1094
5. Schwartzberg AM, Olson TY, Talley CE, Zhang JZ (2006) Synthesis, characterization, and tunable optical properties of hollow gold nanospheres. J Phys Chem B 110:19935–19944
6. Preciado-Flores S, Wang D, Wheeler DA, Newhouse R, Hensel JK, Schwartzberg A et al (2011) Highly reproducible synthesis of hollow gold nanospheres with near infrared surface plasmon absorption using PVP as stabilizing agent. J Mater Chem 21:2344–2350
7. Wheeler DA, Newhouse RJ, Wang H, Zou S, Zhang JZ (2010) Optical properties and persistent spectral hole burning of near infrared-absorbing hollow gold nanospheres. J Phys Chem C 114:18126–18133
8. Kim J, Xia M, Liu H (2005) Extinction coefficients of hemoglobin for near-infrared spectroscopy of tissue. Eng Med Biol Mag IEEE 24:118–121
9. Lu W, Xiong CY, Zhang GD, Huang Q, Zhang R, Zhang JZ et al (2009) Targeted photothermal ablation of murine melanomas with melanocyte-stimulating hormone analog-conjugated hollow gold nanospheres. Clin Cancer Res 15:876–886
10. Zhang JJ (2010) Biomedical applications of shape-controlled plasmonic nanostructures: a case study of hollow gold nanospheres for photothermal ablation therapy of cancer. J Phys Chem Lett 1:686–695
11. Ji X, Shao R, Elliott AM, Stafford RJ, Esparza-Coss E, Bankson JA et al (2007) Bifunctional gold nanoshells with a superparamagnetic iron oxide-silica core suitable for both MR imaging and photothermal therapy. J Phys Chem C 111:6245–6251
12. Melancon MP, Lu W, Yang Z, Zhang R, Cheng Z, Elliot AM et al (2008) In vitro and in vivo targeting of hollow gold nanoshells directed at epidermal growth factor receptor for photothermal ablation therapy. Mol Cancer Ther 7:1730–1739
13. Wang J, Nantz M, Achilefu S, Kang KA (2011) Gold nanoparticle-fluorophore complex for conditionally fluorescing signal mediator. Anal Chim Acta 696:96–104
14. Kang KA, Wang J, Jasinski B, Achilefu S (2011) Fluorescence manipulation by gold nanoparticles: from complete quenching to extensive enhancement. J Nanobiotechnol 9:16–29
15. Wang J, O'Toole M, Massey A; Biswas S, Nantz M, Achilefu S et al (2011) Highly specific, NIR fluorescent contrast agent with emission controlled by gold nanoparticle. Adv Exp Med Biol 701:149–154
16. Duchesne L, Gentili D, Comes-Franchini M, Fernig DG (2008) Robust ligand shells for biological applications of gold nanoparticles. Langmuir 24:13572–13580

Chapter 46
Renal Oxygenation and Function of the Rat Kidney: Effects of Inspired Oxygen and Preglomerular Oxygen Shunting

Christopher S. Wilcox, Fredrik Palm, and William J. Welch

Abstract We investigated the hypothesis that a preglomerular diffusional shunt for O_2 stabilized renal PO_2 and that changes in intrarenal PO_2 determined nephron nitric oxide (NO) availability for blunting of the tubuloglomerular feedback (TGF) response. The inspired O_2 content of anesthetized rats was changed from normal (21%) to low (10%) or high (100%) for 30–45 min. Direct recordings of PO_2 in the lumens of proximal and distal tubules demonstrated significantly ($P<0.05$) lower values at all sites in spontaneously hypertensive rats compared to normotensive Wistar Kyoto (WKY) rats. Low inspired O_2 did not change intratubular PO_2, but high inspired O_2 increased PO_2 modestly (25–50%; $P<0.01$) in both strains and at both sites. Addition of 7-nitroindazole (7-NI; 10^{-4} M) to artificial tubular fluid perfusing the loop of Henle of WKY nephrons to block neuronal (type 1) nitric oxide synthase in the macula densa increased TGF but this increase was less ($P<0.01$) in nephrons of rats breathing high vs. normal inspired O_2 (1.8 ± 0.4 vs. 3.4 ± 0.3 mmHg; $P<0.01$). In conclusion, the PO_2 in the renal tubules was effectively buffered from even extreme changes in arterial PO_2, consistent with a functionally important preglomerular O_2 diffusional shunt. However, high inspired PO_2 increased intratubular PO_2 sufficiently to blunt the effects of NO derived from the macula densa, likely reflecting bioinactivation of NO by reactive oxygen species generated at increased PO_2 levels. Thus, the preglomerular diffusional shunt appeared to stabilize intrarenal PO_2 during changes in arterial oxygen and to protect NO signaling within the kidney.

C.S. Wilcox (✉) • W.J. Welch
Kidney and Vascular Research Center, Georgetown University Hypertension, F 6003 PHC, Washington, DC 20007, USA
e-mail: wilcoxch@georgetown.edu

F. Palm
Kidney and Vascular Research Center, Georgetown University Hypertension, F 6003 PHC, Washington, DC 20007, USA

Department of Medical Cell Biology, Uppsala University, Uppsala, Sweden

Department of Medical and Health Sciences, Linköping University, Linköping, Sweden

W.J. Welch et al. (eds.), *Oxygen Transport to Tissue XXXIV*, Advances in Experimental Medicine and Biology 765, DOI 10.1007/978-1-4614-4989-8_46,

Keywords Kidney • Oxygenation • Shunting

1 Introduction

We used an ultramicro platinum-iridium microelectrode to measure the PO_2 in the outer cortex of the rat's kidney. The PO_2 values averaged 40–42 mmHg in the proximal and distal tubules, 30 mmHg in the outer medulla and 48 mmHg in the efferent arteriole. Since these values were clearly lower than the mean value of 55 mmHg recorded simultaneously in the renal vein [1], we concluded that a preglomerular diffusional O_2 shunt, likely between the arcuate and/or interlobular arteries and veins, limited the supply of O_2 to the renal tissues. However, its functional significance remained unclear. The prime function of a shunt is to stabilize downstream gas tensions. Therefore, we measured the PO_2 within the rat's kidney during short-term (30–45 min) changes in inspired O_2 from normal (21%) to low (10%) or high (100%) values in normotensive Wistar Kyoto (WKY) and spontaneously hypertensive rats (SHR) since the latter have reduced intrarenal PO_2 related to oxidative stress [1].

To further evaluate the functional significance of a shunt that restricted PO_2 in the renal tissues, we investigated intrarenal nitric oxide (NO) signaling at normal and high inspired O_2. NO is bioinactivated by superoxide anion ($O_2^{\bullet-}$). $O_2^{\bullet-}$ production from nicotinamide adenine dinucleotide phosphate (NADPH) oxidase increased with PO_2 [2]. NO produced by neuronal nitric oxide synthase (nNOS) in the macula densa blunted the tubuloglomerular feedback (TGF) response in normal rats, but this effect was lost in SHR because of inactivation of NO by $O_2^{\bullet-}$ [1, 3]. To test the role of inspired O_2 on intrarenal NO, the loop of Henle (LH) was perfused with artificial tubular fluid (ATF) while measuring the proximal stop flow pressure (PSF) upstream from a wax block in the proximal tubule during the addition of vehicle or 7-nitroindazole (7-NI; 10^{-4} M) to block nNOS in the macula densa. The degree to which the TGF was increased by blockade of nNOS was used as a marker of intrarenal NO signaling from the tubules (macula densa) to the afferent arterioles. We compared the effects of nephron perfusion of 7-NI while rats inspired normal O_2 or high O_2 to test the hypotheses that a function of the O_2 shunt was to reduce the normal levels of intrarenal O_2 sufficiently to preserve NO signaling within the kidney.

2 Methods

The methods to measure PO_2 in proximal and distal tubules with an ultramicro, coaxial, platinum/iridium micropipette [1] and to study the role of NO from macula densa nNOS in the regulation of TGF [1, 3] have been published. In brief, adult male Sprague Dawley rats (Harlan, Madison, WI, USA) were anesthetized with thiobarbital (Inactin, 100 mg/kg intraperitoneally; Research Biochemicals Inc., Natick, MA, USA). A catheter was placed in a jugular vein for fluid infusion and in a femoral artery for recordings of mean blood pressure (MAP). The bladder was catheterized and a tracheotomy tube inserted to facilitate breathing. The left kidney was exposed by a flank incision and immobilized in a Lucite cup. After surgery, the rats were infused with saline solution containing 1% bovine serum albumin at a rate of 1.5 mL/h and the studies begun after 60 min of stabilization. Micropuncture measurements to determine intrarenal PO_2 and proximal tubular stop flow pressure (PSF) were conducted while the rats inhaled three different gas mixtures containing 10, 21, and 100% oxygen, respectively. We perfused the loop of Henle (LH) with ATF while measuring the PSF during the addition of vehicle or 7-NI (10^{-4} M) to block nNOS.

3 Results

Values for normal PO_2 in proximal and distal tubules [1] and TGF responses to 7-NI in SHR and WKY rats [1, 3] were similar to our previous publications. PO_2 values were lower in SHR than WKY, confirming a previous report [1] (Fig. 46.1). At 30–45 min after reducing the inspired O_2 from 21% (air) to 10%, there were no significant changes in the PO_2 recorded in the proximal or distal tubules of either WKY or SHR rats. After switching from normal to 100% inspired O_2, there were significant increases at both sites and in both strains. However, despite a fivefold increase in inspired O_2, the intratubular PO_2 values increased on average by only 25–50%.

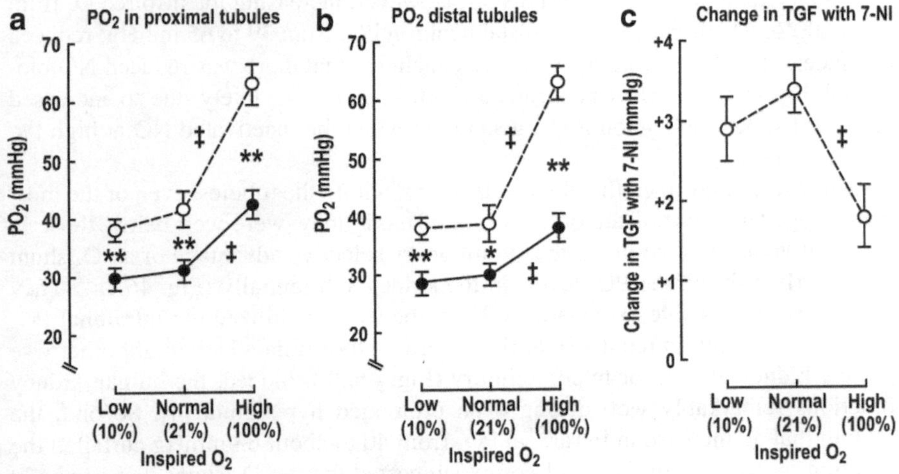

Fig. 46.1 Mean±SEM values (n=4–6) for partial pressure of oxygen (PO_2) in proximal (**a**) and distal tubules (**b**) and absolute change in tubuloglomerular feedback (TGF) response (**c**) before and after intratubular administration of 7-nitroindazole (10^{-4} M) at different inspired oxygen contents. Data are shown for control Wistar Kyoto (WKY) rats (*open circles* and *broken lines*) and hypertensive SHR (*closed circles* and *continuous lines*). Significance of difference between SHR and WKY: *, $P<0.05$; **, $P<0.01$. Significance of difference between values at normal or high inspired O_2: ‡, $P<0.05$

The TGF was quantitated from the reduction in PSF (an index of glomerular capillary pressure) recorded by a pressure pipette upstream from a wax block in the proximal tubule of the test nephron during perfusion of its macula densa segment from the late proximal tubule with ATF. The addition of 7-NI to ATF increased the TGF by 3.4±0.3 mmHg in rats breathing room air similar to prior reports [1, 3]. After changing from normal to low inspired O_2, there were no significant changes in the response to 7-NI. However, 30–45 min after changing from normal to 100% inspired O_2, the increase of TGF with nephron perfusion of 7-NI was reduced significantly by 50% to 1.8±0.4 mmHg ($P<0.01$). This indicated that an increase in intratubular PO_2 above normal values during breathing of 100% O_2 reduced NO signaling within the renal cortex.

4 Discussion

The main new findings of this study were that the directly measured intranephron PO_2 values in the outer cortex of the rat's kidney were sufficiently buffered from contemporary changes in inspired O_2 that there were no significant reductions in PO_2 in the proximal or distal tubules of normal or hypertensive rats when the inspired O_2 was reduced from 21 to 10%. Moreover, during a fivefold increase in inspired O_2 from 20 to 100%, the increases in intranephron PO_2 were limited to 25–50% at both sites. Reducing the inspired O_2 from 21 to 10% did not affect nephron/vascular signaling by NO, as indexed from the increase in TGF responses during blockade of macula densa nNOS. However, increasing the inspired O_2 from 21 to 100%, which increased the distal nephron PO_2 from 39 to 65 mmHg, reduced this index of NO signaling by 50%. We concluded that there was reduced NO bioavailability in the cortical nephrons at high inspired O_2, likely due to increased intrarenal superoxide anion ($O_2^{\bullet-}$) generation which bioinactivated NO at high tissue PO_2 levels.

Our results confirmed that the levels of PO_2 within the tubules, even of the most well-oxygenated part of the outer cortex of the kidney, were well below those in the renal vein. What could be the evolutionary selective advantage of an O_2 shunt that clearly reduced the PO_2 levels in the kidney substantially (Fig. 46.1a)? They point to some possible explanations. First, the shunt stabilized the intrarenal PO_2 and limited the fall in renal PO_2 during arterial hypoxemia which might otherwise put the kidney at risk for hypoxic injury (Fig. 46.2b). Indeed, the human kidney functions remarkably well during even prolonged hypoxemia [4]. Second, the finding that an increase in intrarenal PO_2 from 40 to about 65 mmHg curtailed the function of NO within the renal cortex suggested that an O_2 shunt that normally restricted the PO_2 levels in the kidney may be necessary for fully effective NO signaling. We found that O_2 availability limited NADPH oxidase activity in renal tissues [2]. Thus, a relatively low level of tissue PO_2 may be needed to restrict renal $O_2^{\bullet-}$ generation and to preserve NO signaling. Nevertheless, the kidney is at

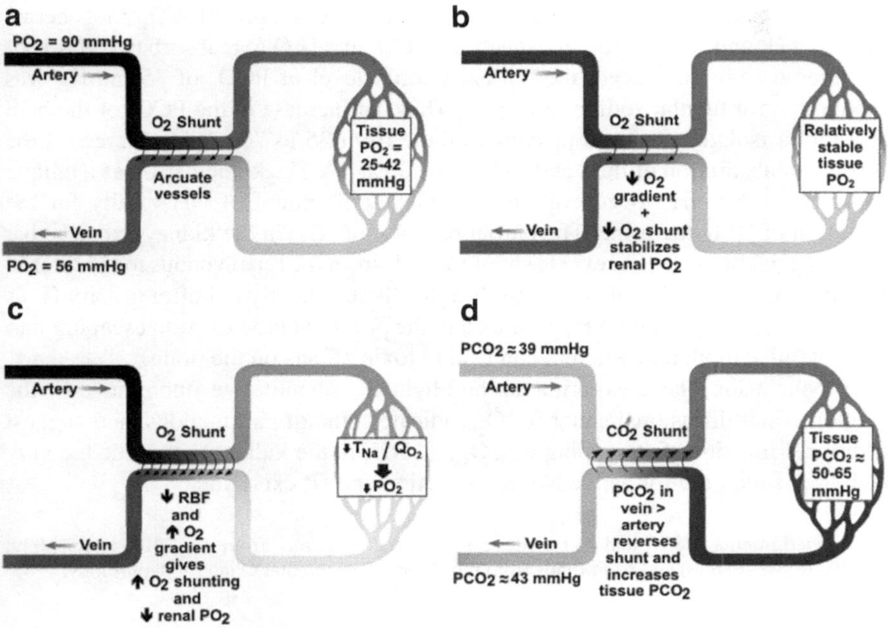

Fig. 46.2 Diagrammatic representation of a preglomerular diffusional gas shunt for O_2 in (**a–c**) or for CO_2 in (**d**). (**a**) Depicts effects in the normal kidney. (**b**) Depicts decreased O_2 shunting at low arterial PO_2 values. (**c**) Depicts increased O_2 shunting during shock. (**d**) Depicts shunting of CO_2 from the preglomerular venous to arterial systems in the normal kidney. For explanation, see Sect. 4

substantial risk of hypoxic/ischemic damage during hypotension and shock. Redfors et al. demonstrated a sharp reduction in oxygenation of the kidney of patients with acute kidney injury after cardiac surgery [5]. Indeed, these findings may explain the extreme vulnerability of the kidney to shock in contrast to its relative immunity from the effects of hypoxemia. A reduction in renal tissue PO_2 due to ROS accumulation during shock should reduce the PO_2 in the venous shunt pathway and enhance the gradient for O_2 diffusion from the artery to the vein. Moreover, a reduction in renal blood flow (RBF) during shock should prolong the time that blood spends in the shunt pathway. These two together would enhance O_2 shunting and further curtail renal tissue PO_2 values (Fig. 46.2c).

A potential advantage of the shunt relates to reversed diffusion of gases that are produced within the kidney and thereby have higher concentrations in venous than arterial blood (Fig. 46.2d). Notably, the PCO_2 levels in the proximal or distal tubule, interstitium, efferent arteriole, or Bowman space are 15–25 mmHg above those in arterial or renal venous blood [6]. This may, in part, represent reverse shunting of CO_2 produced in the kidney from the renal venous to the arterial systems. Renal CO_2 is produced by the reaction of secreted H^+ with filtered HCO_3^- or as a byproduct of cellular respiration. The kidney requires a huge supply of CO_2 to generate sufficient H^+ and HCO_3^- from the reaction of CO_2 and H_2O to reabsorb two-thirds of the filtered sodium. Indeed, the normal arterial level of PCO_2 of 35 mmHg was suboptimal for tubular sodium transport. Thus, an increase in the PCO_2 of the bath of a rabbit isolated perfused proximal tubule from 35 to 70 mmHg increased the absolute reabsorption of fluid and sodium by 30% [7]. The kidney also has a unique requirement for NH_3 to generate approximately 50 mmol of NH_4^+ daily for the excretion of H^+ in the urine. The partial pressure of NH_3 in the kidney exceeds that in the arterial blood [8]. A reverse shunt for NH_3 from the renal venous to the arterial systems could not only enhance renal NH_3 levels sufficiently to buffer urinary H^+ as NH_4^+ but also would limit NH_3 produced in the proximal tubules from escaping into the systemic circulation where it can lead to toxic effects on the brain.

In conclusion, these experiments highlight the quantitative importance of the preglomerular diffusional shunt for O_2, indicate some of its functions, and suggest additional functions for trapping of CO_2 and NH_3 in the kidney to provide the very high fluxes required for active Na^+ reabsorption and H^+ excretion.

Acknowledgments CSW and WJW were supported by grants from the NIH (DK-36079; DK-49870; HL-68686) and from funds from the George E. Schreiner Chair of Nephrology.

References

1. Welch WJ, Baumgärtl H, Lübbers D et al (2001) Nephron pO_2 and renal oxygen usage in the hypertensive rat kidney. Kidney Int 59:230–237
2. Chen Y, Gill PS, Welch WJ (2005) Oxygen availability limits renal NADPH-dependent superoxide production. Am J Physiol Renal Physiol 289:F749–F753
3. Welch WJ, Tojo A, Wilcox CS (2000) Roles of NO and oxygen radicals in tubuloglomerular feedback in SHR. Am J Physiol Renal Physiol 278:F769–F776
4. Wilcox CS, Payne J, Harrison BDW (1982) Renal function in patients with chronic hypoxaemia and cor pulmonale following reversal of polycythaemia. Nephron 30:173–177
5. Redfors B, Bragadottir G, Sellgren J et al (2010) Acute renal failure is NOT an "acute renal success"—a clinical study on the renal oxygen supply/demand relationship in acute kidney injury. Crit Care Med 38:1695–1701
6. DuBose TD Jr, Pucacco LR, Seldin DW et al (1978) Direct determination of PCO_2 in the rat renal cortex. J Clin Invest 62:338–348
7. Jacobson HR (1981) Effects of CO_2 and acetazolamide on bicarbonate and fluid transport in rabbit proximal tubules. Am J Physiol 240:F54–F62
8. Wilcox CS, Granges F, Kirk G et al (1984) Effects of saline infusion on titratable acid generation and ammonia secretion. Am J Physiol 247:F506–F519

Chapter 47
Alteration of the Inflammatory Molecule Network After Irradiation of Soft Tissue

Zhenyu Xiao, Shanmin Yang, Ying Su, Wei Wang, Hengshan Zhang, Mei Zhang, Kunzhong Zhang, Yeping Tian, Yongbing Cao, Liangjie Yin, Lurong Zhang, and Paul Okunieff

Abstract Inflammatory molecules (IMs) play an important role in ionizing radiation (IR)-induced soft tissue damage. The alteration of IMs as a function of time was studied with a protein array containing 62 IMs in mouse cutaneous soft tissues exposed to 30 Gy. The results showed that: (1) 2 days after irradiation, the levels of TGF-β1, MIP-1γ, IL-1α, and sTNF RI increased, while IGFBP-3, CXCL16, and IL-1β decreased in IR skin as compared to control skin; (2) 21 days after IR, TGF-β1, and MIP-1 γ, IL-1α remained high, while CXCL16 and IL-1β remained low; (3) 3 months after IR, the cytokine pattern exhibited reversals. The levels of MIP-1γ decreased, while VCAM-1, IGFBP-3, and TGF-β1 production increased. The data indicated that: (a) IMs change as a function of time after soft tissue irradiation; (b) changing IM levels may reflect the altered balance of the cytokine network, leading to imbalance or homeostasis; and (c) an antibody-based protein array can be used to assess multiple IMs simultaneously, making it useful for bulk screening for changes in tissue cytokine levels.

Keywords Inflammatory molecules • Protein array • Soft tissue • Cytokines • Radiation

1 Introduction

Ionizing radiation (IR)-induced soft tissue toxicity results from the occurrence and progression of an inflammatory molecule (IM) cascade consisting of three surges: the immediate phase (hours to days), the early phase (days to weeks), and the late

---

Z. Xiao • S. Yang • Y. Su • W. Wang • H. Zhang • M. Zhang • K. Zhang • Y. Tian • Y. Cao
L. Yin • L. Zhang • P. Okunieff (✉)
Department of Radiation Oncology, UF Shands Cancer Center, University of Florida, P.O. Box 103633, Gainesville, FL 32610, USA
e-mail: pokunieff@ufl.edu

W.J. Welch et al. (eds.), *Oxygen Transport to Tissue XXXIV*, Advances in Experimental Medicine and Biology 765, DOI 10.1007/978-1-4614-4989-8_47,

phase (months and years) [1–3]. The exact role that each phase of IM expression might play in early inflammation and late fibrosis is not well understood.

IM protein array may be a useful tool to advance understanding of cytokine balance and to screen for patterns of advantageous or deleterious families of cytokine expression. In this study, 62 IMs were measured simultaneously in cutaneous tissues for a better understanding of time-dependent IM alteration patterns.

2 Methods

Female, 6–8-week-old, C57BL/6 mice (National Cancer Institute Mouse Repository, Frederick, MD) were divided into a normal control group ($n = 15$) and an IR group ($n = 15$). In order to model IR-induced soft tissue toxicity, mice received a single 30 Gy dose to the right hind leg via a 137-cesium γ-ray source. At 2 days, 21 days, and 3 months after IR, 5 mice from each group were euthanized and cutaneous tissue samples were collected from the hind legs. Samples were then homogenized in a weight-adjusted volume of lysis buffer. Protein quantification was performed for all samples prior to the array.

The RayBio Mouse Cytokine Antibody Array III & 3.1 (RayBiotech, Inc., Norcross, GA) was used to test 62 mouse cytokines, chemokines, and growth factors. The array membrane contains dots of antigen-specific immobilized antibodies arranged in 14 columns and 10 rows (Fig. 47.1). Six dots are coated with biotin-conjugated IgG (non-cross-reacting positive control antibodies), four are coated with medium (negative controls), and six are uncoated (blank). The protein array was performed according to the manufacturer's instructions. In brief, the membrane was blocked, 1 mL of each sample was incubated with the membrane at room temperature (RT) for 2 h. After washing the unbound proteins, 1 mL of diluted biotin-conjugated antibodies was added at RT for 2 h incubation. Two (2) mL of

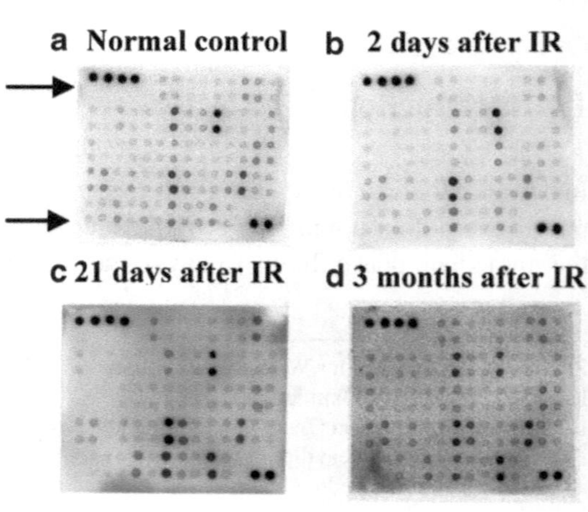

Fig. 47.1 Protein array images of 62 cytokines in pool samples at different time points. (**a**) Normal control; (**b**) 2 days after IR; (**c**) 21 days after IR; and (**d**) 3 months after IR

Table 47.1 The alteration of cytokines at 2 days

Cytokines	Ratio (IR/normal)
MIP-1r	1.429
sT NF RI	1.241
IL-1 alpha	1.088
VCAM-1	1.084
TIMP-1	0.961
MCP1	0.900
IGFBP-6	0.852
CXCL16	0.847
M-CSF	0.803
IL-12 p70	0.767
IGFBP-3	0.480
CTACK	0.367

Table 47.2 The alteration of cytokines at 21 days

Cytokines	Ratio (IR/normal)
IL-1 alpha	1.857
sT NF RI	1.525
MIP-1r	1.194
IGFBP-3	1.056
VCAM-1	0.995
TIMP-1	0.789
CXCL16	0.729
MIP-2	0.282
Cytokines	Ratio (IR/normal)
IL-1 alpha	1.857
sT NF RI	1.525
MIP-1r	1.194

1,000-fold-diluted horseradish peroxidase (HRP)-conjugated streptavidin was then added and incubated at RT for another 2 h. The mixed detection buffer was then added, creating HRP-dependent quantitative florescence; after 2 min at RT, the signal was measured (FluorChem HD Imaging System). In order to normalize the results, each cytokine level was then expressed as a percentage; the average density of the duplicate dots was divided by the average optical density of the six positive controls on each array membrane. This process was completed after the background was subtracted through the formula:

$$\text{Relative cytokine expression level} = \frac{(\text{Density of cytokine} - \text{Density of background})}{(\text{Density of positive control} - \text{Density of background})}$$

Because the protein array detects only relative expression levels and not absolute values, there is no defined lower detection limit. However, for the sake of specificity, we limited the sensitivity of the test by ignoring expression levels <10% of the positive controls. Tables 47.1, 47.2, and 47.3 list the high-expression proteins under study.

Table 47.3 The alteration of cytokines at 3 months

Cytokines	Ratio (IR/normal)
VCAM-1	1.889
RANTES	1.566
TIMP-1	1.311
IGFBP-3	1.274
Lymphotactin	1.152
TCA-3	1.121
CXCL16	1.014
s TNFRI	0.953
LIX	0.93
MIP-1r	0.77
IL-1 alpha	0.505
L-selection	0.344
Eotaxin	0.335
IL-3 Rb	0.245

3 Results

The irradiated cutaneous soft tissue from the hind leg was harvested, weighed to 500 mg, and homogenized in 2 mL of lysis buffer containing protease inhibitor cocktail. After spinning down the tissue debris, we measured the soluble protein supernatant for protein concentration; 300 μg/sample in 600 μL buffer was used to determine the level of 62 IMs with 62 antibody-coated arrays. This high-throughput, semiquantitative assay utilizes a small sample that would be inadequate for a conventional ELISA that requires 100 μL per IM. A densitometer obtained the relative quantitative value for 62 IM levels at 2 days (Fig. 47.1 and Tables 47.1, 47.2, and 47.3). The levels of cytokines were easily followed using the array technology and demonstrate results consistent with the time-dependent cutaneous changes (erythema, indurations, and fibrosis) after irradiation.

Typically, changes in cytokine levels in excess of twofold are defined to be significant. However, with this approach, false positives occur for low expressing proteins due to background signal noise; thus, cytokines expressed near the detection limit were ignored. As a result, some of the cytokines on the membrane are of limited value. The following IMs were considered relevant for the study of IR-induced IM expression: interleukin-1 alpha (IL-1α), soluble tumor necrosis factor receptor 1 (sTNF-RI), insulin-like growth factor-binding protein 3 (IGFBP-3), chemokine ligand 16 (CXCL16), vascular cell adhesion protein 1 (VCAM-1), and metallopeptidase inhibitor 1 (TIMP-1) (Tables 47.1, 47.2, and 47.3). We found at least 4 IM patterns after IR: (1) cytokines that were initially elevated 2 days after IR and then declined at 21 days and 3 months (e.g., macrophage inflammatory protein-1γ, MIP-1γ); (2) IGFBP-3 was low at 2 days and then incrementally increased; (3) sTNF-RI peaked at 21 days after IR; and (4) other IMs continuously increased over the 3 months (e.g., CXCL16, VCAM-1, and TIMP-1) (Fig. 47.2).

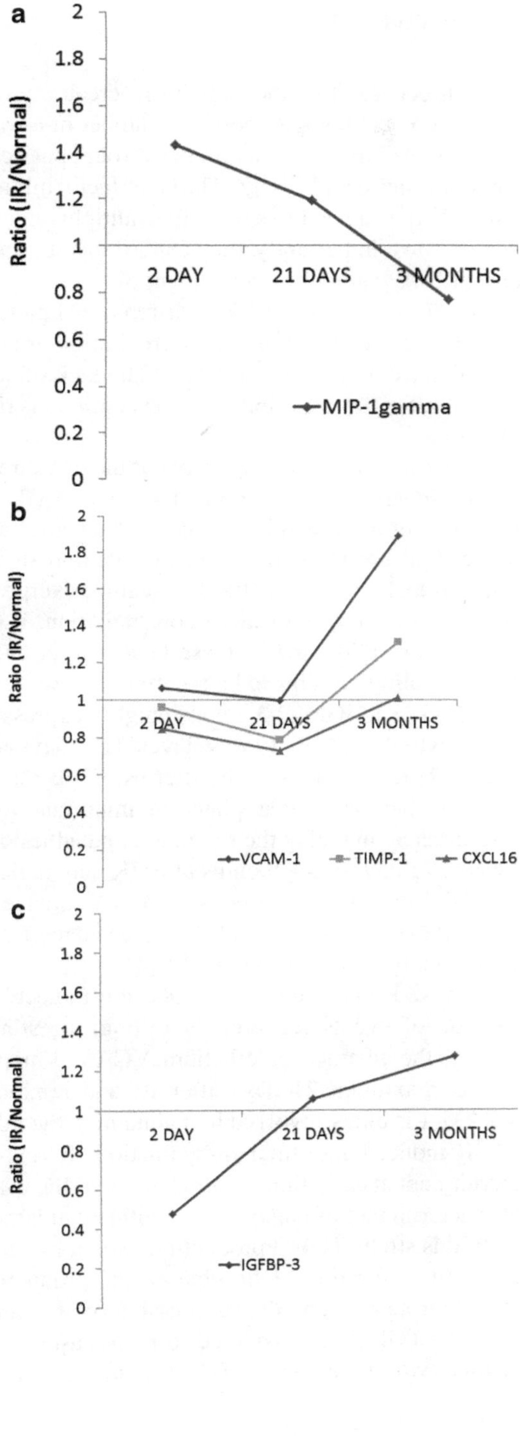

Fig. 47.2 Different alteration patterns of cytokines in mouse soft tissue after irradiation were analyzed using protein array (**a–c**)

4 Discussion

As a direct result of their ability to create multiplex reactions, emerging protein microarray technologies confer a number of advantages over conventional ELISA-based procedures, such as higher throughput screening and reduced reagent consumption and sample usage. The latter factor minimizes the risk of sample exhaustion when there is a need to screen for multiple target analytes. Arrays also have weaknesses. Most importantly, they feature specific antibody epitopes that, like commercial ELISA, can require refinement.

We discovered at least four different IM patterns. Certain IMs increased. These are either directly responding to irradiation or are attempting to balance other IMs that directly respond to radiation. The lack of subsequent regulation of these IMs months after IR, as exhibited by the continuous rise in TGF-β, will likely contribute to fibrosis.

IL-1α was the most responsive of the measured cytokines following IR. Altering IL-1α signaling by interleukin-1 receptor 1 (IL-1R1) knockout reduces cutaneous radiation fibrosis. Similarly, agents that decrease IL-1 production following IR reduce both early inflammation and late fibrosis of soft tissues in mice. The reverse is also true. For example, transgenic mice overproducing IL-1α develop debilitating spontaneous fibrosis even without radiation. IL-1Ra is a competitive inhibitor of IL-1. Arrays allow for all these factors to be measured and permit the balance of IL-1 signaling proteins to be ascertained simultaneously.

MIP-1γ and IGFBP-3 are also highly expressed, secondly only to IL-1α among the 62 cytokines. MIP-1γ was elevated at 2 days and then showed continuous decline both at 21 days and 3 months after IR. IGFBP-3 exhibited the reverse pattern. MIP-1γ is a chemokine that plays an important role in immune and inflammatory responses by inducing the migration and adhesion of leukocytes. Lower concentrations at 21 days and 3 months after IR may indicate an abnormal cytokine balance. IGFBP-3 increases in response to DNA damage and hypoxia. The steady increase in IGFBP-3 is consistent with the expected reduced reparative angiogenesis expected to worsen over time following IR [4].

VACM-1 and intercellular adhesion molecule-1 (ICAM-1) are involved in the cascade of events resulting in rolling, arresting, and transmigrating leukocytes through the inflamed endothelium. VCAM-1 maintained a high level after IR, reaching its maximum 21 days after IR and remaining above normal for 3 months. VCAM-1 is likely involved in maintaining the inflammation process. In experimental IR-induced intestinal inflammation, ICAM-1 plays a key role in leukocyte recruitment at early times after abdominal IR, whereas VCAM-1 is the main molecular determinant of leukocyte recruitment at later time points [5, 6].

In this study, TGF-β production, which has an important effect on the proliferation and maturation of fibroblasts, the production of collagen, the regulation of metalloproteases, and the suppression of aberrant angiogenesis [7], was very high after IR. TGF-β is considered to be a master switch for the fibrotic program [8]. In this cytokine network, TGF-β could act in conjunction with other cytokines.

For example, the anti-proliferative signaling of IGFBP-3 appears to require an active TGF-β signaling pathway, and IGFBP-3 stimulates phosphorylation of the TGF-β signaling intermediates Smad2 and Smad3 [9].

Taken together, using a high-throughput array to study multiple IMs may enhance our understanding of the complex IM network in IR-induced soft tissue toxicity and provide clues to possible interventions.

5 Conclusion

IR-induced IMs change as a function of time, and different cytokines act at different stages to upregulate or downregulate fibrosis, a characteristic of the IM alteration network. Limitations of arrays parallel to those of ELISA. High affinity and highly specific antibodies must be printed on the membrane to be useful. We found that antibody-based protein arrays are a powerful tool for screening cytokine patterns after radiation.

Acknowledgments This project is supported in part by U19 AI067733. We thank Kate Casey-Sawicki for editing this manuscript.

References

1. Rubin P, Johnston CJ, Williams JP et al (1995) A perpetual cascade of cytokines postirradiation leads to pulmonary fibrosis. Int J Radiat Oncol Biol Phys 33:99–109
2. Burmeister BH, Smithers BM, Davis S et al (2002) Radiation therapy following nodal surgery for melanoma: an analysis of late toxicity. ANZ J Surg 72:344–348
3. Okunieff P, Augustine E, Hicks JE et al (2004) Pentoxifylline in the treatment of radiation-induced fibrosis. J Clin Oncol 22:2207–2213
4. Grimberg A, Coleman CM, Burns TF et al (2005) p53-Dependent and p53-independent induction of insulin-like growth factor binding protein-3 by deoxyribonucleic acid damage and hypoxia. J Clin Endocrinol Metab 90:3568–3574
5. Molla M, Gironella M, Miquel R et al (2003) Relative roles of ICAM-1 and VCAM-1 in the pathogenesis of experimental radiation-induced intestinal inflammation. Int J Radiat Oncol Biol Phys 57:264–273
6. Epperly MW, Sikora CA, DeFilippi SJ et al (2002) Pulmonary irradiation-induced expression of VCAM-I and ICAM-I is decreased by manganese superoxide dismutase-plasmid/liposome (MnSOD-PL) gene therapy. Biol Blood Marrow Transplant 8:175–187
7. Rajalalitha P, Vali S (2005) Molecular pathogenesis of oral submucous fibrosis—a collagen metabolic disorder. J Oral Pathol Med 34:321–328
8. Martin M, Lefaix J, Delanian S (2000) TGF-beta1 and radiation fibrosis: a master switch and a specific therapeutic target? Int J Radiat Oncol Biol Phys 47:277–290
9. Baxter RC (2001) Signalling pathways involved in antiproliferative effects of IGFBP-3: a review. Mol Pathol 54:145–148

Chapter 48
Imaging the Redox States of Human Breast Cancer Core Biopsies

H.N. Xu, J. Tchou, B. Chance, and L.Z. Li

Abstract Currently, the gold standard to establish benign vs. malignant breast tissue diagnosis requires an invasive biopsy followed by tissue fixation for subsequent histopathological examination. This process takes at least 24 h resulting in tissues that are less suitable for molecular, functional, or metabolic analysis. We have recently conducted redox scanning (cryogenic NADH/flavoprotein fluorescence imaging) on snap-frozen breast tissue biopsy samples obtained from human breast cancer patients at the time of their breast cancer surgery. The redox state was readily determined by the redox scanner at liquid nitrogen temperature with extraordinary sensitivity, giving oxidized flavoproteins (Fp) an up to tenfold discrimination of cancer to non-cancer of breast in our preliminary data. Our finding suggests that the identified metabolic parameters could discriminate between cancer and non-cancer breast tissues without subjecting tissues to fixatives. The remainder of the frozen tissue is available for additional analysis such as molecular analysis and conventional histopathology. We propose that this novel redox scanning procedure may assist in tissue diagnosis in ex vivo tissues.

This article is dedicated to the memory of Dr. Britton Chance, who devoted himself to the research process with sleepless nights and his profound insights into science as well as his great attention to detail until the last moment of his life.

H.N. Xu • L.Z. Li (✉)
Department of Radiology, University of Pennsylvania, Philadelphia, PA, USA
e-mail: linli@mail.med.upenn.edu

J. Tchou
Department of Surgery, University of Pennsylvania, Philadelphia, PA, USA

Rena Rowan Breast Center, Abramson Cancer Center, Perelman School of Medicine, University of Pennsylvania, Philadelphia, PA, USA

B. Chance
Department of Biochemistry and Molecular Biophysics, Johnson Research Foundation, University of Pennsylvania, Philadelphia, PA, USA

W.J. Welch et al. (eds.), *Oxygen Transport to Tissue XXXIV*, Advances in Experimental Medicine and Biology 765, DOI 10.1007/978-1-4614-4989-8_48,

Keywords Mitochondrial redox ratio • Fluorescence imaging • Metabolism • Tumor aggressiveness • Oxygenation

1 Introduction

Biochemical analysis of cancer core biopsies has little diagnostic value in the clinic so far. A novel approach to utilize the biochemistry of biopsies is afforded by our redox scanner [1, 2]. This diagnostic tool requires only a small tissue sample (1 mm in 2 dimensions) to obtain tissue mitochondrial redox states at a high spatial resolution down to 50 μm. In this study, we have discovered that redox scanning may provide a novel biochemical approach to identify breast cancer tissue in clinical samples. The differential redox state in breast cancer vs. normal tissues is presumably due to tissue reoxygenation which reactivates the electron transport activity in the mitochondria. In our previous studies [3–6], we discovered that the redox state of tumor tissue was accurately determined by the redox ratio, Fp/(Fp+NADH), a sensitive indicator of the tumor metabolic activity which correlates with the degree of invasiveness. In another study, we also showed that the premalignant pancreatic tissue is more heterogeneous in mitochondrial redox state than the normal one [7].

Currently, the gold standard to establish benign vs. malignant breast tissue diagnosis requires an invasive biopsy followed by tissue fixation for subsequent histological examination. The required tissue fixation step renders the tissue less suitable for additional molecular, functional, and metabolic analyses. So far the histological examination and redox scanning cannot be correlated as redox scanning requires viable tissue with a preserved metabolic state while histology evaluation often requires tissue fixation resulting in nonviable tissues. These two seemingly incompatible requirements are now met by our simple observation that indicates that tissue obtained from normal adjacent tissue in the periphery of the surgical specimen and from the centre of the tumor specimen obtained by core biopsy needle are well oxygenated and the mitochondrial electron transport metabolism is reactivated.

In this paper, we report the preliminary data on imaging the redox state of breast biopsies to discriminate the cancer to non-cancer tissues. Upon further confirmation with more patient samples and standardization of tissue sample collection procedures, this novel redox scanning procedure may assist in real-time tissue diagnosis in ex vivo tissues at the time of surgery.

2 Methods

Tissue collection was performed according to a protocol approved by the Internal Review Board of our institution. Our procedure involves excising a thin section from normal adjacent breast tissue at the periphery of the surgical specimen at the time of breast cancer surgery. Tissue from the central portion of the surgical specimen containing tumor tissue was collected using a 20-gauge core biopsy needle (~$1 \times 1 \times 5$ mm^3). The redox scan requires mounting the snap-frozen tissue slice or core biopsies in the frozen mounting buffer (H_2O:ethanol: glycerol = 10:30:60, freezing point −30°C), followed by redox scanning in two or three dimensions for the redox states of NADH and Fp. Unused portion of both normal and tumor tissues were preserved in the frozen mounting buffer and stored in liquid nitrogen tank for histopathology evaluation.

Breast tissues (normal adjacent tissue collected as a small tissue block and tumor tissue collected by core needle) collected from patients 1 and 2 were placed on saline-moistened paper at room temperature in the operating room. Both tissues were then brought outside the surgery room where they were immediately wrapped with aluminum foil and dipped into liquid N_2. The estimated time interval between removing tissue from the body and snap-freezing was 5–10 min. The core needle samples of cancerous tissue from various locations in the surgical specimen from patient 1 were dipped into liquid nitrogen temporally in the following order: a, b, and d with a time interval of 30 ~ 60 s. For samples from patient 3, both cancerous and normal tissues were given as a thin tissue block of ~$0.3 \times 2 \times 2$ cm^3. The estimated time interval between tissue removal from the patient and snap-freeze was ~5 min. All collected tissues, both normal and cancerous tissues, were from the affected cancer-bearing breasts.

The detailed procedures for embedding tissue samples for redox scanning have been reported elsewhere [5, 7–9]. Small portions of the frozen core biopsies were embedded. For tissue blocks, thin pieces were first sliced off from the snap-frozen tissue blocks using a handsaw on dry-ice-chilled metal surface. The thin piece was embedded in the mounting buffer in such a way that the starting layer for redox scanning was the surface exposed to the air before snap-freeze. NADH and Fp standards were first embedded with chilled glycerol/ethanol/water mounting buffer. The tissue was then placed adjacently using a pair of chilled forceps. Cold mounting buffer slush was added on top of the tissues to secure.

The embedded samples were carefully milled flat under liquid N_2 by shaving off the top surfaces, ~50 μm from tumor tissue (core biopsy) and ~100–300 μm from normal tissue. The samples were then scanned under liquid N_2 using the redox scanner. Two to four sections with spacing between 40 and 120 μm (depth range 0–240 μm) were scanned for each core biopsy sample. Two sections with spacing ~200–300 μm were scanned for each tissue block sample.

The acquired NADH and Fp signals were analyzed using Matlab® software which constructed all images. Both the NADH and Fp images were displayed as concentration maps, where concentration was in the unit of μM calibrated to the fluorescence from the corresponding standard. The redox ratio images were displayed as the concentration ratio of Fp/(Fp + NADH) in the range of 0 ~ 1.

For multiple tissue samples from patient 1, the redox scanning indices (Fp, NADH, and redox ratio) were averaged across tissue sections and then across tissue samples from various locations. Single factor ANOVA analysis was performed to compare the differences among the mean values of three core biopsies from different locations.

3 Results and Discussion

Figures 48.1 and 48.2 illustrate the typical findings of the redox images of breast tumor tissue and normal breast tissue. The results of redox scans of the human biopsy samples are shown in Table 48.1. The first and most notable finding is the high values of oxidized flavoprotein Fp, clearly indicating that the electron transfer chain to oxygen was operating efficiently prior to snap-freezing. Fp nominal concentration in cancerous tissue ($Fp_t = 652 \pm 93$ μM) is significantly higher than that in normal tissue ($Fp_n = 118 \pm 89$ μM) with $p = 0.027$. Their ratio, Fp_t/Fp_n, averaged over three patients is 7.7 ± 4.4.

In comparing the cancerous and normal tissues, we find a still greater difference, the redox ratio is different for all three normal (0.40 ± 0.09) and three cancerous tissues (0.61 ± 0.10), despite the borderline statistical difference which might be partially due to the very different disease histories. Patient 1 and 2 received no neo-adjuvant therapy, while patient 3 had been treated with Arimidex for 6 months before she had the surgery. In spite of the disparities in patient disease history and sample collection procedures, the cancerous tissue has strikingly higher Fp (almost tenfold higher). This cannot be fully accounted for by slightly higher O_2 diffusion coefficient in tumor tissue than that in normal one [10–14]. The oxygen diffusion distance is more than 500 μm for a time window of 5 min assuming a diffusion constant of 10^{-5} cm²/s. The reported redox scanning data were from the top layers of both the tissue blocks (depth ~0–300 μm) and the core biopsy samples

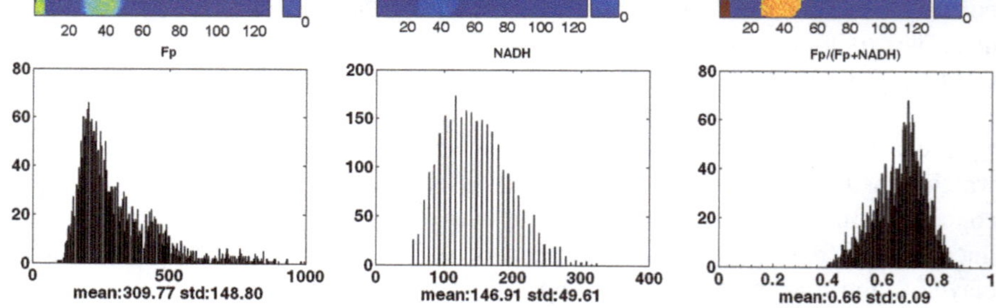

Fig. 48.1 Typical redox images of breast tumor tissue (patient 1) (image matrix 128 × 128, resolution 40 μm). The redox ratio ranges between 0 and 1; the Fp or NADH images are in the unit of μM. The *x* axes of the corresponding histograms represent the redox ratio or concentration. The y axes represent the number of pixels in the tumor section having a specific value of redox ratio or concentration

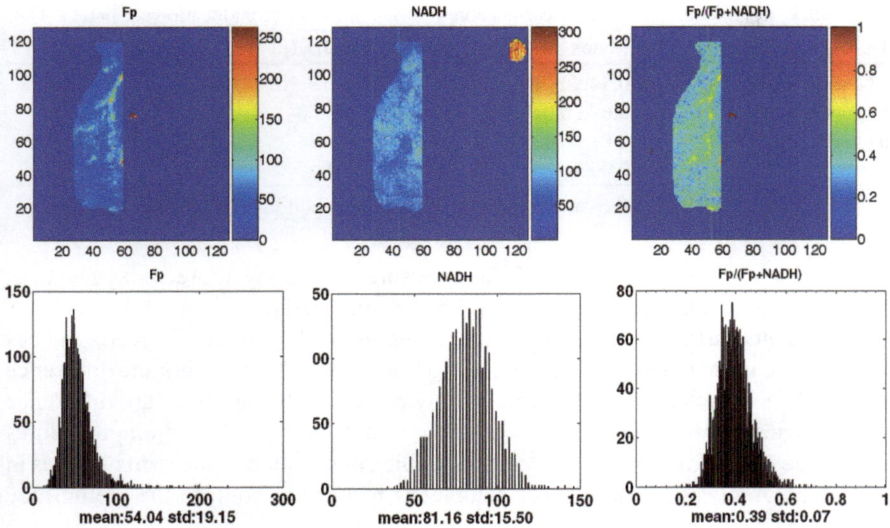

Fig. 48.2 Typical redox images of normal breast tissue (patient 1) (image matrix 128 × 128, resolution 100 μm)

Table 48.1 Redox indices of the cancerous and normal breast tissues from three patients

	Cancerous			Normal			
Patient	Redox ratio$_t$	NADH$_t$ (μM)	Fp$_t$	Redox ratio$_n$	NADH$_n$ (μM)	Fp$_n$ (μM)	Fp$_t$/Fp$_n$
1[a]	0.71	153	569	0.42	86	65	9.2
2[b]	0.52	699	759	0.31	153	69	11.1
3[c]	0.59	295	601	0.48	174	221	2.7
Mean	0.61	384	652	0.40	138	118	7.7[d]
SD	0.10	281	93	0.09	46	89	4.4
p value[e]	0.062	0.248	0.027				

[a]2.1 cm, invasive ductal carcinoma (IDC), triple negative (TN) receptors, negative nodes, poorly differentiated, 3 core biopsies, and 1 normal tissue block
[b]3 cm, BRCA1 mutation, TN, negative nodes, poorly differentiated, 1 core biopsy, and 1 normal tissue block
[c]10 cm, ER+, positive node, Arimidex treated for 6 months, 1 cancer and 1 normal tissue block
[d]Average of Fp$_t$/Fp$_n$ over the three patients
[e]Comparison of the redox indices between cancerous and normal tissue of the three patients

(depth ~0–240 μm). Thus, it is reasonable to assume that the reoxygenation due to oxygen diffusion should be similar for both normal and cancerous tissue samples.

For patient 1, several cancerous tissue core biopsies from different locations were collected and redox-scanned. Table 48.2 summaries the highly significant heterogeneity in the redox ratio, NADH, and Fp in her tumor tissue obtained by core needle in the following temporal order a→b→d. The highly significant difference in the redox indices is readily recognized. The temporal order of the sample snap-freezing a→b→d coincides with the increase of both NADH and Fp concentrations.

Table 48.2 Redox indices of the cancerous core biopsies from different locations of patient 1

Tissue location	Redox ratio	$NADH_t$ (μM)	Fp_t (μM)
a (2 sections)	0.54±0.01	45±13	57±20
b (4 sections)	0.69±0.03	142±30	351±37
d (5 sections)	0.81±0.03	218±34	1,044±337
p value[a]	0.00001	0.0004	0.0017

[a]Single factor ANOVA analysis was used to show the statistical difference between locations

Delayed snap-freeze led to longer air-exposure and likely more reoxygenation, which should result in increasing Fp and decreasing NADH.

It warrants further investigation whether the marked difference in redox indices between the normal and cancerous tissues of the three patients and/or the difference among the samples of patient 1 were mainly caused by the air-exposure difference or the intrinsic biological difference in these tissues. Sample collection procedures need to be standardized to minimize its possible contribution to the redox indices in the future. Analysis of more tissue samples is planned to confirm the preliminary results reported here.

4 Conclusions

In summary the mitochondrial redox states of human breast cancer tissue samples can be measured fluorometrically and show marked differences between cancer and non-cancer tissues. The redox scanning results clearly indicate aerobiosis of the samples. We report here, for the first time, the feasibility of redox cryo-imaging of breast biopsies which may open up new avenues to explore the correlation of histo-pathology and redox detection of cancer and perhaps to monitor efficacy of thera-peutics. We propose that this novel redox scanning procedure may be used as a real-time tissue diagnosis tool in ex vivo tissues.

Acknowledgments This work was supported by the Susan G. Komen Foundation Grant KG081069 (L.Z. Li), the Center of Magnetic Resonance and Optical Imaging (CMROI)—an NIH supported research resource P41RR02305 (R. Reddy), the Small Animal Imaging Program (SAIR) 2U24-CA083105 (J. Glickson & L. Chodosh), and the Abramson Cancer Center Pilot Grant funded by the NCI Cancer Center Support Grant (J. Tchou)

References

1. Chance B, Schoener B, Oshino R et al (1979) Oxidation-reduction ratio studies of mitochondria in freeze-trapped samples. NADH and flavoprotein fluorescence signals. J Biol Chem 254:4764–4771
2. Quistorff B, Haselgrove JC, Chance B (1985) High spatial resolution readout of 3-D metabolic organ structure: an automated, low-temperature redox ratio-scanning instrument. Anal Biochem 148:389–400

3. Li LZ, Zhou R, Zhong T et al (2007) Predicting melanoma metastatic potential by optical and magnetic resonance imaging. Adv Exp Med Biol 599:67–78
4. Li LZ, Zhou R, Xu HN et al (2009) Quantitative magnetic resonance and optical imaging biomarkers of melanoma metastatic potential. Proc Natl Acad Sci U S A 106:6608–6613
5. Xu HN, Nioka S, Glickson JD et al (2010) Quantitative mitochondrial redox imaging of breast cancer metastatic potential. J Biomed Opt 15:036010
6. Li LZ, Xu HN, Ranji M et al (2009) Mitochondrial redox imaging for cancer diagnostic and therapeutic studies. J Innov Opt Health Sci 2:325–341
7. Xu HN, Nioka S, Chance B et al (2011) Heterogeneity of mitochondrial redox state in premalignant pancreas in a PTEN null transgenic mouse model. Adv Exp Med Biol 201:207–213
8. Xu HN, Wu B, Nioka S et al (2009) Quantitative redox scanning of tissue samples using a calibration procedure. J Innov Opt Health Sci 2:375–385
9. Xu HN, Wu B, Nioka S et al (2009) Calibration of redox scanning for tissue samples. Proceedings of Biomedical Optics in San Jose, CA, Jan. 24, Ed. SPIE 7174:71742F
10. Evans NTS, Naylor PFD, Quinton TH (1981) The diffusion coefficient of oxygen in respiring kidney and tumour tissue. Respir Physiol 43:179
11. Groebe K, Vaupel P (1988) Evaluation of oxygen diffusion distances in human breast cancer xenografts using tumor-specific in vivo data: role of various mechanisms in the development of tumor hypoxia. Int J Radiat Oncol Biol Phys 15:691
12. Krogh A (1919) The rate of diffusion of gases through animal tissues, with some remarks on the coefficient of invasion. J Physiol 52:391–408
13. Macdougall JDB, McCabe M (1967) Diffusion coefficient of oxygen through tissues. Nature 215:1173
14. Vaupel P, Mayer A, Briest S et al (2005) Hypoxia in breast cancer: role of blood flow, oxygen diffusion distances, and anemia in the development of oxygen depletion. Adv Exp Med Biol 566:333–342

Chapter 49
Early Life Hypoxic or Hypoxic/Hypercapnic Stress Alters Acute Ventilatory Sensitivity in Adult Mice

Kui Xu, Solomon Raju Bhupanapadu Sunkesula, Pengjing Huang, Constantinos P. Tsipis, Thomas Radford, Gerald Babcock, Walter F. Boron, and Joseph C. LaManna

Abstract In this study we investigated the effect of early life conditioning (hypoxia ± hypercapnia) on adult acute ventilatory sensitivity to hypoxia and hypercapnia. Mice were exposed to either hypoxia (5% O_2) or hypoxia/hypercapnia (5% O_2/8% CO_2) in a normobaric chamber for 2 h at postnatal day 2 (P2), and then returned to normoxia. At 3 months of age, hypoxic ventilatory response (HVR) and hypercapnic ventilatory response (HCVR) were measured using a plethysmograph system. Results showed that HVR was significantly decreased in the P2-hypoxia mice but not in the P2 hypoxia/hypercapnia mice as compared to the P2-normoxic mice, respectively. However, HCVR was significantly decreased in the P2 hypoxia–hypercapnia group but not in the P2-hypoxia group. These data suggest early postnatal hypoxic stress vs. hypoxic/hypercapnic stress plays different roles in fetal programming of the respiratory control system as shown by altered adult acute ventilatory sensitivity.

Keywords Hypoxia • Hypercapnia • Ventilatory response • Respiratory control

1 Introduction

Brief hypoxic exposure in early life has led to persistent structural and functional changes in the lung [1], heart [2], and brain [3]. For example in the lung, these mechanisms may involve mitochondrial pore opening [4]. It was shown that even

K. Xu (✉)
Departments of Neurology, School of Medicine, Case Western Reserve University, 10900 Euclid Avenue, Robbins E732, Cleveland, OH, USA
e-mail: kxx@case.edu

S.R.B. Sunkesula • P. Huang • C.P. Tsipis • T. Radford • G. Babcock • W.F. Boron • J.C. LaManna
Departments of Physiology and Biophysics, School of Medicine, Case Western Reserve University, Cleveland, OH, USA

W.J. Welch et al. (eds.), *Oxygen Transport to Tissue XXXIV*, Advances in Experimental Medicine and Biology 765, DOI 10.1007/978-1-4614-4989-8_49,

brief exposure to postnatal hypoxia had lung effects that persisted to the adult phenotype [5]. In humans, pulmonary circulation is especially vulnerable to perinatal hypoxia exposed to the ambient oxygen tension of altitude (4,559 m) [6]. Studies conducted in rats showed that perinatal hypoxia led to long-term pulmonary changes [7] most likely due to endothelial cell dysfunction [8]. The roles of epigenetic mechanisms in fetal programming [7, 9] have become the latest concepts to support Barker's hypothesis [10, 11]. Furthermore, hypoxic/ischemic events during infancy may, through epigenetic mechanisms, result in greater vulnerability to subsequent challenges in adulthood, such as hypoxic exposure or ischemic insult.

The goal of this study was to assess the degree of future risk in mice exposed to early life hypoxic ± hypercapnic events. We have developed a novel model to induce changes in the adult structure/function of the central nervous system, in mice by giving brief, severe, nonlethal hypoxia at an early stage of postnatal development (e.g., postnatal day 2, P2). In this study we investigated the effect of P2 hypoxia (±hypercapnia) conditioning on acute ventilatory sensitivity to hypoxia or hypercapnia in 3-month-old mice.

2 Methods

The experimental protocol employed by this study was approved by the Institutional Animal Care and Use Committee (IACUC) at Case Western Reserve University.

2.1 P2 Hypoxic or Hypoxic/Hypercapnic Exposure

Postnatal (P2) mice were exposed to either hypoxia (5% oxygen in nitrogen) or hypoxia/hypercapnia (5% O_2, 8% CO_2 in N_2) in a chamber where the inflow of gases is automatically controlled (Oxycycler model A44XO, Biospherix, Redfield, NY) for 2 h. Mice were then returned to normoxic environment. Littermates were maintained normoxic in the same room and used as controls. Ambient temperature and humidity of gases is monitored and maintained at 22–24°C and 40–50%, respectively.

2.2 Measurement of Hypoxic Ventilatory Response and Hypercapnic Ventilatory Response

Hypoxic ventilatory response (HVR) and hypercapnic ventilatory response (HCVR) were measured in mice at P90 using a plethysmograph system (Buxco Electronics, Troy, NY). Conscious, unrestrained mice were placed individually in a precalibrated 0.6 L barometric chamber and continuously ventilated with humidified room air or gas mixtures passing through the chamber at a rate of 0.5 L/min. Pressure changes in the chamber due to animal inspiration and expiration were measured by a high gain differential pressure transducer. Analog signals were amplified, continuously digitized and analyzed by the Buxco software, with a rejection algorithm induced in the breath-by-breath analysis routine which allows accurate rejection of motion-induced artifact. The room air, 8% O_2 in N_2 and 8% CO_2 in air, were ventilated for 10 min each to determine the normoxic baseline, hypoxic and hypercapnic ventilation, respectively. Tidal volume, breath frequency, and minute volume (MV) were recorded for subsequent analysis. HVR and HCVR were calculated by the following:

$$\text{HVR} = \text{hypoxic}(8\%\text{O}_2)\text{MV} / \text{normoxic baseline MV}$$

$$\text{HCVR} = \text{hypercapnic}(8\%\ \text{CO}_2)\text{MV} / \text{normoxic baseline MV}$$

2.3 *Statistical Analysis*

Data are expressed as mean ± SD. Statistical analyses were performed using SPSS V19.0 for Windows. The comparison between any two groups was analyzed with a *t*-test, and significance was considered at the level of $P<0.05$.

3 Results

3.1 *Physiological Variables*

Body weights and packed red cell volume (hematocrit) were compared amongst the P2 conditioning (normoxia, hypoxia, and hypoxia/hypercapnia) groups in 3-month-old mice. There was no significant difference between any two groups in body weight (~26 g on average) or hematocrit (48% on average).

3.2 *Acute Ventilatory Responses to Hypoxia or Hypercapnia*

HVR and HCVR were measured in 3-month-old adult mice with different P2 conditions (normoxia, hypoxia, and hypoxia/hypercapnia), as shown in Fig. 49.1. HVR was decreased significantly in the P2 hypoxia group (1.7 ± 0.4, mean ± SD, $n=26$) as compared to the P2 normoxia group (2.0 ± 0.4, $n=40$). However, the P2 hypoxia/hypercapnia group had similar values of HVR to that of the P2 normoxia group (Fig. 49.1a). HCVR was decreased significantly in the P2 hypoxia/hypercapnia group (1.8 ± 0.2, mean ± SD, $n=26$) but not in the P2-hypoxia group (2.4 ± 0.4, $n=25$) compared to the P2 normoxia group (2.3 ± 0.4, $n=28$).

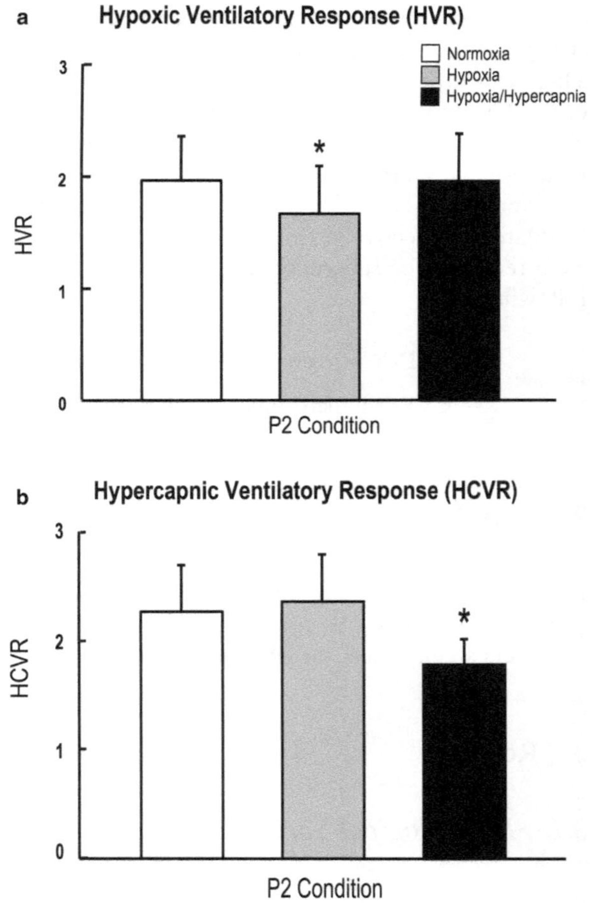

Fig. 49.1 Hypoxic ventilatory response (**a**) and hypercapnic ventilatory response (HCVR) (**b**) in 3-month-old mice with different P2 conditioning, normoxia, hypoxia or hypoxia/hypercapnia. Values are mean ± SD, *asterisk* indicates significant difference ($P < 0.05$) compared to the P2-normoxia group

4 Discussion

In this study, we have shown that brief postnatal (P2) exposure to hypoxia (±hypercapnia) resulted in altered acute ventilatory sensitivities to hypoxia or hypercapnia in adult mice. P2-hypoxic conditioning decreased HVR in adult, but co-exposure with hypercapnia seemed to reverse this effect. A hypercapnic-induced increase in blood flow is potentially a rationale for our finding. On the other hand, hypoxic conditioning alone at P2 did not affect the HCVR in adulthood, but the addition of hypercapnia decreased the HCVR in the future. Therefore hypoxic and hypercapnic exposure plays different roles in the fetal programming of the respiratory control system. Though our data support Barker's hypothesis, the mechanisms responsible for the altered adult ventilatory sensitivity induced by P2-conditioning need to be further elucidated. Possible mechanisms may involve the epigenetic change of hypoxia-inducible genes and the structural and functional changes in chemoreflex pathways, such as in brainstem neurons.

Acknowledgments This study was supported by NIH grant NS 38632.

References

1. Tang JR, Le Cras TD, Morris KG Jr et al (2000) Brief perinatal hypoxia increases severity of pulmonary hypertension after reexposure to hypoxia in infant rats. Am J Physiol Lung Cell Mol Physiol 278:L356–L364
2. Del DD, Tadevosyan A, Karbassi F et al (2010) Hypoxia in early life is associated with lasting changes in left ventricular structure and function at maturity in the rat. Int J Cardiol. doi:10.1016/j.ijcard.2010.10.135
3. Akers KG, Nakazawa M, Romeo RD et al (2006) Brief perinatal hypoxia increases severity of early life modulators and predictors of adult synaptic plasticity. Eur J Neurosci 24:547–554
4. Hagberg H, Mallard C, Rousset CI et al (2009) Apoptotic mechanisms in the immature brain: involvement of mitochondria. J Child Neurol 24:1141–1146
5. Massaro GD, Olivier J, Massaro D (1989) Short-term perinatal 10% O_2 alters postnatal development of lung alveoli. Am J Physiol 257:L221–L225
6. Sartori C, Allemann Y, Trueb L et al (1999) Augmented vasoreactivity in adult life associated with perinatal vascular insult. Lancet 353:2205–2207
7. Rexhaj E, Bloch J, Jayet PY et al (2011) Fetal programming of pulmonary vascular dysfunction in mice: role of epigenetic mechanisms. Am J Physiol Heart Circ Physiol 301(1):H247–H252. Epub 2011 May 2
8. Rexhaj E, Garcin S, Rimoldi SF et al (2011) Reproducibility of acute mountain sickness in children and adults: a prospective study. Pediatrics 127:e1445–e1448
9. Tarry-Adkins JL, Ozanne SE (2011) Mechanisms of early life programming: current knowledge and future directions. Am J Clin Nutr 94:1765S–1771S
10. Barker DJ, Winter PD, Osmond C et al (1989) Weight in infancy and death from ischaemic heart disease. Lancet 2:577–580
11. Feldt K, Raikkonen K, Eriksson JG et al (2008) Childhood growth and cardiovascular reactivity to psychological stressors in late adulthood. J Intern Med 264:72–82

Chapter 50
3D Analysis of Intracortical Microvasculature During Chronic Hypoxia in Mouse Brains

Kouichi Yoshihara, Hiroyuki Takuwa, Iwao Kanno, Shinpei Okawa, Yukio Yamada, and Kazuto Masamoto

Abstract The purpose of this study is to determine when and where the brain microvasculature changes its network in response to chronic hypoxia. To identify the hypoxia-induced structural adaptation, we longitudinally imaged cortical microvasculature at the same location within a mouse somatosensory cortex with two-photon microscopy repeatedly for up to 1 month during continuous exposure to hypoxia (either 8 or 10% oxygen conditions). The two-photon microscopy approach made it possible to track a 3D pathway from a cortical surface arteriole to a venule up to a depth of 0.8 mm from the cortical surface. The network pathway was then divided into individual vessel segments at the branches, and their diameters and lengths were measured. We observed 3–11 vessel segments between the penetrating arteriole and the emerging vein over the depths of 20–460 μm within the 3D reconstructed image ($0.46 \times 0.46 \times 0.80$ mm^3). The average length of the individual capillaries (<7 μm in diameter) was 67 ± 46 μm, which was not influenced by hypoxia. In contrast, 1.4 ± 0.3 and 1.2 ± 0.2 fold increases of the capillary diameter were observed 1 week after exposure to 8 % and 10% hypoxia, respectively. At 3 weeks

K. Yoshihara • S. Okawa • Y. Yamada
Department of Mechanical Engineering and Intelligent Systems, University of Electro-Communications, Chofu, Tokyo, Japan

H. Takuwa • I. Kanno
Molecular Imaging Center, National Institute of Radiological Sciences, Chiba, Japan

K. Masamoto (✉)
Molecular Imaging Center, National Institute of Radiological Sciences, Chiba, Japan

Center for Frontier Science and Engineering, University of Electro-Communications, 1-5-1 Chofugaoka, Chofu, Tokyo 182-8585, Japan
e-mail: masamoto@mce.uec.ac.jp

W.J. Welch et al. (eds.), *Oxygen Transport to Tissue XXXIV*, Advances in Experimental Medicine and Biology 765, DOI 10.1007/978-1-4614-4989-8_50,

from the exposure, the capillary diameter reached 8.5 ± 1.9 and 6.7 ± 1.8 μm in 8% and 10 % hypoxic conditions, respectively, which accounted for the 1.8 ± 0.5 and 1.4 ± 0.3 fold increases relative to those of the prehypoxic condition. The vasodilation of penetrating arterioles (1.4 ± 0.2 and 1.2 ± 0.2 fold increases) and emerging veins (1.3 ± 0.2 and 1.3 ± 0.2 fold increases) showed relatively small diameter changes compared with the parenchymal capillaries. These findings indicate that parenchymal capillaries are the major site responding to the oxygen environment during chronic hypoxia.

Keywords Two-photon microscopy • Oxygen transport • Hypoxic adaptation • Somatosensory cortex

1 Introduction

The brain is known to be an organ vulnerable to a lack of oxygen. Only 2–3 min without an adequate oxygen supply leads to irreversible damage of central nervous system (CNS) functions. To maintain an adequate supply, the brain vasculature is well organized to allow the diffusion of oxygen to every region of the tissue [1]. Previous studies have shown that the density of brain microvessels adaptively increased in response to chronic hypoxia [2, 3]. Angiogenesis and degeneration of the microvasculature were also reported in response to a progress of brain tumors and Alzheimer's disease, respectively [4, 5]. These studies strongly indicate that close interaction exists between brain microvasculature and tissue activity to maintain an adequate supply of oxygen. However, the mechanism that regulates the brain microvasculature and its effect on tissue oxygen homeostasis remains mostly unknown.

Because one of the reasons for the lack of knowledge about the structural adaptations of brain microvasculature is a lack of longitudinal studies at the capillary network scale, here, we longitudinally tracked the spatial and temporal adaptation of the microvasculature during chronic hypoxia in a mouse somatosensory cortex with two-photon microscopy. The same three-dimensional (3D) microvasculature was imaged every week for up to 1 month during continuous exposure to hypoxia (at either 8 or 10 % oxygen). Two-photon microscopy is based on the principle of nonlinear optics [6–8]. In addition, two-photon microscopy allows excitation with near-infrared light, which has the advantage of providing a long penetration depth in biological tissues [9]. In the present study, volume images of the cortical microvasculature were captured up to a depth of 0.8 mm from the cortical surface with a step size of 0.005 mm and reconstructed with 3D rendering. The 3D pathways of the parenchymal microvasculature were identified and divided into individual segments at the branches. The diameters and lengths of the segments were measured, and their temporal changes were longitudinally traced over the periods of the measurements. Finally, the surface areas and volumes of the vessel segments were calculated to evaluate the adaptation of the microvasculature.

2 Materials and Methods

2.1 Image Analysis

The 3D image (field of view: $0.46 \times 0.46 \times 0.80$ mm^3) consists of approximately 160 2D images in the *x-y* plane parallel to the brain surface with a depth step of 5 μm, and each 2D image has $1{,}024 \times 1{,}024$ pixels with an individual pixel size of 0.45 μm. The image analysis performed in each step is described below.

1. Identification of parenchymal arteriovenous pathway.
   One target pathway between a surface artery and a vein through parenchymal microvasculature was identified within the reconstructed image (Fig. 50.1a, b),

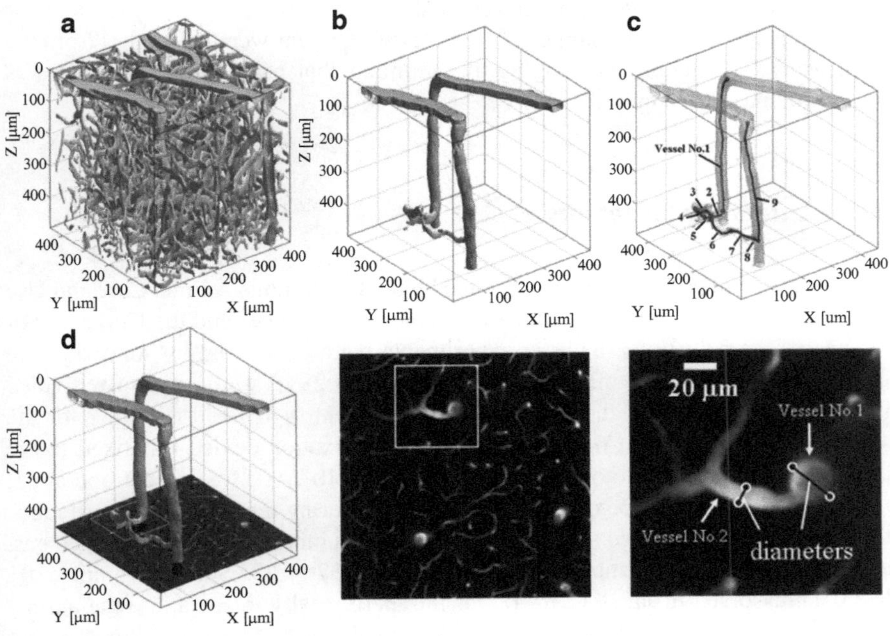

Fig. 50.1 Methods for measuring vessel diameter and length. (**a**) Reconstructed volume image of parenchymal microvasculature; binarized from 3D image that was reconstructed by rendering obtained 2D *x-y* images. (**b**) Selection of a single vessel pathway; the selection criterion was to have the entire connection from the surface arteriole to the surface venule visible within the image. (**c**) Segmentation and measurement of the length of individual vessel segments. (**d**) Manual measurement of vessel diameter in the image

by tracking all branches from the artery to the vein including penetrating arterioles, parenchymal capillaries (<7 μm in diameter), and emerging veins.

2. Segmentation of vessels.
   The vessel segments in the pathway were sequentially numbered (Fig. 50.1c). One vessel segment was defined as the vessel having two-branches at both ends. The sequential numbers were assigned, beginning from the penetrating artery and ending with the emerging vein.
3. Measurement of the length of a vessel segment.
   The vessel length was measured as the distance from one end to the other end along the centerline of the vessel segment. Several measurement points were manually placed on the centerline to account for the curvature of the vessel. By summing all the distances between the neighboring two measuring points, the vessel length was obtained.
4. Measurement of the vessel diameter.
   The vessel diameter was measured as the average of the diameters at ten different positions in the vessel segment. Because the intensity differences (i.e., contrast) in the images were sufficient to discriminate the vessel edges relative to the tissue, it was not difficult to manually assign the edges of the vessels to measure the vessel diameters (Fig. 50.1d).
5. Calculation of the vessel surface area and volume.
   The surface areas and volumes of each vessel segment were calculated from the measured diameters and lengths by assuming that each vessel segment was approximated as a cylinder.

2.2 *Animal Experiments*

All experimental protocols were approved by the Institutional Animal Care and Use Committee of the National Institute of Radiological Sciences and the University of Electro-Communications, and the experiments were conducted by following the approved protocols. Ten male C57BL/6J mice (21–23 g) were randomly divided into two groups: one for the 8 % oxygen experiment ($n=5$) and the other for the 10 % oxygen experiment ($n=5$). A custom-made fixation device was fixed to the animal's head for two-photon microscopy imaging [10]. At 2 weeks post-operation, the experiments with hypoxic exposure initiated. During the imaging experiments, the mice were anesthetized with isoflurane (1 %) in a mixture of air and oxygen gas, and their rectal temperature was maintained at 37 ± 1°C. Sulforhodamine 101 (SR101) dissolved in saline (5 mM) was intraperitoneally injected into the animal (8 mL/kg) for labeling blood plasma, and cortical vasculature was imaged with a two-photon microscope (TCS SP5MP, Leica Microsystems, Germany) equipped with a Ti:Sapphire laser (MaiTai HP, Spectra-Physics, CA). The excitation wavelength was 900 nm (average power 2.0 W output), and the emission signal was detected through a bandpass filter (610/75 nm).

At day 0 (start of exposure to hypoxia), a reference image of the cortical surface vasculature was obtained with reflection acquisition mode excited at 633 nm with a low magnification object lens (field of view: 3.6×3.6 mm^2). The cortical arteries and veins were distinguished in the reflection image; the arteries and veins exhibited different light intensities because of their different light absorption characteristics [9]. Because the pattern of the cortical surface vessels was preserved over the period of the longitudinal experiments, the reference image was used to identify the measurement locations at different imaging experiments. Except while undergoing imaging, the animals were kept in either an 8 or 10 % oxygen room in which the oxygen level was monitored with an oxygen sensor. The imaging experiments were repeatedly performed on days 0, 7, 14, 21, and 31 after exposure to hypoxia. Because three out of five animals were dead after 3 weeks in the 8 % oxygen experiment, the analytical results were only compared for the first 3 weeks of measurements. Data were represented as mean ± standard deviation.

3 Results and Discussion

Reconstructed 3D images of cortical microvasculature are represented in Fig. 50.2. The images show the apparently same 3D structures over the period of 1 month during continuous exposure to hypoxia. The 3D pathways were identified from penetrating arteriole (A) to emerging vein (V) through parenchymal capillaries (C),

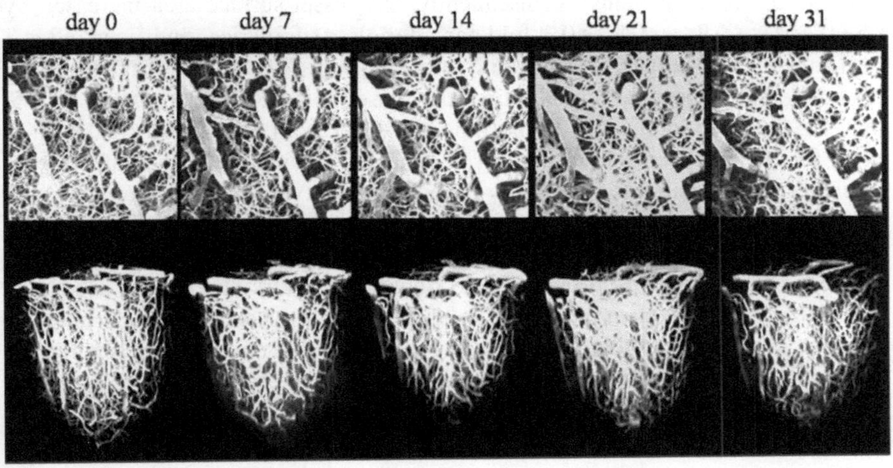

Fig. 50.2 The 3D imaging of cortical microvasculature with two-photon microscopy. Volume images (0.46×0.46×0.80 mm^3) were obtained from day 0 (start of exposure to 10 % oxygen) today 31 for the identical region within the cortex. The upper panels show the maximum intensity projected images in the *x*-*y* plane (0.46×0.46 mm^2), and the lower panels show the 3D reconstructed images

Table 50.1 Vessel diameter and length measurements of single pathways from a penetrating arteriole (A) to the emerging vein (V) through parenchymal capillaries (C, <7 μm in diameter) during chronic hypoxia

	Diameter (μm)				Length (μm)			
(total #)	day 0	day 7	day 14	day 21	day 0	day 7	day 14	day 21
8 % O_2 (n=5)								
A (32)	11±5	14±5	15±5	16±6	77±41	78±39	76±41	77±38
C (61)	4.9±0.9	6.8±1.3	7.5±1.7	8.5±1.9	68±48	69±49	67±46	67±46
V (23)	17±9	21±10	23±12	23±13	91±69	88±66	86±61	81±58
10 % O_2 (n=5)								
A (42)	14±6	17±7	18±7	18±7	126±108	123±108	122±106	123±106
C (69)	5.0±1.0	5.9±1.4	6.4±1.5	6.7±1.8	61±37	62±36	60±36	61±35
V (21)	17±8	19±8	20±9	21±10	136±127	141±122	134±121	133±119

which had 5–13 branches between the surface artery and the vein over depths of 20–460 μm within the image. The total length from the beginning of a penetrating arteriole to the end point of the emerging vein was 703±278 μm (311–1,221 μm, n=30 networks), which was unchanged during chronic hypoxia over 3 weeks (Table 50.1). In contrast, the parenchymal capillary diameter was predominantly increased by 1.4±0.3 and 1.2±0.2 fold at 1 week after continuous exposure to 8 % and 10 % oxygen, respectively (Table 50.1). The vasodilation of parenchymal capillaries continued over 3 weeks: 1.8±0.5 and 1.4±0.3 fold relative to that of the prehypoxic condition. This increase was significantly higher than both the penetrating arterioles (1.4±0.2 times) and emerging veins (1.3±0.2 times) under the 8 % oxygen condition but not the 10 % oxygen condition (1.2±0.2 fold for arterioles and 1.3±0.2 fold for veins). Consequently, the vessel surface area increased by 1.5±0.4, 1.8±0.7, and 1.2±0.3 fold for arterioles, capillaries, and veins after 3 weeks under 8 % oxygen exposure and by 1.2±0.3, 1.4±0.5, and 1.3±0.3 fold for 10 % oxygen, respectively. In addition, the vessel volume was 2.1±0.9, 3.5±2.5, and 1.7±0.6 times larger for 8 % oxygen and 1.6±0.5, 2.1±1.6, and 1.6±0.6 times larger for 10 % oxygen. These results show that depending on the level of hypoxia, vascular restructuring was induced for different vessel types. In particular, the lower oxygen level (8 %) caused a predominant increase in the parenchymal capillary volumes, indicating that several mechanisms may be participating in the hypoxia-induced restructuring of cortical microvasculature. Future studies are needed to investigate the mechanism of oxygen level sensing and the driving force leading to a change in the capillary diameter to further understand the mechanism involved in maintaining tissue oxygen homeostasis in the brain.

4 Conclusions

The present study quantified hypoxia-induced restructuring of cerebral microvasculature based on 3D network pathways measured with two-photon microscopy at the same locations within mouse brains over 1 month. We found that hypoxia-induced vasodilation occurred predominantly in the parenchymal capillaries, indicating that parenchymal capillaries are the major site responding to the chronic hypoxic environment.

Acknowledgments The authors thank Mr. Ryutaro Asaga and Mr. Ryota Sakamoto for their help in the preparation of the experiments. This work was partially supported by Special Coordination Funds for Promoting Science and Technology (K.M.).

References

1. Masamoto K, Tanishita K (2009) Oxygen transport in brain tissue. J Biomech Eng 131:074002
2. Boero JA, Ascher J, Arregui A et al (1999) Increased brain capillaries in chronic hypoxia. J Appl Physiol 86:1211–1219
3. Xu K, LaManna JC (2006) Chronic hypoxia and the cerebral circulation. J Appl Physiol 100:725–730
4. Wsseling P, Ruiter DJ, Burger PC (1997) Angiogenesis in brain tumors; pathobiological and clinical aspects. J Neurooncol 32:253–265
5. Meyer EP, Ulmann-Schuler A, Staufenbiel M et al (2008) Altered morphology and 3D architecture of brain vasculature in a mouse mode for Alzheimer's disease. Proc Natl Acad Sci U S A 105:3587–3592
6. Denk W, Strickler JH, Webb WW (1990) Two-photon laser scanning fluorescence microscopy. Science 248:73–76
7. So PT, Dong CY, Masters BR et al (2000) Two-photon excitation fluorescence microscopy. Annu Rev Biomed Eng 2:399–429
8. Helmchen F, Denk W (2005) Deep tissue two-photon microscopy. Nat Methods 2:932–940
9. Yamada Y (1995) Light-tissue interaction and optical imaging in biomedicine. Ann Rev Heat Transfer 6:1–59
10. Takuwa H, Autio J, Nakayama H et al (2011) Reproducibility and variance of a stimulation-induced hemodynamic response in barrel cortex of awake behaving mice. Brain Res 1369:103–111

Chapter 51
Contribution of Brain Glucose and Ketone Bodies to Oxidative Metabolism

Yifan Zhang, Youzhi Kuang, Joseph C. LaManna, and Michelle A. Puchowicz

Abstract Ketone bodies are an alternative energy substrate to glucose in brain. Under conditions of oxidative stress, we hypothesize that ketosis stabilizes glucose metabolism by partitioning glucose away from oxidative metabolism towards ketone body oxidation. In this study we assessed oxidative metabolism in ketotic rat brain using stable isotope mass spectrometry analysis. The contribution of glucose and ketone bodies to oxidative metabolism was studied in cortical brain homogenates isolated from anesthetized ketotic rats. To induce chronic ketosis, rats were fed either a ketogenic (high-fat, carbohydrate restricted) or standard rodent chow for 3 weeks and then infused intravenously with tracers of [U-^{13}C] glucose or [U-^{13}C] acetoacetate for 60 min. The measured percent contribution of glucose or ketone bodies to oxidative metabolism was analyzed by measuring the ^{13}C-label incorporation into acetyl-CoA. Using mass spectrometry (gas-chromatography; GC-MS, and liquid-chromatography; LCMS) and isotopomer analysis, the fractional amount of substrate oxidation was measured as the M+2 enrichment (%) of acetyl-CoA relative to the achieved enrichment of the infused precursors, [U-^{13}C]glucose or [U-^{13}C] acetoacetate. Results: the percent contribution of glucose oxidation in cortical brain in rats fed the ketogenic diet was 71.2 ± 16.8 (mean% ± SD) compared to the standard chow, 89.0 ± 14.6. Acetoacetate oxidation was significantly higher with ketosis

Y. Zhang
Departments of Biomedical Engineering, School of Medicine,
Case Western Reserve University, Cleveland, OH, USA

Y. Kuang • J.C. LaManna
Departments of Physiology and Biophysics, School of Medicine,
Case Western Reserve University, Cleveland, OH, USA

M.A. Puchowicz (✉)
Departments of Nutrition, School of Medicine, School of Medicine,
Case Western Reserve University, 10900 Euclid Avenue, Biomedical Research Bldg,
Room 927, Cleveland, OH, USA
e-mail: map10@case.edu

W.J. Welch et al. (eds.), *Oxygen Transport to Tissue XXXIV*, Advances in Experimental Medicine and Biology 765, DOI 10.1007/978-1-4614-4989-8_51,

compared to standard chow, 41.7±9.4 vs. 21.9±10.6. These data confer the high oxidative capacity for glucose irrespective of ketotic or non-ketotic states. With ketosis induced by 3 weeks of diet, cortical brain utilizes twice as much acetoacetate compared to non-ketosis.

Keywords Ketosis • Mass spectrometry • Rat cortex • Stable isotopes

1 Introduction

This study was developed on the basis that ketones are effective against pathology associated with altered glucose metabolism, such as with ischemia reperfusion injury and seizure disorders. Ketosis can be induced by prolonged fasting or ketogenic (KG) diet. We had previously reported neuroprotection by ketosis following recovery from transient focal ischemia [1]. Using a rat model of ketosis, we hypothesized that the cerebral metabolic rate for glucose (CMRglc) decreases with increasing ketosis. Thus, a shift of oxidative metabolism away from glucose towards ketone bodies may result in neuroprotection, irrespective of the mechanistic link. It has been described in humans and rodents that CMRglc decreases with ketosis [2, 3]. To show the partitioning of glucose metabolism towards ketone oxidation, one would need to simultaneously measure CMRglc under ketotic and non-ketotic conditions. CMRglc can be readily measured by a metabolic trapping mechanism using ^{18}Fluorodeoxyglucose (^{18}FDG) tracer and Positron Emission Tomography (PET) imaging system [2]. However, the CMRket cannot be reliably measured due to various constraints. These include (1) costly use of PET imaging systems, (2) lack of a trapping tracer for accurate measurements of ketone body utilization, (3) short half-life of the currently available tracers, and (4) relatively low sensitivity of nuclear magnetic resonance (NMR) [3–5].

To test in the cortical brain the partitioning of ketone utilization during ketosis, we designed a study using stable isotope tracers and mass spectrometry to estimate the fractional contribution of glucose or ketone bodies (acetoacetate) to oxidative metabolism. This enabled the implementation of a relatively inexpensive method (compared to PET), using an in vivo rat model of ketosis to study cortical brain glucose and ketone body metabolism. GC-MS and LC-MS systems are used for investigating intermediary metabolism, as they have high sensitivities to many analytes and metabolites. Compared to NMR methods, the use of small sample size therefore allows a smaller blood volume to sample. Stable isotopes of ^{13}C-labeled tracers were infused into anesthetized rats and assayed by mass spectrometry. This approach assumes that the ^{13}C-label incorporated acetyl-CoA is from the oxidation of the precursors, ^{13}C-glucose or ^{13}C-acetoacetate. Using isotopomer analysis, the M+2 enrichment of acetyl-CoA was measured and the fractional percent contribution of substrates (glucose or ketone bodies) to oxidative metabolism was calculated as the mole percent enrichment (MPE) of acetyl-CoA [6, 7].

2 Methods

The experimental protocol employed in this study was approved by the Institutional Animal Care and Use Committee (IACUC) at Case Western Reserve University.

2.1 *Animal Preparation and Diets*

Adult male Wistar rats (final weight: 310–440 g; $n=20$) were purchased from Charles River and were allowed to acclimatize in the CWRU Animal Resource Center (ARC) for at least 1 week before feeding their respective diets. Rats were then fed either the ketogenic (KG) or Standard (STD) diets for 3 weeks prior to the experimental day [1, 2]. The KG diet was purchased from Research Diet (New Brunswick, NJ, USA) and the standard rodent chow (Teklab 8664) was provided by CWRU ARC. All procedures were performed with approval from the Case Western Reserve University IACUC. On the experimental day, both diet groups (KG and STD) underwent the same surgical procedures for the placement of jugular and arterial catheters and tracer infusions [2]. Rats were morning fasted for 4 h prior to tracer infusions prior to infusions. Anesthesia was induced with isofluorane balanced with a mixture of N_2/O_2 and the rats were maintained under light anesthesia during the tracer infusions. The flow rates of the gases were manually adjusted to maintain breath rates (60–80 breath/min). Arterial blood gases were measured (ABL-5, Radiometer, Copenhagen) to confirm stable arterial blood pH.

2.2 *Experimental Design, Tracer Preparation, and Infusions*

Four study groups were implemented: rats were infused with tracers of [U-^{13}C]glucose or [U-^{13}C]AcAc and fed either standard (STD) or ketogenic (KG) diets. [U-^{13}C] glucose (99.8 %) was solved in 0.9 % NaCl solution to a final concentration of 38.7 mM. [U-^{13}C]AcAc was derived from [U-^{13}C]ethyl-acetoacetate, as previously described [8] and concentrated to 137 mM. All chemicals were purchased from Sigma-Aldrich. Tracers were infused via the jugular vein catheter (0.50 or 1.0 mmol/kg/h) (Harvard Apparatus syringe pump-11 Plus) for 60 min. To verify steady-state conditions, blood samples (100–200 μL) were taken from the tail artery at time point 0 (pre-infusion), and at 15, 30, 40, 50, and 60 min (during infusion), immediately centrifuged and the plasma frozen for GC-MS analysis of the [U-^{13}C] precursor enrichments and concentrations of glucose and acetoacetate. At the end of infusion, the rats were decapitated; the brains were dissected immediately, frozen in liquid nitrogen, and stored at −80 °C. Cortical sections (~200 mg tissue) were then dissected under frozen conditions and homogenized using a specific organic solvent mixture designed for isolation of acyl-CoAs and related metabolites [6, 7].

2.3 *Estimation of the Contribution of Acetoacetate and Glucose to Oxidative Metabolism*

Cortical brains were processed for ^{13}C-acetyl-CoA (M+2) enrichments (MPEs) using LC-MS, a similar method as previously described [6, 7]. The plasma MPE of ^{13}C-glucose and ^{13}C-AcAc was measured using GC-MS methods [6, 7]. After background correction, the MPEs of the precursor ^{13}C-substrates and the oxidative product (acetyl-CoA), were calculated from the measured ion masses (M+4, [U-^{13}C] AcAc; M+6, [U-^{13}C]glucose; M+2, [U-^{13}C]acetyl-CoA) to the unlabeled (MO, endogenous intermediate); e.g., acetyl-CoA (M+2) MPE=[M2/(M2+M0)×100]. The percent fractional contribution of glucose or AcAc to oxidative metabolism in cortical brain was estimated from the MPE of acetyl-CoA relative to the plasma MPE of the ^{13}C-infused substrates and calculated: Substrate contribution to oxidative metabolism (%)=[(brain acetylCoA MPE×2)/(plasma glucose or AcAc MPE)]×100. All data are expressed as mean±SD. Statistical analyses were performed using a two sample *t*-test. Significance was considered at the level of $p < 0.05$.

3 Results and Discussion

The fractional contribution of glucose or AcAc to cortical brain oxidative metabolism was estimated in anesthetized ketotic rats using stable isotope mass spectrometry analysis. The plasma MPE tracer dilution profiles of ^{13}C-glucose and ^{13}C-AcAc reached steady-state conditions by 50 min (time course not shown). Ketosis induced by KG diet did not significantly affect plasma ^{13}C-glucose or ^{13}C-AcAc MPE compared to STD groups (9.8±1.0 % vs. 9.2±0.5 % and 20.9±5.5 vs. 24.7±3.3, respectively) (Fig. 51.1). Cortical oxidative metabolism was significantly altered by

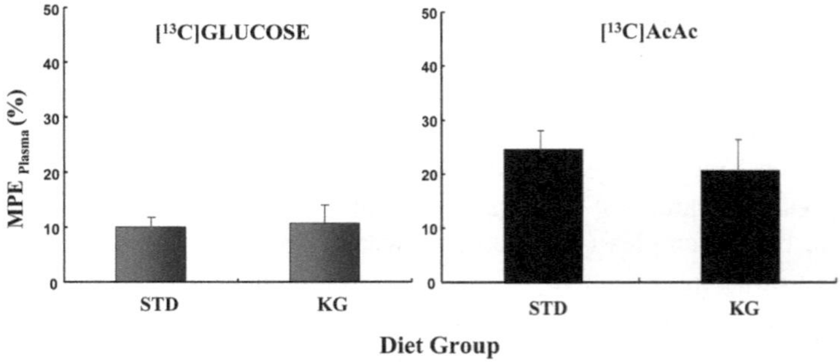

Fig. 51.1 Plasma molar enrichment (MPE %) at $t=50$ min. Tracers of [U-^{13}C]glucose and [U-^{13}C] AcAc (acetoacetate) were infused in two diet groups, standard (STD) and ketogenic (KG). Steady-state MPE was achieved by $t=50$ min (time course not shown). ^{13}C-glucose infusions resulted in a 10 % plasma MPE in both diet groups. As a result of an increase in infusion rate of [U-^{13}C]AcAc in the KG diet group compared to STD diet, a twofold increase in the ^{13}C-AcAc plasma MPE was observed (mean±SD; $*p < 0.05$)

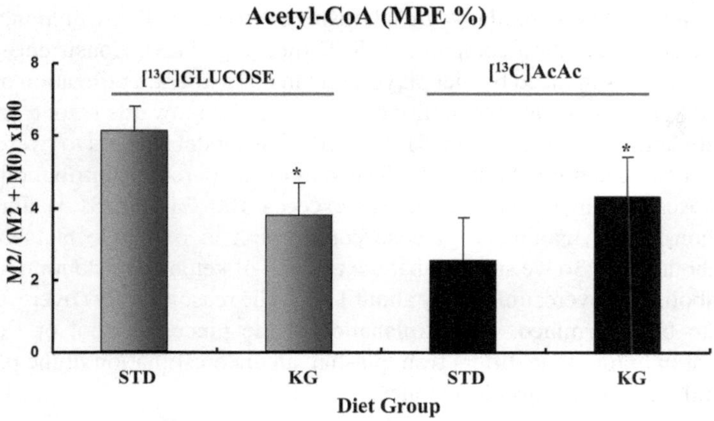

Fig. 51.2 Acetyl-CoA MPE in cortical brain. Rats fed STD or KG diets were infused with either [U-^{13}C]glucose or [U-^{13}C]AcAc tracers. Ketosis resulted in decreased glucose MPE with a parallel increase in AcAc MPE

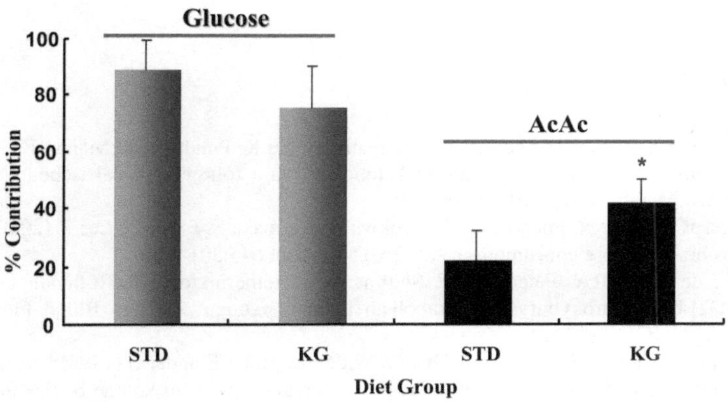

Fig. 51.3 Contribution of glucose and AcAc to oxidative metabolism. Percent contribution of glucose oxidation in cortical brain decreased with ketosis. A significant increase in percent contribution to AcAc oxidation with ketosis was also observed

ketosis (Figs. 51.2 and 51.3). With glucose oxidation, a 30 % decrease in acetyl-CoA MPE was observed (Fig. 51.2, see STD and KG groups given tracer infusions of [U-^{13}C]glucose), whereas with AcAc oxidation (see [U-^{13}C]AcAc), acetyl-CoA MPE increased about 40 % with ketosis (STD vs. KG groups). These data show a partitioning of brain glucose oxidation towards ketone body oxidation with chronic ketosis. When estimating the percent contribution of glucose to oxidative metabolism, ketosis (KG) resulted in a decrease in glucose oxidation which was not significantly different from the STD diet group (Fig. 51.3). These data confirm the high oxidative capacity for glucose in cortical brain irrespective of ketosis or non-ketosis. With respect to percent contribution of ketone body oxidation, ketosis resulted in an increase in oxidative metabolism, as shown by the twofold increase in AcAc percent contribution compared to STD diet (Fig. 51.3). Consistent with our hypothesis, ketosis induced by diet plays a role in cortical brain utilization of AcAc. These findings demonstrate the ability of brain to switch towards ketone body oxidation with ketosis (Figs. 51.2 and 51.3) [4, 5]. This model appears to overestimate oxidative metabolism by about 15 %. The sum of the percent contribution of glucose and AcAc to oxidative metabolism exceeds 100 % (Fig. 51.3). Indeed, in healthy non-ketotic mammals, glucose contribution to oxidative metabolism in brain is about 90 %. So we suspect that the fraction of ketone contribution to oxidation metabolism is overestimated by about 15 %. The reason for this overestimation remains to be determined. One explanation is the precursor pool of ^{13}C-AcAc enrichment in brain tissue differs from plasma; an underestimation of the precursor MPE could account for this discrepancy.

Acknowledgments The authors would like to thank the CASE MMPC, affiliated staff and faculty, for their technical assistance and helpful discussions on mass isotopomer analysis. This research was supported by the National Institutes of Health, R01 HL092933-01, R21 NS062048-01 and Mouse Metabolic Phenotyping Center, MMPC U24 DK76169.

References

1. Puchowicz MA, Zechel J, Valerio J, Emancipator D, Xu K, Pundik S, LaManna JC, Lust WD (2008) Neuroprotection in diet induced ketotic rat brain following focal ischemia. J Cereb Blood Flow Metab 28(12):1907–1916
2. LaManna JC, Salem N, Puchowicz M, Erokwu B, Koppaka S, Flask C, Lee Z (2009) Ketone suppress brain glucose consumption. Adv Exp Med Biol 645:301–306
3. Pan JW, de Graaf RA, Petersen KF, Shulman GI, Hetherington HP, Rothman DL (2002) [2,4-13C2]-beta-hydroxybutyrate metabolism in human brain. J Cereb Blood Flow Metab 22:890–898
4. Blomqvist G, Thorell JO, Ingvar M, Grill V, Widen L, Stone-Elander S (1995) Use of R-β-[1-11C]hydroxybutyrate in PET studies of regional cerebral uptake of ketone bodies in humans. Am J Physiol 269(5):E948–E959
5. Bentourkia M, Tremblay S, Pifferi F, Rousseau J, Lecomte R, Cunnane S (2009) PET study of 11C-acetoacetate kinetics in rat brain during dietary treatment affecting ketosis. Am J Physiol 296:E796–E801
6. Gu L, Zhang GF, Kombu RS, Allen F, Kutz G, Brewer WU, Roe CR, Brunengraber H (2010) Parenteral and enteral metabolism of anaplerotic triheptanoin in normal rats. II. Effects on lipolusis, glucose production, and liver acyl-CoA profiles. Am J Physiol 298:E362–E371
7. Deng S, Zhang GF, Kasumov T, Roe CR, Brunengraber H (2009) Interrelations between C4 ketogenesis, C5 ketogenesis, and anaplerosis in the perfused rat liver. J Biol Chem 284(41):27799–27807
8. Des Rosiers C, Montgomery JA, Desrochers S, Garneau M, David F, Mamer OA, Brunengraber H (1988) Interference of 3-hydroxyisobutyrate with measurements of ketone body concentration and isotopic enrichment by gas chromatography-mass spectrometry. Anal Biochem 173:96–105

Chapter 52
Alteration of Circulating Mitochondrial DNA Concentration After Irradiation

Mei Zhang, Bingrong Zhang, Yansong Guo, Lei Zhang, Shanmin Yang, Liangjie Yin, Sadasivan Vidyasagar, David Maguire, Steve Swarts, Zhenhuan Zhang, Amy Zhang, Lurong Zhang, and Paul Okunieff

Abstract Mitochondrial DNA (mtDNA) is maternally inherited and controls the oxygen-related production of adenosine-5′-triphosphate, which is transported from the mitochondria to other cellular compartments and used as energy for cellular activities. The mtDNA is physically separated from nuclear DNA (nDNA). Ionizing radiation (IR) causes the release of both mtDNA and nDNA into circulation. Our previous study demonstrated that nDNA has potential to be a biodosimeter. In this study, branched DNA technology was used to explore the alteration pattern of mtDNA after IR. C57BL/6 mice were exposed to 0, 1.5, 3, 6, 8, or 10 Gy total body irradiation; thereafter, plasma mtDNA was assessed with samples collected at 3, 6, 9, 15, 24, 48, 72, or 168 h. We found that: (1) the designed probesets were specific for mtDNA extracted from the liver, and they recognized the small amount of mtDNA mixed in the nDNA; (2) plasma mtDNA exhibited a statistically significant increase only at 6 h after 8 Gy irradiation. The alteration of mtDNA was not dose-dependent or time-dependent; hence, it is unlikely to be an effective biodosimeter.

Keywords Circulating mitochondrial DNA • Irradiation

M. Zhang (✉) • B. Zhang • Y. Guo • L. Zhang • S. Yang • L. Yin • S. Vidyasagar
D. Maguire • S. Swarts • Z. Zhang • A. Zhang • L. Zhang • P. Okunieff
Department of Radiation Oncology, University of Florida, P.O. Box 100385, Gainesville, FL 32610, USA
e-mail: mzhang14@ufl.edu

W.J. Welch et al. (eds.), *Oxygen Transport to Tissue XXXIV*, Advances in Experimental Medicine and Biology 765, DOI 10.1007/978-1-4614-4989-8_52,

1 Introduction

Mitochondria are responsible for the production of adenosine-5′-triphosphate (ATP), the "energy currency" for cellular activity. Mitochondrial DNA (mtDNA) contains 37 genes that play important roles in the regulation of cellular metabolism, apoptosis, and oxidative stress control.

Since mtDNA has a less robust DNA repair capacity than nuclear DNA (nDNA), it is particularly susceptible to reactive oxygen species (ROS), which act as mediators of ionizing radiation (IR)-induced cellular damage [1]. Radiation induces a relatively high incidence of mtDNA mutation, deletion, or depletion and is likely to trigger programmed cell death and the release of mtDNA [2, 3]. In addition, our previous studies and others revealed that mtDNA copy number increased in the tissues and blood cells of irradiated mice [4–6]. These facts might indicate an increased release of mtDNA into circulation after IR; however, there are few reports regarding circulating mtDNA copy number after IR [6, 7].

Identification of rapid and sensitive biodosimetry is of great importance for radiation triage. Currently, there is no satisfactory biodosimetry to guide medical intervention after nuclear incidents [8]. Our previous study demonstrated that plasma nDNA increased in a radiation dose-dependent and time-dependent manner; thus, it could serve as a radiation biodosimeter for triage, especially within 1–2 days after IR exposure [9, 10]. Using state-of-the-art branched DNA (bDNA) technology, this current study seeks to determine if plasma mtDNA has an alteration pattern similar to that of plasma nDNA.

2 Methods

Eight-week-old, male, C57BL/6 mice were divided into groups of 5. Mice were housed in a pathogen-free barrier facility in accordance with United States Public Health Service guidelines, and the use of animals was approved by the Institutional Animal Use Committee. Groups received a single dose of 1.5, 3, 6, 8, or 10 Gy total body irradiation (TBI); control mice were not irradiated. Animals were immobilized with a plastic restraint during exposure to a cesium-137 γ-ray source at a dose rate of 1.84 Gy/min. At 3, 6, 9, 15, 24, 48, 72, or 168 h after TBI, mice were euthanized; 500–800 μL of blood was collected in ethylenediaminetetraacetic acid (EDTA)-treated polypropylene tubes, as described previously [11]. The blood was centrifuged ($200 \times g$ for 10 min at 4°C), followed by separation of plasma into fresh tubes with care taken to leave the cellular portion undisturbed. The samples were then stored at −80°C until analysis.

Control murine livers were resected. Total DNA was extracted from these tissues by standard proteolytic digestion followed by phenol/chloroform/isoamyl alcohol purification. For mtDNA extraction, tissues were finely chopped and placed in ice-cold buffer containing 250 mM sucrose, 1 mM ethylene glycol tetraacetic acid, and 10 mM 4-(2-hydroxyethyl)-1-piperazineethanesulfonic acid (HEPES) (pH 7.4).

The chopped tissues were homogenized with a glass/teflon potter tissue grinder and then centrifuged for 10 min at 500 × *g* at 4°C. Supernatants were centrifuged for 10 min at 12,000 × *g* at 4°C. The sediments were resuspended with HEPES buffer (250 mM sucrose and 10 mM HEPES, pH 7.4) and centrifuged again for 10 min at 12,000 × *g* at 4°C. Afterwards, the isolated mitochondria were prepared for extraction by phenol/chloroform/isoamyl alcohol purification.

The mitochondrial probesets, including four capture extenders, ten label extenders, and six blocking probes were designed by DiaCarta, LLC (Hayward, CA, USA) and were synthesized by Integrated DNA Technologies, Inc. (Coralville, IA, USA). The assay was performed according to instructions, as described previously [9].

To further confirm the assay's specificity, we used the QuantiGene 2.0 assay (DiaCarta, LLC), which targets the murine B1 and mitochondrial sequences, to parallel test the total nDNA and isolated mtDNA from the livers.

Analysis of variance with Dunnett's post hoc testing was used to analyze any potential difference between the irradiated groups and the control group. A P value <0.05 was regarded as statistically significant.

3 Results

The specificity of the mitochondrial probesets was tested. Total DNA extracted from the liver showed that more nDNA bound to the murine B1 probes than to the mtDNA probeset. When the same amount of nDNA and mtDNA were measured with the mitochondrial probes, we found that mtDNA bound more to its own probes while total DNA binding was negligible. However, the B1 probes also could detect the total DNA in the mtDNA extracted from isolated livers (Fig. 52.1a).

Quantitative real-time polymerase chain reaction (RT-PCR) detected a 177-bp amplicon in the murine mitochondrial genome and was performed to confirm the specificity of the bDNA-based mitochondrial assay. The PCR primers and protocol used in the experiment were reported in a previous publication [5].

Quantitative RT-PCR showed a significant increase in the relative ratio of mtDNA to nDNA in mtDNA extracted from isolated hepatic mitochondria as compared to the total DNA extracted from liver tissue (Fig. 52.1b). This result indicates that the mtDNA extracted from isolated hepatic mitochondria was essentially pure mtDNA with little nDNA.

To determine the precision of the bDNA-based mitochondrial assay, serial dilutions of murine DNA were prepared in Tris-EDTA buffer and used to construct a standard curve expressed as the relative luminescence unit (RLU) vs. the input DNA concentrations in the range of 6.25–400 ng/mL. The assay yielded an excellent linearity with $R^2=0.985$ (Fig. 52.2). This experiment was repeated at least three times; similar linearity data were obtained each time.

Plasma samples collected at different times with different radiation doses were assayed with the bDNA-based mtDNA assay to quantify the circulating mtDNA concentration. At 3 and 6 h after TBI, mtDNA had a tendency to increase, but this

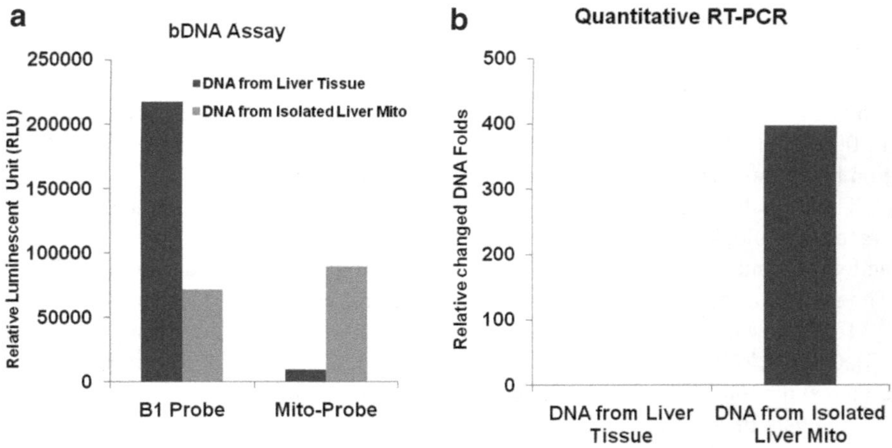

Fig. 52.1 Total DNA and mtDNA from isolated tissues tested with a bDNA-based assay with B1 and mitochondrial probes showed specific binding of mtDNA to the mitochondrial probes (**a**). The RT-PCR assay tested total DNA and mtDNA from tissues to determine the mtDNA to nDNA ratio (**b**). It indicated that the isolated mtDNA was essentially pure and contained little nDNA

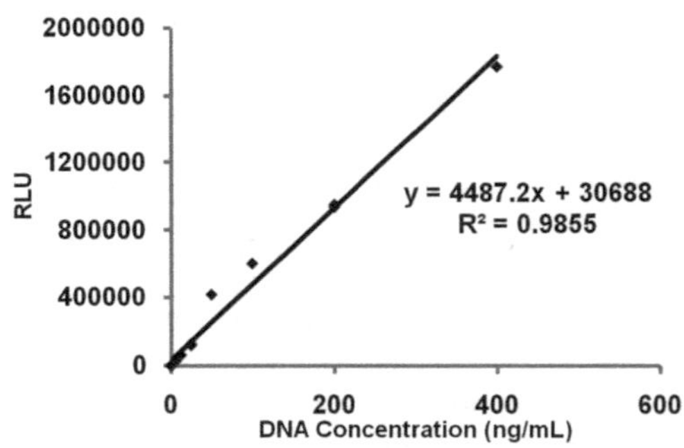

Fig. 52.2 Standard curve for bDNA-based mitochondrial assay with serial dilutions of murine DNA. The assay was linear with $R^2=0.985$

increase was only statistically significant at 6 h after 8 Gy (Fig. 52.3a, b). At 9 h after TBI, mtDNA in the 1.5 Gy group was 212.6 ± 103.9 ng/mL as compared to 32.60 ± 10.41 ng/mL in the 0 Gy control group. A similar tendency was observed in 3 and 6 Gy mice 15 h after TBI (Fig. 52.3c, d). However, considering the variation between individual animals, the alteration was not statistically significant. At 24 and 48 h after TBI, the mean mtDNA concentrations in the control and irradiated groups were similar. At 72 and 168 h, the 6–10 Gy groups presented decreased mtDNA concentrations as compared to the 0 Gy control group (data not shown).

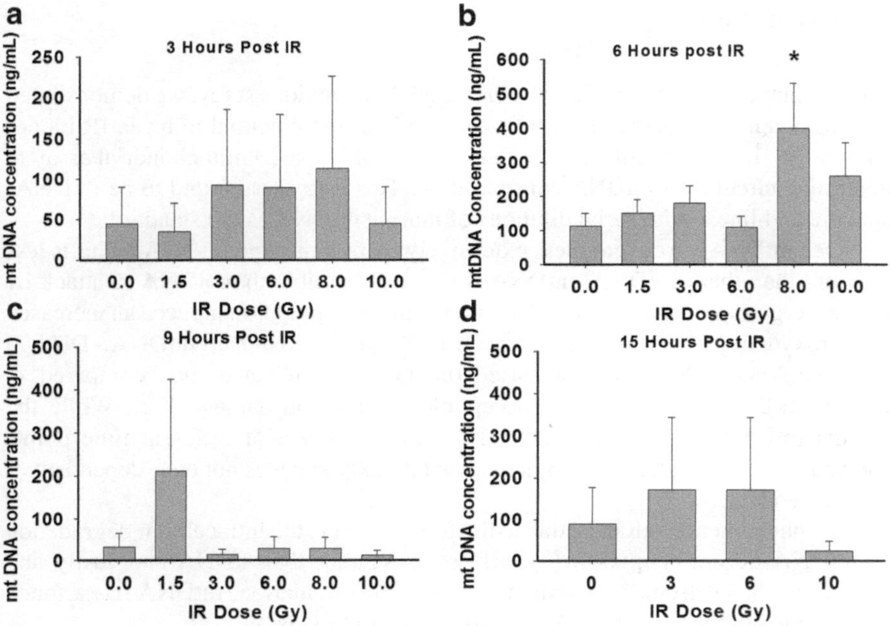

Fig. 52.3 Plasma mtDNA patterns at different radiation doses (**a–d**) (*asterisk* represent $P<0.05$ compared to 0 Gy). The data are presented as mean ± standard error of the mean (SEM) for 4–5 mice/group

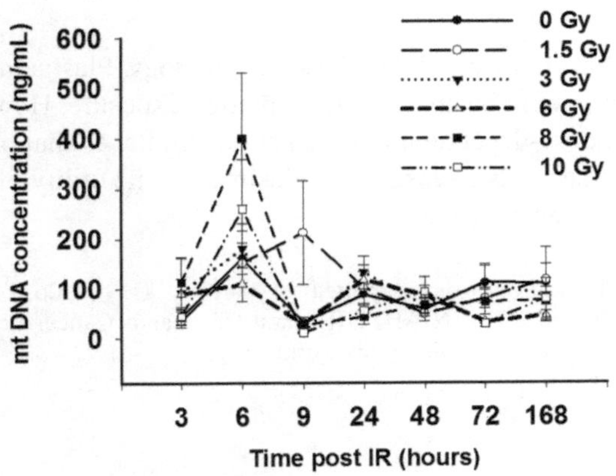

Fig. 52.4 Plasma mtDNA concentration in different radiation-dose groups at different time points. Each point represents the mean mtDNA value at indicated times after radiation. The *error bars* represent SEM

When the mtDNA concentration in each group was plotted against different time points after irradiation, the higher dose radiation groups exhibited a clear trend of elevated mtDNA concentrations at 6 h (Fig. 52.4). However, this elevation was not dose dependent or time dependent.

4 Discussion

The hallmark of IR damage is DNA breakage. In a previous study, we demonstrated that plasma nDNA increases with IR dose and has the potential to be an IR biodosimeter [9]. In our current study, we utilized a bDNA-based mitochondrial assay to determine circulating mtDNA concentration. This assay was found to be mtDNA-specific and linear with serial dilutions of murine mtDNA as the standard.

Since mtDNA are covered less extensively by proteins and mtDNA repair is less efficient than that of nDNA, mtDNA is more susceptible than nDNA to attack by active oxygen species. Richter et al. demonstrated that radiation induced an increased 8-hydroxydeoxyguanosine (oh^8dG) level in the order of isolated mtDNA>DNA in mitochondria>nDNA. This increased oh^8dG level indicated that, compared to nDNA, mtDNA is much more susceptible to radiation damage [12]. While the plasma mtDNA from mice irradiated at different doses at different time points showed an increased trend at 6 h, it declined quickly and was not dose dependent or time dependent.

This phenomenon might be due to the following: (1) the intracellular degradation of mtDNA during apoptosis; (2) mtDNA is smaller than nDNA and easily and quickly excluded from the circulation; (3) in the circulation, mtDNA has a much lower copy number than nDNA as measured by nDNA assay.

5 Conclusion

MtDNA can be measured with bDNA-based technology. Plasma mtDNA increases after irradiation and might be an index for radiation exposure. However, it does not exhibit a radiation dose-dependent release pattern similar to that of plasma nDNA; therefore, it cannot serve as a sensitive biodosimeter. The utility of plasma mtDNA remains to be further investigated.

Acknowledgments This project is supported in part by U19 AI067733, RC1AI078519, RC2-AI-087580, RC1-AI081274 (NIAID/NIH), and UF Shands Cancer Center startup funds. We thank Kate Casey-Sawicki for editing this work.

References

1. Morales A, Miranda M, Sanchez-Reyes A, Biete A, Fernandez-Checa JC (1998) Oxidative damage of mitochondrial and nuclear DNA induced by ionizing radiation in human hepatoblastoma cells. Int J Radiat Oncol Biol Phys 42:191–203
2. Wilding CS, Cadwell K, Tawn EJ, Relton CL, Taylor GA et al (2006) Mitochondrial DNA mutations in individuals occupationally exposed to ionizing radiation. Radiat Res 165:202–207

3. Murphy JE, Nugent S, Seymour C, Mothersill C (2005) Mitochondrial DNA point mutations and a novel deletion induced by direct low-LET radiation and by medium from irradiated cells. Mutat Res 585:127–136
4. Zhang H, Maguire DJ, Zhang M, Zhang L, Okunieff P (2011) Elevated mitochondrial DNA copy number and POL-gamma expression but decreased expression of TFAM in murine intestine following therapeutic dose irradiation. Adv Exp Med Biol 701:201–206
5. Zhang H, Maguire D, Swarts S, Sun W, Yang S et al (2009) Replication of murine mitochondrial DNA following irradiation. Adv Exp Med Biol 645:43–48
6. Evdokimovskii EV, Patrushev MV, Ushakova TE, Gaziev AI (2007) Sharp changes in the copy number of mtDNA and its transcription in the blood cells of X-ray irradiated mice are observed, and mtDNA fragments appear in the blood serum. Radiats Biol Radioecol 47:402–407
7. Evdokimovsky EV, Ushakova TE, Kudriavtcev AA, Gaziev AI (2011) Alteration of mtDNA copy number, mitochondrial gene expression and extracellular DNA content in mice after irradiation at lethal dose. Radiat Environ Biophys 50:181–188
8. Flynn DF, Goans RE (2006) Nuclear terrorism: triage and medical management of radiation and combined-injury casualties. Surg Clin North Am 86:601–636
9. Zhang L, Zhang M, Yang S, Cao Y, Bingrong Zhang S et al (2010) A new biodosimetric method: branched DNA-based quantitative detection of B1 DNA in mouse plasma. Br J Radiol 83:694–701
10. Zhang H, Zhang SB, Sun W, Yang S, Zhang M et al (2009) B1 sequence-based real-time quantitative PCR: a sensitive method for direct measurement of mouse plasma DNA levels after gamma irradiation. Int J Radiat Oncol Biol Phys 74:1592–1599
11. Vasilyeva IN (2001) Low-molecular-weight DNA in blood plasma as an index of the influence of ionizing radiation. Ann N Y Acad Sci 945:221–228
12. Richter C, Park JW, Ames BN (1988) Normal oxidative damage to mitochondrial and nuclear DNA is extensive. Proc Natl Acad Sci U S A 85:6465–6467

Erratum to: Oxygen Transport to Tissue XXXIV

William J. Welch, Fredrik Palm, Duane F. Bruley and David K. Harrison

Erratum to:
W.J. Welch et al. (eds.), *Oxygen Transport to Tissue XXXIV*, Advances in Experimental Medicine and Biology 765, DOI 10.1007/978-1-4614-4989-8

The below listed chapters were originally published © Springer Science+Business Media New York, but have now been made available Open Access and © The Authors under a CC BY 4.0 license.

The front matter and cover have also been updated with the copyright information.

Chapter 4 titled "Canonical Correlation Analysis in the Study of Cerebral and Peripheral Haemodynamics Interrelations with Systemic Variables in Neonates Supported on ECMO".

Chapter 10 titled "Normobaric Hyperoxia Does Not Change Optical Scattering or Pathlength but DoesIncrease Oxidised Cytochrome *c* Oxidase Concentration in Patients with Brain Injury".

The updated online version of this book can be found at
https://dx.doi.org/10.1007/978-1-4614-4989-8

The updated online versions of the chapters can be found at
https://dx.doi.org/10.1007/978-1-4614-4989-8_4
https://dx.doi.org/10.1007/978-1-4614-4989-8_10
https://dx.doi.org/10.1007/978-1-4614-4989-8_13
https://dx.doi.org/10.1007/978-1-4614-4989-8_15
https://dx.doi.org/10.1007/978-1-4614-4989-8_28

W.J. Welch et al. (eds.), *Oxygen Transport to Tissue XXXIV*, Advances in Experimental Medicine and Biology 765, DOI 10.1007/978-1-4614-4989-8_53,

Chapter 13 titled "Modelling Cerebrovascular Reactivity: A Novel Near-Infrared Biomarker of Cerebral Autoregulation?"

Chapter 15 titled "Oscillations in Cerebral Haemodynamics in Patients with Falciparum Malaria".

Chapter 28 titled "Wavelet Cross-Correlation to Investigate Regional Variations in Cerebral Oxygenation in Infants Supported on Extracorporeal Membrane Oxygenation".

Author Index

W.J. Welch et al. (eds.), *Oxygen Transport to Tissue XXXIV*, Advances in Experimental Medicine and Biology 765, DOI 10.1007/978-1-4614-4989-8,

Subject Index

W.J. Welch et al. (eds.), *Oxygen Transport to Tissue XXXIV*, Advances in Experimental Medicine and Biology 765, DOI 10.1007/978-1-4614-4989-8,

Kolibri 360 GmbH
Ferdinand-Jühlke-Straße 7
99095 Erfurt, Deutschland
produktsicherheit@kolibri360.de